Nanotechnology in Agriculture

Ramesh S. Chaughule • Amruta V. Shelar
Editors

Nanotechnology in Agriculture

Pioneering Progress and Challenges

Springer

Editors
Ramesh S. Chaughule
Ramnarain Ruia Autonomous College
Mumbai, Maharashtra, India

Amruta V. Shelar
Savitribai Phule Pune University
Pune, India

ISBN 978-3-031-97543-1 ISBN 978-3-031-97544-8 (eBook)
https://doi.org/10.1007/978-3-031-97544-8

This Springer imprint is published by the registered company Springer Nature Switzerland AG
The registered company address is: Gewerbestrasse 11, 6330 Cham, Switzerland

Foreword 1

As our world faces the dual challenges of ensuring food security and protecting the environment, innovative agricultural solutions are more crucial than ever. *Nanotechnology in Agriculture: Pioneering Progress and Challenges* brings together diverse perspectives and cutting-edge research to emphasize nanotechnology's significant impact on modern farming practices.

This volume guides readers from foundational concepts to complex applications. It begins by establishing nanotechnology as a driver of sustainable agricultural practices—a central theme throughout this book. The early chapters explore the molecular mechanisms behind nanofertilizers, nanopesticides, and nanofungicides, illustrating how these innovations enable precise nutrient delivery and effective pest control. This precision enhances crop productivity and reduces environmental impact by minimizing chemical runoff and waste.

Building on this foundation, this book further explores advanced applications such as plasmonic nanosensors for real-time detection of agrotoxins, nanopriming methods to enhance seed germination, and the implementation of nanobiofertilizers and nanocoatings to improve plant health and postharvest management. Additionally, sustainable systems like hydroponics and aeroponics are examined, highlighting how nanotechnology is poised to revolutionize water and nutrient management in controlled environments.

Notably, the volume also addresses the challenges that accompany these innovations. Detailed discussions on the toxicological effects of nanomaterials, along with comprehensive frameworks for risk assessment and regulatory policies, remind us that technological progress must always align with strict safety standards. Furthermore, the integration of nanotechnology with advanced tools such as CRISPR-Cas genome editing opens new avenues for enhancing plant defense mechanisms and genetic resilience, paving the way for transformative advancements in crop protection.

The preface establishes a visionary tone by highlighting the dual promise of improved agricultural efficiency and environmental sustainability. Organized into five comprehensive sections, this book provides a well-structured road map that addresses key research areas, practical applications, and vital policy considerations. This holistic approach reflects the complex nature of nanotechnology in agriculture and emphasizes the collaborative spirit necessary to address global challenges.

By bringing together the expertise of top researchers and practitioners, the editors have developed a forward-looking resource aimed at inspiring scientists, industry leaders, and policymakers alike. We genuinely hope the insights presented in these pages will foster further innovation and contribute to building a resilient, sustainable future for agriculture. I extend my heartfelt congratulations to Professor Chaughule and Ms. Shelar as well as the publisher for their significant contributions to advancing knowledge in this dynamic area of research.

Sincerely,

Department of Agricultural and Biosystems Engineering, South Dakota State University, Brookings, South Dakota, USA
Kasiviswanathan Muthukumarappan
March 30, 2025

Foreword 2

I am delighted to write a foreword for the book, *Nanotechnology in Agriculture: Pioneering Progress and Challenges*. At a time when food security and environmental sustainability are pressing global issues, this book offers valuable insights into how nanotechnology is reshaping modern agriculture.

The chapters contained in this book, authored by distinguished experts, investigate the transformative impact of nanotechnology in sustainable agriculture. This book comprises chapters that explore diverse applications, including the enhanced nutrient delivery systems, optimized nanopesticides delivery systems, the enhancement of plant defenses, and the innovation of hydroponic and aeroponic methodologies. The authors scrutinize scientific advancements, exemplified by the molecular mechanisms underlying nano-fertilizers and nanopesticides, and practical innovations, such as nanopriming techniques promoting seed development.

What makes this collection truly unique is its balanced approach. While it highlights groundbreaking progress, it also thoughtfully addresses the risks and challenges of using nanomaterials in agriculture. By exploring risk assessments and safe implementation practices, this book ensures that advancements are innovative and responsibly managed.

This work serves as both a comprehensive resource and an inspiring call to action for researchers, practitioners, and policymakers. It celebrates creativity and dedication, key ingredients in fostering a sustainable future for agriculture.

I invite you to explore this book and join in the journey of blending technology with traditional agricultural practices. May the insights within spark new ideas and collaborations that pave the way for a healthier, more sustainable future in agricultural science.

My congratulations to the editors and Springer Nature Publishers for their significant contributions to advancing knowledge in this dynamic area of research.

Vishwakarma University – VU, Pune,
India
Savitribai Phule Pune University, Pune,
India

Wasudeo Gade

Preface

Nanotechnology offers transformative solutions for sustainable agriculture by enhancing productivity and reducing environmental impact. It enables the development of smart delivery systems, such as nanofertilizers and nanopesticides, which precisely release nutrients and active agents to plants, minimizing wastage and runoff. Nanobiosensors facilitate real-time monitoring of soil health, pest detection, and agrochemical residues, ensuring efficient resource utilization. Additionally, nanomaterials like zinc oxide and silver nanoparticles promote plant growth and protect against pathogens. In seed technology, nanopriming enhances germination rates and improves stress tolerance. Collectively, these advancements support sustainable farming by reducing chemical inputs, increasing crop yields, and preserving ecological balance.

We are pleased to present the fifth book published by Springer Nature, *Nanotechnology in Agriculture: Pioneering Progress and Challenges*, a dynamic field made by the researchers. This book highlights the transformative role of nanotechnology in modern agriculture, with a focus on nanofertilizers, nanopesticides, nanobiosensors, and other advancements, catering to both aspiring and practicing scientists. It examines the potential of nanotechnology across diverse areas, including food production, precision farming, enhanced nutrient delivery, pest control, and water management. By leveraging the unique properties of nanomaterials, researchers and industry leaders are driving innovative solutions for sustainable agriculture and resource optimization.

This book is organized into five comprehensive sections: The first section establishes the need for nanotechnology and explains the fundamental molecular mechanisms. The second section focuses on pest management and detection, as well as plant defense mechanisms. The third section explores the applications of nanotechnology in seed development and fertilizers and highlights the significance of hydroponics and aeroponics. The fourth section introduces sustainability-focused materials, postharvest technologies, and cutting-edge genetic advancements. Finally, the fifth section concludes the book by addressing toxicology, risk assessment, and policy considerations.

1. Nanotechnology revolutionizes sustainable agriculture by addressing resource depletion, environmental degradation, and food security. Nanofertilizers and nanopesticides enable precise nutrient delivery and targeted crop protection, minimizing waste and contamination. Nanomaterials improve soil health, while nanosensors optimize resource use through real-time monitoring of plants, soil, and water. Efficient water management and innovations like nanocoatings reduce postharvest losses. By lowering reliance on conventional agrochemicals, nanotechnology minimizes ecological harm and ensures cost-effective solutions for farmers. Its integration enhances productivity and supports sustainable farming, essential for global food security. The chapter by Shelar et al. discusses technologies for sustainable agriculture, water management, and soil health and their implications for sustainability and environmental impact.

   Nanofertilizers, nanopesticides, and nanofungicides play a crucial role in enhancing crop productivity and plant protection. Nanofertilizers optimize nutrient delivery through controlled release and improved plant absorption, thereby regulating metabolic processes and boosting biomass production. Nanopesticides and nanofungicides effectively target pests and pathogens through oxidative stress mechanisms and advanced encapsulation techniques, ensuring both environmental safety and efficacy. These innovations address critical food security and sustainability challenges by improving agricultural efficiency and minimizing ecological impact. The chapter by Amisha Rani et al. explores the molecular mechanisms underlying the action of nanofertilizers, nanopesticides, and nanofungicides, highlighting their ability to penetrate the plant cuticle and induce transcriptional reprogramming for enhanced metabolome regulation.
2. Researchers have focused on the development and application of nanofertilizers, nanopesticides, and nanofungicides. Nanofertilizers enhance nutrient uptake and metabolic regulation through targeted delivery, increasing biological yields and complementing conventional methods. Nanopesticides and nanofungicides utilize metallic nanoparticles and encapsulation technologies to achieve precise pest and pathogen control while minimizing ecological impact. These innovations highlight the transformative potential of nanotechnology in advancing sustainable agricultural practices. The chapter by M. G. Kassa et al. provides a comprehensive overview of the mechanisms, applications, and future prospects of nanofertilizers, nanopesticides, and nanofungicides, emphasizing their critical role in promoting sustainability in agriculture.

   The integration of plasmonic nanosensors in precision farming, the challenges associated with scalability, and the transition toward eco-friendly agricultural practices are key areas of focus. The chapter by Shelar and Didwal explores the cutting-edge role of plasmonic nanosensors in revolutionizing agrotoxin detection for sustainable agriculture. It highlights the advancements in sensitivity, specificity, and real-time monitoring capabilities enabled by plasmonic nanostructures.

   Integrating nanotechnology with CRISPR-Cas systems revolutionizes plant immunity by enabling precise gene editing. Nanomaterials such as lipid-based nanoparticles and carbon nanotubes facilitate targeted delivery, protect CRISPR components, and enhance cellular uptake. This approach supports multiplexed edits and controlled activation, optimizing plant immune responses under specific conditions. The chapter by Dabhade and Kaur delves into the transformative role of nanotechnology in enhancing plant defense mechanisms by triggering immune responses at the molecular level. It examines the interaction of nanoparticles with plant signaling pathways, their role in activating systemic acquired resistance, and their potential to combat biotic stressors. The chapter highlights innovative strategies for sustainable crop protection and improved agricultural resilience by integrating molecular insights and nano-technological advancements.

3. Exploring novel approaches, nanopriming technology has emerged as a promising method to enhance seed germination and seedling development, providing sustainable solutions for improved crop resilience. The chapter by Salve et al. examines nanopriming technology as a groundbreaking approach to boosting seed germination and seedling development. It discusses the role of nanoparticles in enhancing seed vigor, stress tolerance, and nutrient uptake, addressing key agricultural challenges. The chapter underscores its potential to revolutionize crop production and resilience across diverse environmental conditions by analyzing the mechanisms of nanoparticle-mediated priming and its practical applications.

   The foliar application of nanofertilizers is emerging as a transformative approach in sustainable agriculture, addressing challenges like declining soil fertility, overuse of chemical fertilizers, and environmental degradation. By directly spraying nutrients onto plant leaves, this method enhances nutrient delivery efficiency, reduces losses, and minimizes environmental impacts. The chapter by Chaughule and Shelar explores the mechanisms of nutrient uptake via foliar nanofertilizers, highlighting their potential to improve nutrient use efficiency, mitigate abiotic stresses, and boost crop productivity. It also discusses the advantages of this approach over traditional methods, including reduced fertilizer usage and enhanced environmental sustainability, while addressing challenges related to formulation, scalability, and field adoption.

   Nanotechnology enhances hydroponics by optimizing nutrient delivery, improving uptake efficiency, and increasing crop resilience to stress. Nanosensors enable real-time monitoring for precision control, while nanoparticles enhance substrate quality, boosting growth and yield. Aeroponics involves growing plants in air or mist, delivering nutrients directly to the roots, ensuring water efficiency, faster growth rates, and effective space utilization. The chapter by Mahajan et al. examines the advantages and disadvantages of hydroponics and aeroponics systems, highlighting resource efficiency, reduced chemical dependency, and sustainability as benefits of hydroponics, while noting challenges like high initial costs and technical complexity. For aeroponics, it discusses benefits such as minimal resource use and high yield potential, alongside drawbacks like susceptibility to power failures and high maintenance costs.

4. Nanobiochar (NBC), a nanoscale biochar derivative, is gaining attention in sustainable agriculture and environmental management due to its high surface area, porosity, and functional groups. Produced through advanced methods like

ball milling and hydrothermal synthesis, NBC serves as a soil amendment, slow-release fertilizer, and contaminant remediation agent. The chapter by Popat et al. explores NBC production techniques and their role in enhancing soil health, nutrient retention, and plant resilience under stress. Challenges and safety considerations for effective use are also discussed.

Postharvest technologies aim to preserve agricultural produce quality through methods like packaging, storage, and distribution. Nanotechnology offers innovative solutions for postharvest management, including nanoencapsulation to protect against stressors, nanocoatings to regulate ripening and reduce water loss, and advanced packaging to extend shelf life. The chapter by Poorvaja and Preethi discusses how these approaches improve product quality, safety, and farmer revenue while minimizing waste. However, careful evaluation of nanoparticle migration, toxicity, and permissible limits is essential for safe application in edible packaging.

CRISPR-Cas has revolutionized plant genome editing, but efficient delivery of its components remains a challenge. Nanotechnology offers innovative solutions using materials like metal nanoparticles, carbon-based nanomaterials, and polymeric nanoparticles to enhance delivery. These nanocarriers improve uptake, stability, and specificity, reducing off-target effects and increasing editing efficiency. The chapter by Shende and Patil highlights the synergy between nanotechnology and CRISPR-Cas, showcasing its transformative potential in advancing plant genome editing.

5. Nanotechnology is revolutionizing agriculture by boosting productivity with nano-agrochemicals like nanopesticides and nanofertilizers. Its unique properties hold promise, and studies also reveal potential toxic effects on soil microbiota, fungi, and bacteria, raising environmental concerns. The chapter by Sharbidre examines the physiological, biochemical, and genotoxic impacts of nanomaterials, emphasizing the need for safer and sustainable practices in agroecosystems and food chains, offering insights for safer and more sustainable use.

   Nanotechnology offers transformative potential in agriculture and food industries through nano-agrochemicals, but their environmental and health risks remain a concern. Insufficient data on toxicity, bioaccumulation, and impacts on soil microorganisms create uncertainties in consumer safety and regulations. The chapter by Raut and Sharbidre discusses the protocols for toxicity testing, life-cycle analysis, and risk assessment, alongside strict regulations for production and disposal. Public awareness and nano-ecotoxicology studies are crucial for ensuring safe and sustainable applications in the agri-food sector.

   As this interdisciplinary science continues to evolve, further innovations in agriculture are expected to increase crop production to face the global population.

   The editors wish to thank all the distinguished and expert contributors for their enthusiastic participation in this endeavor. We are confident that this book will serve as a valuable guide for researchers and students involved in nanotechnology, environmental science, food science, and agriculture. Dr. Ramesh Chaughule wishes to thank his family members and the staff of Ramnarain Ruia Autonomous College for all their support. Ms. Amruta Shelar extends her heartfelt gratitude to the Department of Technology and the Department of Biotechnology, Savitribai Phule Pune University (SPPU), for their invaluable support and unwavering encouragement.

Last but not least, the editors sincerely thank the staff of Springer Nature for their support from time to time.

Mumbai, Maharashtra, India Ramesh S. Chaughule
Pune, India Amruta V. Shelar

Contents

Nanotechnology: A Catalyst for Sustainable Agricultural Practices

Amruta V. Shelar, Ramesh S. Chaughule, and Rajendra Patil

Contents

A. V. Shelar
Department of Technology, Savitribai Phule Pune University, Pune, India

R. S. Chaughule
Department of Chemistry, Ramnarain Ruia Autonomous College, Matunga, Mumbai, India

R. Patil (✉)
Department of Biotechnology, Savitribai Phule Pune University, Pune, India
e-mail: rpatil@unipune.ac.in

R. S. Chaughule, A. V. Shelar (eds.), *Nanotechnology in Agriculture*,
https://doi.org/10.1007/978-3-031-97544-8_1

1 Introduction

Agriculture has long been the foundation of human civilization, providing food, fiber, and fuel for a growing global population. However, the modern agricultural landscape faces unprecedented challenges, including soil degradation, water scarcity, climate change, and the escalating demand for food driven by population growth [1]. By 2050, the global population is projected to reach nearly 10 billion, necessitating a 70% increase in agricultural productivity to meet demand [2]. Achieving this level of productivity sustainably is critical to avoid exacerbating environmental degradation and resource depletion. Sustainable agriculture, therefore, has emerged as a key paradigm to address these issues. It emphasizes the optimization of resource use, the reduction of environmental impact, and the promotion of ecological balance while maintaining economic viability for farmers. Sustainable agriculture is not merely about maintaining productivity but also ensuring food security, environmental health, and resilience against climate extremes [3]. Conventional farming practices, while effective in boosting yields, have often relied heavily on chemical inputs such as synthetic fertilizers and pesticides, contributing to issues like soil erosion, water pollution, and biodiversity loss. Transitioning to sustainable practices requires innovative approaches that integrate advanced technologies to optimize resource use and minimize ecological footprints. Among the emerging tools to support this transition, nanotechnology has shown tremendous potential.

1.1 Nanotechnology and Its Potential Applications in Agriculture

Nanotechnology, the manipulation and application of materials at the nanoscale (1–100 nanometers), has revolutionized several sectors, from medicine to energy. In agriculture, its application is relatively nascent but rapidly expanding, promising transformative impacts across various domains. At the nanoscale, materials exhibit unique physical, chemical, and biological properties, including increased reactivity, improved solubility, and enhanced mechanical strength [4]. As shown in Fig. 1, these attributes have enabled the development of novel solutions tailored to address the specific challenges of modern agriculture.

One of the most promising applications of nanotechnology lies in *nanofertilizers and nutrient management.* Unlike traditional fertilizers, which are often inefficiently utilized by plants, nanofertilizers offer controlled and targeted nutrient delivery. Their high surface area and reactivity enable precise release mechanisms, reducing nutrient loss due to leaching or volatilization [5]. This not only improves nutrient use efficiency but also minimizes the environmental impacts associated with excessive chemical fertilizer use, such as waterway eutrophication and soil contamination.

Similarly, nanopesticides represent a significant advancement in crop protection. Conventional pesticides often require repeated applications and result in off-target effects, harming nontarget organisms and ecosystems. Nanopesticides, by contrast, provide enhanced efficacy and specificity, ensuring that active ingredients are delivered precisely where needed [6]. This reduces overall chemical usage, mitigates resistance development in pests, and lowers risks to beneficial organisms and human health.

Nanomaterials also play a vital role in *soil health and fertility management.* They can be engineered to improve soil structure, enhance microbial activity, and facilitate the release of nutrients in a manner that promotes long-term fertility [7]. Furthermore, nanotechnology offers innovative solutions for *water management,* such as nanofiltration systems for irrigation, which improve water quality by removing contaminants while conserving resources.

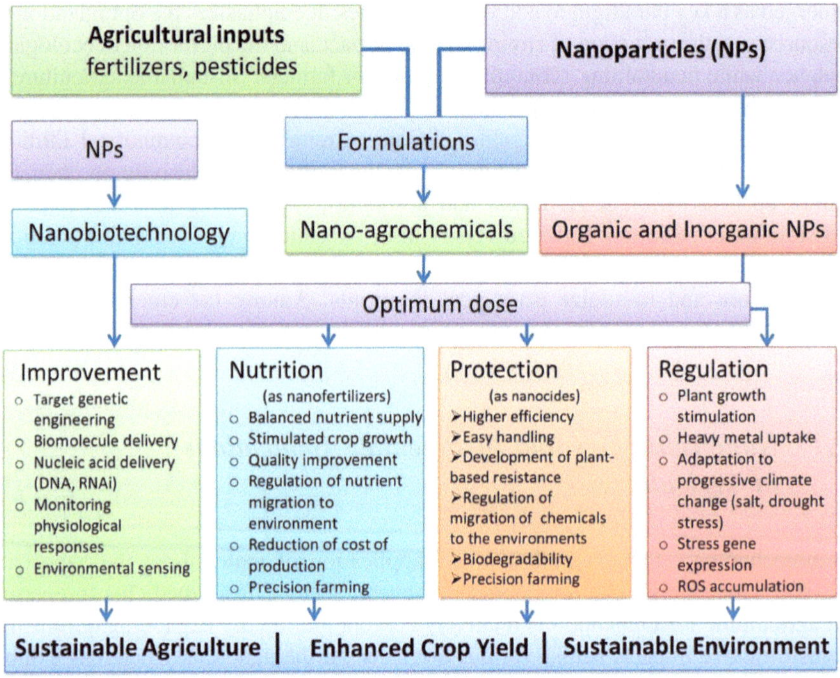

Fig. 1 Nanomaterials also play a crucial role in boosting crop yields through nanofertilizers and nanopesticides, while promoting plant growth and resilience to the challenges posed by progressive climate change. This is a reprinted image of Ref. [83] with permission (Licensed under CC BY)

Another critical application of nanotechnology is in *postharvest management.* Innovations such as *nanocoatings* and *smart packaging* are revolutionizing the preservation and transportation of agricultural produce [8]. Nanocoatings can extend the shelf life of fruits and vegetables by preventing microbial spoilage and moisture loss. Smart packaging, embedded with nanosensors, allows real-time monitoring of food quality, enabling timely interventions to reduce postharvest losses [9]. These advancements align closely with the goals of sustainable agriculture by reducing waste and enhancing resource efficiency.

The integration of *nanosensors* into agricultural systems further underscores the potential of nanotechnology. These sensors provide precise, real-time data on soil moisture, nutrient levels, and pest activity, enabling farmers to make informed decisions and optimize resource use. This "smart agriculture" approach enhances productivity while minimizing input costs and environmental impacts [9].

This chapter aims to provide a comprehensive exploration of nanotechnology's role in advancing sustainable agriculture. By examining the current state of nanotechnology applications and their potential to address critical challenges, this chapter highlights the transformative potential of nanotechnology in fostering a more sustainable and resilient agricultural system. The specific objectives of this chapter include:

1.2 Global Need for Sustainable Agriculture

This chapter underscores the pressing need to transition to sustainable agricultural practices in response to global challenges such as climate change, resource depletion, and population growth. It provides a contextual understanding of why sustainability in agriculture is not merely desirable but essential.

1.2.1 Exploring Nanotechnology's Capabilities

This section delves into the unique properties of nanomaterials and their potential applications across key areas of agriculture, including nutrient management, pest control, soil health, water conservation, and postharvest management. It emphasizes how these technologies differ from conventional solutions and their contributions to achieving sustainability goals.

1.2.2 Evaluating Environmental and Economic Benefits

This chapter examines how nanotechnology reduces dependency on traditional chemical inputs, thereby minimizing environmental impacts. It also explores the cost-effectiveness of nano-based solutions, addressing concerns related to affordability and accessibility for farmers, particularly in developing regions.

1.2.3 Discussing Safety, Regulation, and Public Perception

While nanotechnology holds immense promise, its adoption is not without challenges. This section discusses the regulatory frameworks required to ensure the safe use of nanomaterials in agriculture and addresses concerns related to potential environmental and human health risks. The importance of public awareness and acceptance of these technologies is also emphasized.

1.2.4 Identifying Future Directions and Research Needs

This chapter concludes by outlining areas where further research and development are needed to maximize the benefits of nanotechnology in agriculture. This includes improving the scalability and affordability of nano-based solutions, addressing regulatory gaps, and exploring innovative applications to meet emerging challenges.

By addressing these objectives, the chapter aims to provide a holistic understanding of how nanotechnology can revolutionize agriculture, making it more efficient, sustainable, and resilient. The scope of this chapter is broad, encompassing fundamental concepts, technological advancements, and practical implications. It is intended to serve as a valuable resource for researchers, policymakers, agricultural professionals, and anyone interested in the intersection of nanotechnology and sustainable farming practices.

The introduction of nanotechnology into agriculture marks a paradigm shift in how we approach food production and resource management. Its ability to address critical challenges while aligning with sustainability goals positions it as a key enabler of the agricultural systems of the future. By fostering innovation and collaboration, nanotechnology has the potential to reshape the agricultural landscape, ensuring that it meets the demands of a growing population without compromising environmental integrity. This chapter seeks to illuminate this potential, offering insights into how nanotechnology can be harnessed to create a sustainable and thriving agricultural ecosystem.

2 Role of Nanotechnology in Agriculture

Nanotechnology has emerged as a transformative tool in agriculture, providing innovative solutions to longstanding challenges. Its unique ability to manipulate materials at the nanoscale allows for unprecedented precision, efficiency, and sustainability in agricultural practices. This section delves into the critical applications of nanotechnology in agriculture, with a focus on nanofertilizers, nanopesticides, and nanomaterials for soil health.

2.1 Advantages of Nanofertilizers Over Conventional Fertilizers

Conventional fertilizers, while pivotal in boosting agricultural productivity, suffer from inefficiencies and environmental drawbacks. A significant portion of nutrients from these fertilizers is lost due to leaching, volatilization, or runoff, contributing to water pollution and greenhouse gas emissions. Nanofertilizers offer a revolutionary alternative, addressing these limitations through enhanced nutrient efficiency and reduced ecological impacts (Fig. 2) [10].

Nanofertilizers, composed of nanoparticles or nanoscale coatings, possess high surface-area-to-volume ratios and improved solubility, allowing them to deliver nutrients with precision. This targeted delivery minimizes nutrient wastage and ensures that plants receive essential nutrients in optimal quantities. Moreover, the slow-release mechanisms of nanofertilizers extend nutrient availability over longer periods, reducing the frequency of application and labor costs [10]. By enhancing nutrient use efficiency, nanofertilizers can significantly decrease the amount of fertilizers required, mitigating their environmental footprint.

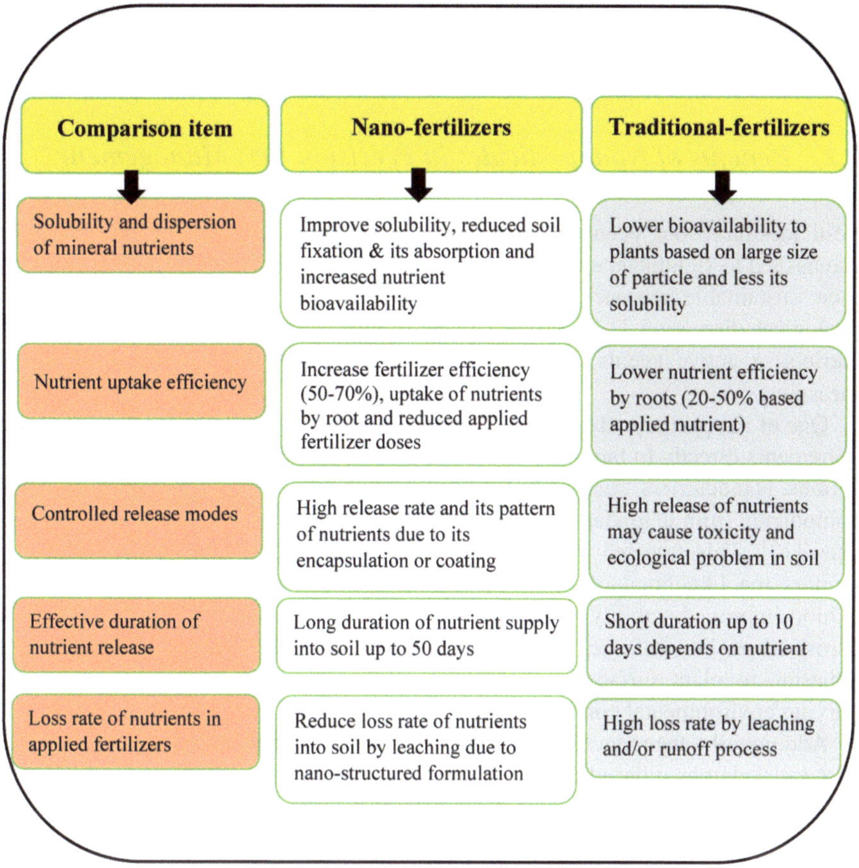

Fig. 2 A comparative analysis of traditional fertilizers and nanofertilizers. This is a reprinted image of Ref. [85] with permission (Licensed under CC BY)

2.1.1 Mechanisms of Controlled Nutrient Release and Uptake

The efficacy of nanofertilizers lies in their advanced mechanisms of nutrient delivery and uptake. Nanoparticles can be engineered to encapsulate essential nutrients within biodegradable coatings, which dissolve gradually in response to environmental triggers such as moisture, pH changes, or microbial activity. This controlled release ensures a steady supply of nutrients, preventing nutrient overloads or deficiencies common with conventional fertilizers. Additionally, the small size of nanoparticles facilitates better penetration through plant cuticles and cell walls, improving nutrient absorption at the cellular level [11]. For instance, nanostructured zinc and iron fertilizers have demonstrated superior uptake efficiency compared to their bulk counterparts, addressing micronutrient deficiencies in crops. Furthermore, some nanofertilizers are designed to interact with plant root exudates, enhancing nutrient bioavailability in the rhizosphere. These mechanisms collectively enable precise nutrient management, leading to improved crop yields and quality.

2.2 Benefits of Nanopesticides in Precision Pest Management

Pesticides have long been a cornerstone of crop protection, but their indiscriminate use has led to significant environmental and health concerns. Nanopesticides offer a more sustainable approach by providing targeted pest management with minimal ecological disruption [12]. These advanced formulations utilize nanoparticles as carriers or active agents to enhance the efficacy and precision of pest control measures.

One of the primary advantages of nanopesticides is their ability to deliver active ingredients directly to target pests, reducing the need for excessive chemical applications. Nanocarriers, such as liposomes or polymeric nanoparticles, protect active compounds from degradation and facilitate their controlled release at pest-specific sites [12]. This targeted delivery not only minimizes off-target effects but also reduces the likelihood of pesticide resistance development. Nanopesticides also exhibit improved stability and dispersion properties, ensuring uniform application across crops. For instance, silica-based nanopesticides have demonstrated superior adhesion to plant surfaces, enhancing their persistence and effectiveness under varying environmental conditions [13].

Additionally, the reduced particle size of nanopesticides allows them to penetrate pest exoskeletons more efficiently, increasing their lethality. The successful application of nanopesticides has been demonstrated in various agricultural contexts. For example, nanosilver-based pesticides have shown remarkable efficacy against fungal pathogens in crops like wheat and rice, reducing disease incidence while maintaining crop safety [14]. Similarly, chitosan nanoparticles loaded with botanical insecticides have been used to control aphid populations in vegetable crops, achieving higher mortality rates compared to conventional sprays [15]. Another notable example is the use of copper oxide nanoparticles as fungicides. These nanoformulations have been effective in managing fungal diseases such as powdery mildew and downy mildew in grapevines, demonstrating prolonged activity and reduced application rates [16]. Such case studies highlight the potential of nanopesticides to revolutionize pest management practices by combining efficacy, safety, and sustainability.

2.3 Role of Nanoparticles in Improving Soil Structure and Microbial Activity

Healthy soils are fundamental to sustainable agriculture, serving as a reservoir for nutrients, water, and beneficial microorganisms. However, intensive farming practices often degrade soil structure and deplete its nutrient reserves. Nanotechnology

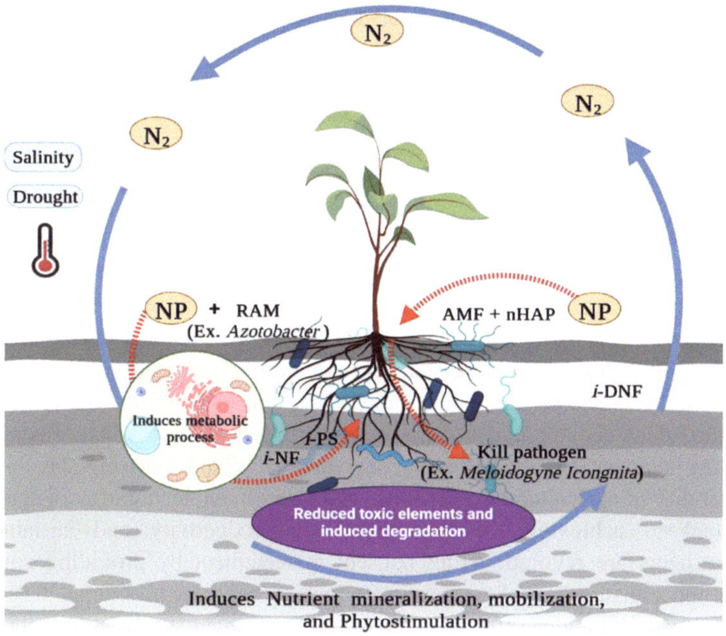

Fig. 3 Nanotechnology promotes plant growth under challenging soil conditions by enhancing the functionality of root-associated microbes, optimizing nutrient cycling, and facilitating the phytostimulation process. This is a reprinted image of Ref. [84] with permission (Licensed under CC BY)

offers innovative solutions to restore and enhance soil health through the application of nanomaterials [17].

Nanoparticles can improve soil structure by modifying its physical and chemical properties (Fig. 3). For instance, clay-based nanoparticles have been used to enhance soil aggregation and porosity, promoting better water retention and aeration [18]. These improvements create a favorable environment for root growth and nutrient uptake, contributing to increased crop productivity.

Moreover, certain nanomaterials can stimulate microbial activity in the soil, a critical factor for nutrient cycling and organic matter decomposition. For example, iron oxide nanoparticles have been shown to enhance the activity of nitrogen-fixing bacteria, improving soil fertility [19]. Similarly, silica nanoparticles can interact with soil enzymes, boosting their efficiency in breaking down organic matter and releasing nutrients [20]. These interactions underscore the potential of nanotechnology to support regenerative agricultural practices that prioritize soil health.

2.3.1 Remediation of Contaminated Soils

Soil contamination is a pressing issue in many agricultural regions, stemming from the accumulation of heavy metals, pesticides, and industrial pollutants.

Nanotechnology provides powerful tools for soil remediation, leveraging the unique properties of nanoparticles to remove or neutralize contaminants.

Nanoparticles, such as zerovalent iron and titanium dioxide, are highly effective in immobilizing heavy metals like lead, cadmium, and arsenic, preventing their uptake by plants [21]. These materials work by adsorbing or chemically transforming contaminants into less toxic forms, reducing their mobility and bioavailability. Additionally, carbon-based nanomaterials like graphene oxide have been used to absorb organic pollutants, such as pesticide residues, from contaminated soils [22]. Phytoremediation, the use of plants to clean up contaminated soils, can also be enhanced with nanotechnology. By applying nanoparticles that boost plant metabolism and stress tolerance, the efficiency of pollutant uptake and degradation can be significantly increased. For example, silver nanoparticles have been shown to enhance the phytoremediation of heavy metals by improving plant growth and antioxidant activity under stress conditions [23]. These advancements offer promising avenues for rehabilitating degraded soils and ensuring the long-term sustainability of agricultural ecosystems. The role of nanotechnology in agriculture is pivotal to achieving the twin goals of productivity and sustainability. Nanofertilizers are revolutionizing nutrient management by providing controlled release and enhanced uptake, addressing inefficiencies in traditional fertilization methods. Nanopesticides offer a targeted and environmentally friendly approach to pest management, reducing chemical usage and associated risks. Meanwhile, nanomaterials are transforming soil health practices by improving structure, boosting microbial activity, and enabling effective remediation of contaminated soils. By integrating these nanotechnological innovations into agricultural systems, we can address critical challenges such as resource depletion, environmental degradation, and food security.

As research and development in this field continue to advance, the potential for nanotechnology to reshape agriculture becomes increasingly evident, paving the way for a more sustainable and resilient future.

3 Advanced Technologies Enabled by Nanotechnology

The advent of nanotechnology has ushered in a new era of innovation in agriculture, introducing advanced technologies that address critical challenges in farming. From precision monitoring to efficient resource utilization, nanotechnology has enhanced the scope of agricultural practices. This section explores the cutting-edge advancements enabled by nanotechnology, focusing on nanosensors, water management, and postharvest management.

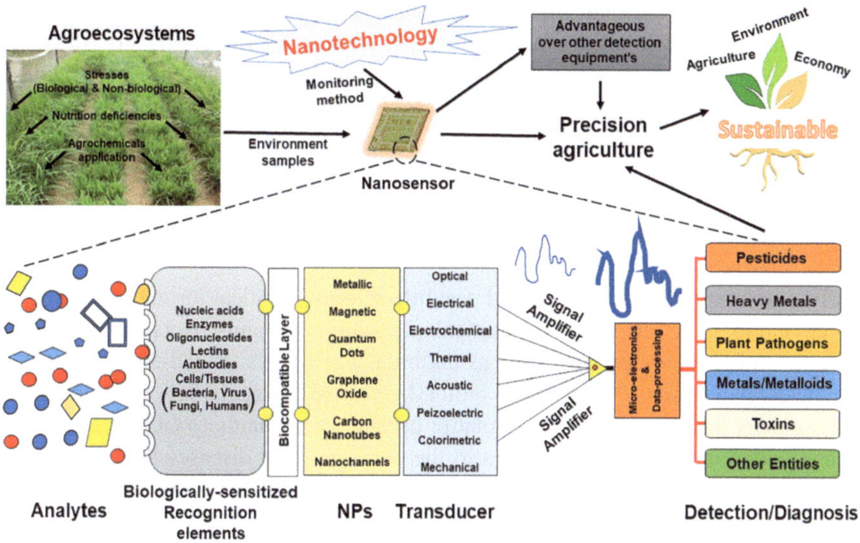

Fig. 4 A simplified diagram showcasing the components of nanosensors for monitoring agroecosystems. This is a reprinted image of Ref. [80] with permission (Licensed under CC BY)

3.1 Nanosensors for Sustainable Agriculture

Nanosensors are one of the most transformative applications of nanotechnology in agriculture, offering unparalleled precision and real-time monitoring capabilities (Fig. 4) [24]. These miniature devices, often embedded with nanomaterials like carbon nanotubes, graphene, or metal nanoparticles, are designed to detect and measure specific parameters crucial for optimizing agricultural practices.

3.1.1 Real-Time Monitoring of Soil Moisture, pH, and Nutrient Levels

Traditional methods for monitoring soil properties often involve periodic sampling and laboratory analysis, which can be time-consuming and labor-intensive. Nanosensors revolutionize this process by providing real-time, on-site data about soil conditions, enabling farmers to make timely and informed decisions [24]. Nanosensors designed for soil moisture detection are particularly useful in regions facing water scarcity. These sensors can measure minute changes in soil moisture content with exceptional accuracy, ensuring that irrigation systems are activated only when necessary.

This precision reduces water waste and prevents over-irrigation, which can lead to soil degradation. In addition to moisture monitoring, nanosensors capable of detecting soil pH and nutrient levels play a crucial role in maintaining soil health. For example, sensors embedded with metal oxide nanoparticles can measure pH changes and detect nutrient deficiencies, such as nitrogen or phosphorus, in real-time [25]. This information allows farmers to apply fertilizers more effectively, reducing nutrient loss and environmental pollution. By integrating nanosensors into smart farming systems, precision agriculture becomes a reality, optimizing resource use while minimizing costs and environmental impacts.

3.1.2 Disease and Pest Detection Systems

Early detection of plant diseases and pest infestations is critical for minimizing crop losses and reducing the need for chemical interventions. Nanosensors equipped with specific recognition elements, such as antibodies, aptamers, or molecular imprints, can detect pathogens or pest-related biomarkers at the molecular level. For instance, nanosensors designed to identify volatile organic compounds (VOCs) released by plants under stress can alert farmers to the early onset of diseases [26]. Similarly, biosensors integrated with gold or silver nanoparticles can detect fungal, bacterial, or viral pathogens by recognizing their DNA or protein signatures. These systems enable targeted interventions, reducing the widespread use of chemical pesticides and mitigating their environmental impact. Furthermore, nanosensors can be deployed in pest monitoring systems to track pest populations and predict outbreaks [27]. Such technologies empower farmers to adopt precision pest management strategies, ensuring crop protection with minimal chemical use. By enabling real-time monitoring and early detection, nanosensors significantly enhance the resilience and productivity of agricultural systems.

3.2 Nanotechnology in Water Management

Water is a vital resource in agriculture, but its availability is increasingly threatened by climate change, population growth, and over-extraction. Nanotechnology offers innovative solutions for water management, ensuring efficient use and sustainability of this precious resource.

Nanofiltration and Water Purification Systems for Irrigation Nanofiltration systems have revolutionized water purification for agricultural irrigation [28]. These systems utilize membranes embedded with nanomaterials, such as carbon nanotubes, graphene oxide, or titanium dioxide, to filter contaminants from water [29]. Nanofiltration membranes are highly effective in removing salts, heavy metals, organic pollutants, and microorganisms, making even saline or contaminated water suitable for irrigation.

One of the key advantages of nanofiltration systems is their high selectivity and efficiency. Unlike conventional filtration methods, nanofiltration requires less energy and can operate at lower pressures, reducing operational costs. Additionally, these systems can be tailored to target specific contaminants, ensuring that irrigation water meets the quality requirements for various crops. In arid and semiarid regions, where freshwater resources are scarce, nanofiltration systems can be integrated with desalination technologies to convert seawater into usable irrigation water [30]. This advancement not only addresses water scarcity but also supports sustainable agricultural practices in challenging environments.

Nanotechnology also contributes to water efficiency in agriculture through the development of water-retentive nanomaterials. For example, hydrogel nanoparticles can be incorporated into soils to enhance water retention capacity. These hydrogels absorb and store large quantities of water, releasing it gradually as needed by plants. This reduces the frequency of irrigation and ensures a steady water supply during dry periods. Additionally, nanotechnology-enabled smart irrigation systems integrate nanosensors to monitor soil moisture and weather conditions in real time [31]. These systems automate irrigation schedules based on precise data, preventing water wastage and optimizing crop growth. By coupling nanosensors with nanofiltration systems, farmers can achieve a holistic approach to water management, addressing both quality and efficiency.

3.3 Nanotechnology for Postharvest Management

Postharvest losses account for a significant proportion of global food waste, particularly in developing regions. Nanotechnology offers innovative solutions to reduce these losses and improve the shelf life and quality of agricultural produce.

3.3.1 Nanocoatings for Food Preservation

Nanocoatings are thin, nanomaterial-based layers applied to the surface of fruits, vegetables, and other perishable items to enhance their shelf life (Fig. 5). These coatings create a protective barrier that prevents moisture loss, microbial spoilage, and oxidation, thereby preserving the freshness and nutritional value of produce. Edible nanocoatings, often made from biocompatible materials such as chitosan, cellulose nanocrystals, or silk fibroin, have gained popularity for their safety and effectiveness [32]. These coatings can be infused with antimicrobial nanoparticles, such as silver or zinc oxide, to inhibit the growth of pathogens on produce surfaces. Additionally, nanocoatings with oxygen-scavenging properties can delay ripening and reduce spoilage caused by oxidation [32]. The use of nanocoatings has been particularly beneficial for high-value crops like berries, avocados, and apples, where even minor losses can result in significant economic impacts. By extending shelf life and reducing spoilage, nanocoatings contribute to food security and minimize waste along the supply chain.

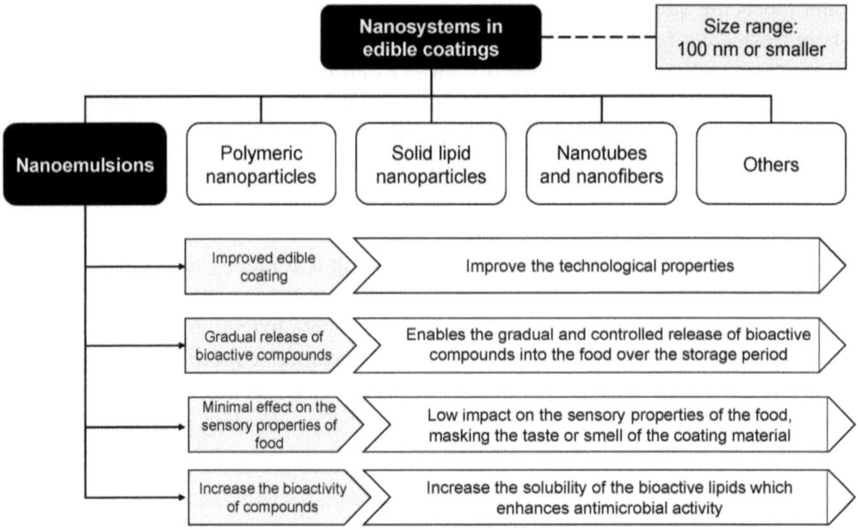

Fig. 5 Advancements in the development of nanosystems integrated with food-grade ingredients have enabled the exploration of functional modifications in food coating materials. These include nanoemulsions, polymeric nanoparticles, nanostructured lipid carriers, nanotubes, nanocrystals, nanofibers, and more. This is a reprinted image of Ref. [86] with permission (Licensed under CC BY)

3.3.2 Smart Packaging for Reducing Postharvest Losses

Smart packaging, enabled by nanotechnology, represents a breakthrough in postharvest management. These packaging systems are designed to monitor and maintain the quality of agricultural produce during storage and transportation, reducing losses and ensuring consumer satisfaction. Nanosensors integrated into packaging materials can detect changes in temperature, humidity, gas composition, or microbial activity, providing real-time information about the condition of the packaged produce [33]. For example, colorimetric nanosensors can change color to indicate spoilage or contamination, alerting stakeholders to take corrective actions. Such systems enhance transparency and traceability in the supply chain, improving food safety and reducing waste. In addition to monitoring, smart packaging can actively regulate the storage environment to extend shelf life. For instance, nanomaterials with moisture-absorbing or ethylene-scavenging properties can maintain optimal conditions for fresh produce. Active packaging solutions, such as antimicrobial films embedded with silver or copper nanoparticles, can further protect against microbial spoilage [34]. The cost-effectiveness and scalability of smart packaging technologies have made them increasingly accessible to farmers and food distributors. By reducing postharvest losses and improving the quality of produce, these innovations contribute to economic sustainability and consumer satisfaction.

Advanced technologies enabled by nanotechnology are transforming agriculture, addressing critical challenges related to resource efficiency, crop protection, and food preservation. Nanosensors offer real-time monitoring capabilities that empower farmers with data-driven insights, enhancing precision and productivity [24]. Nanotechnology-driven water management solutions, including nanofiltration systems and water-retentive materials, ensure sustainable use of water resources in agriculture. Postharvest management innovations, such as nanocoatings and smart packaging, reduce food waste and improve supply chain efficiency.

These advancements not only enhance the sustainability of agricultural practices but also contribute to global food security by optimizing resource use and minimizing losses. As research and development in nanotechnology continue to progress, the integration of these technologies into mainstream agriculture holds the promise of a more resilient, efficient, and sustainable future [35].

4 Environmental and Economic Impacts of Nanotechnology in Agriculture

Nanotechnology has emerged as a game-changer in agriculture, providing innovative solutions to long-standing challenges while ensuring sustainability. The integration of nano-based solutions not only revolutionizes agricultural practices but also has profound environmental and economic implications [36]. This section explores the dual benefits of reducing chemical usage and enhancing cost-effectiveness through nanotechnology in agriculture.

4.1 Reduction in Chemical Usage and Its Environmental Benefits

The overuse of chemical fertilizers, pesticides, and herbicides has been a cornerstone of conventional agriculture. While these chemicals have contributed significantly to increasing crop yields, they have also created a plethora of environmental problems, including soil degradation, water pollution, loss of biodiversity, and greenhouse gas emissions. Nanotechnology offers an opportunity to mitigate these issues by reducing chemical inputs without compromising agricultural productivity [37].

4.2 Minimized Fertilizer Runoff and Pollution

Traditional fertilizers often exhibit low nutrient use efficiency, with a significant proportion of nutrients lost to the environment through leaching, volatilization, or runoff. These losses not only reduce the efficacy of fertilizers but also lead to the eutrophication of water bodies, harming aquatic ecosystems. Nanofertilizers, on the other hand, are designed for controlled and targeted nutrient delivery [38]. By encapsulating nutrients in nanomaterials or creating slow-release formulations, nanofertilizers ensure that plants absorb a higher percentage of applied nutrients. This reduces the quantity of fertilizer required and minimizes nutrient losses, thereby preventing waterway contamination and promoting sustainable farming practices. For example, studies have shown that nanofertilizers containing zinc, nitrogen, or phosphorus can enhance nutrient uptake by up to 40% compared to conventional fertilizers [39]. This improvement not only boosts crop yields but also significantly reduces the environmental footprint of agricultural practices.

4.3 Targeted Pest and Disease Management

The indiscriminate use of chemical pesticides and herbicides has led to the contamination of soils, groundwater, and nontarget organisms. Additionally, pests and weeds have developed resistance to many conventional chemicals, necessitating higher application rates and further exacerbating environmental damage. Nanopesticides provide a sustainable alternative by enabling precision pest and disease management. Nanoparticles can be engineered to deliver active ingredients directly to target pests or pathogens, reducing off-target effects and the overall quantity of chemicals required. For instance, nanoencapsulated pesticides can penetrate pest exoskeletons more effectively, enhancing their efficacy at lower doses [40]. Similarly, silver and copper nanoparticles have demonstrated strong antimicrobial properties, making them effective against fungal and bacterial infections in crops. By reducing chemical usage and improving targeting, nanopesticides contribute to healthier ecosystems and lower environmental contamination (Fig. 6).

4.4 Enhanced Soil Health and Biodiversity

Excessive chemical use in agriculture has degraded soil health, disrupting its natural structure and microbial communities. Nanotechnology addresses this challenge by promoting practices that enhance soil quality. Nanomaterials can improve nutrient cycling, soil aeration, and microbial activity, all of which contribute to soil fertility [41]. Furthermore, the reduced reliance on conventional chemicals ensures that beneficial soil organisms, such as nitrogen-fixing bacteria and earthworms, thrive, maintaining the ecological balance of agricultural soils.

The cumulative environmental benefits of reduced chemical usage extend beyond individual farms. By mitigating pollution, preserving biodiversity, and enhancing soil health, nanotechnology supports the global transition toward regenerative and sustainable agricultural systems.

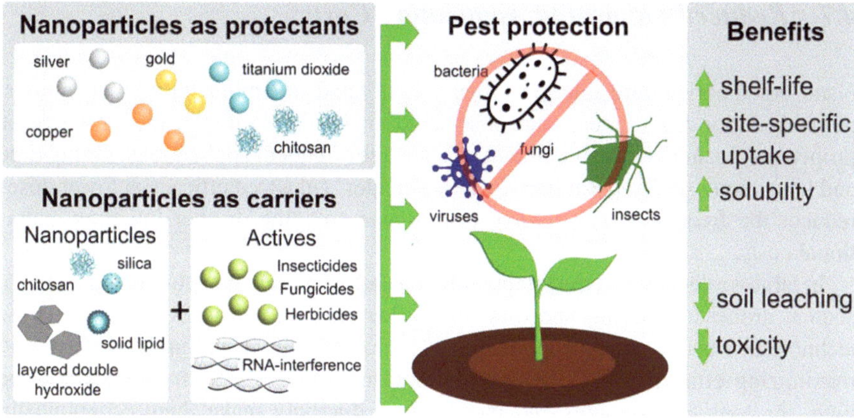

Fig. 6 Nanomaterials as protectants or carriers for crop protection: This schematic illustrates nanomaterials delivering actives like insecticides, fungicides, herbicides, or RNAi molecules, targeting pests and pathogens. Benefits include enhanced shelf life, targeted uptake, increased solubility, and reduced soil leaching and toxicity. This is a reprinted image of Ref. [87] with permission (Licensed under CC BY)

4.5 Cost-Effectiveness of Nano-Based Agricultural Solutions

The economic viability of agricultural technologies is a critical factor influencing their adoption by farmers. Nano-based solutions, despite their advanced nature, have demonstrated remarkable cost-effectiveness, making them accessible and appealing to a broad spectrum of stakeholders [42].

4.6 Improved Efficiency and Resource Optimization

One of the primary economic advantages of nanotechnology lies in its ability to optimize resource use. Nanofertilizers and nanopesticides, with their controlled-release mechanisms and targeted delivery systems, ensure that inputs are utilized more efficiently. This reduces the quantity of fertilizers and pesticides needed, translating to cost savings for farmers. For instance, the use of nanofertilizers has been reported to lower input costs by up to 30% while maintaining or enhancing crop yields. Additionally, nanosensors play a pivotal role in resource optimization [36]. By providing real-time data on soil conditions, nutrient levels, and pest activity, these sensors enable precision agriculture practices. Farmers can make informed decisions about irrigation, fertilization, and pest control, avoiding unnecessary expenditures and maximizing returns on investment. The cost savings associated with such precision techniques make nanotechnology an economically viable option, particularly for resource-constrained farmers.

4.7 Reduced Labor and Application Costs

Nano-based agricultural solutions often simplify and streamline farming operations, reducing labor requirements and associated costs. For example, the slow-release properties of nanofertilizers eliminate the need for frequent applications, saving time and labor expenses [43]. Similarly, nanopesticides' enhanced efficacy at lower doses reduces the frequency and volume of pesticide spraying, further lowering operational costs.

In large-scale farming, the integration of nanotechnology into automated systems, such as drones and robotic sprayers, has further enhanced cost-effectiveness. These technologies ensure uniform application of nanoformulations, minimizing waste and maximizing efficiency [44]. While the initial investment in such systems may be high, the long-term savings and increased productivity make them economically attractive.

4.8 Postharvest Economic Benefits

Postharvest losses represent a significant economic burden on the agricultural sector, particularly in developing countries. Nanotechnology offers solutions to reduce these losses, enhancing the profitability of farming operations. Nanocoatings and smart packaging technologies play a crucial role in preserving the quality and shelf life of agricultural produce, reducing spoilage during storage and transportation [8]. For instance, edible nanocoatings can protect perishable fruits and vegetables from microbial contamination and moisture loss, extending their marketability. Smart packaging, embedded with nanosensors, provides real-time information about the condition of stored produce, enabling timely interventions to prevent spoilage [45]. These innovations not only reduce economic losses but also improve supply chain efficiency, benefiting farmers, distributors, and consumers alike.

4.9 Scalability and Accessibility

The scalability of nano-based solutions is a key factor in their economic impact. Advances in nanotechnology have made it possible to produce nanomaterials at lower costs, ensuring that these innovations are accessible to smallholder farmers as well as large-scale agricultural enterprises [46]. Governments and agricultural organizations have also recognized the potential of nanotechnology, providing subsidies and incentives to promote its adoption. These measures further enhance the affordability and accessibility of nano-based solutions, ensuring their widespread implementation.

4.10 Market Potential and Value Addition

Nanotechnology's ability to improve crop yields, quality, and sustainability has significant implications for market potential and value addition. For instance, crops grown using nanofertilizers and nanopesticides often exhibit superior nutritional content, pest resistance, and esthetic appeal, commanding higher prices in domestic and international markets. Similarly, produce preserved using nanocoatings or smart packaging maintains its quality over extended periods, reducing waste and increasing profitability. The growing consumer demand for sustainably produced food further enhances the market potential of nano-enabled agricultural products [47]. By reducing chemical residues and environmental impact, nanotechnology aligns with global trends favoring eco-friendly and health-conscious food choices. This creates new opportunities for farmers and agribusinesses to differentiate their products and capture premium market segments.

4.11 Environmental-Economic Synergy

The environmental and economic benefits of nanotechnology in agriculture are deeply interconnected, creating a synergistic effect that amplifies its impact. By reducing chemical usage, nanotechnology not only mitigates environmental damage but also lowers input costs for farmers. Improved soil health and water efficiency enhance long-term agricultural productivity, ensuring economic stability while conserving natural resources. Moreover, the cost savings achieved through nano-based solutions enable farmers to reinvest in sustainable practices, such as organic farming or regenerative techniques. This creates a positive feedback loop where economic gains drive environmental stewardship, fostering a resilient and sustainable agricultural sector.

5 Challenges of Nanotechnology for Sustainable Agriculture

Despite its immense potential, the adoption of nanotechnology in agriculture faces challenges that must be addressed to maximize its environmental and economic benefits (Fig. 7).

Regulatory frameworks for nano-based products are still evolving, and concerns about potential risks to human health and the environment need to be thoroughly addressed [48]. Additionally, the initial costs of nanotechnology-based systems may deter adoption among resource-constrained farmers, highlighting the need for subsidies, training programs, and public–private partnerships.

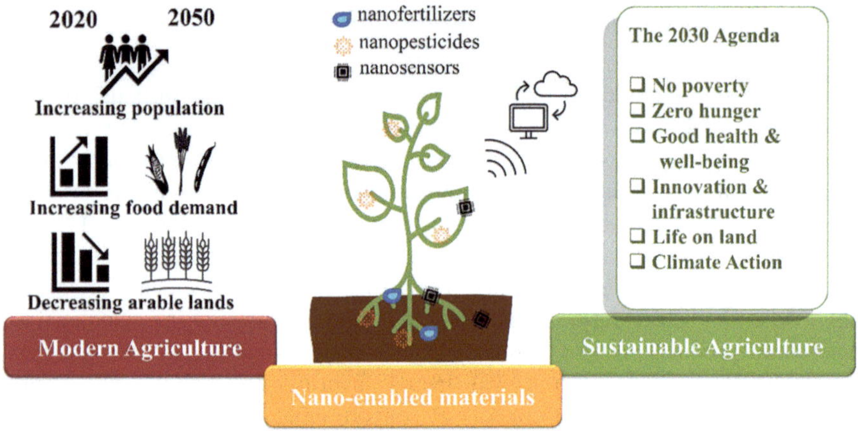

Fig. 7 Challenges in adopting nanotechnology in agriculture must be addressed. This is a reprinted image of Ref. [88] with permission (Licensed under CC BY)

Future research should focus on improving the scalability, affordability, and safety of nano-based solutions. Innovations such as biodegradable nanomaterials, multifunctional nanosensors, and integrated nano-agricultural systems hold promise for further enhancing the environmental and economic impact of nanotechnology [49]. Collaborative efforts among researchers, policymakers, and industry stakeholders will be essential to realizing the full potential of this transformative technology.

Nanotechnology has redefined the landscape of agriculture, offering solutions that address both environmental and economic challenges. By reducing chemical usage, nano-based agricultural technologies promote sustainability and minimize ecological harm. Their cost-effectiveness, driven by resource optimization, labor savings, and value addition, makes them accessible and appealing to farmers worldwide [50].

As the global demand for sustainable food production grows, the integration of nanotechnology into agricultural systems will play an increasingly vital role. By fostering environmental-economic synergy, nanotechnology not only enhances agricultural productivity but also ensures the long-term viability of farming practices [51]. With continued innovation and collaboration, nanotechnology has the potential to transform agriculture into a more sustainable, resilient, and prosperous sector.

6 Limitations of Nanotechnology for Sustainable Agriculture

The integration of nanotechnology into agriculture has transformed the sector by enhancing productivity, resource efficiency, and sustainability. Despite its numerous advantages, nanotechnology is not without challenges and limitations. From potential toxicity and environmental risks to scalability and affordability for small-holder farmers, as well as regulatory and ethical concerns, these barriers must be addressed to ensure the responsible and equitable implementation of nanotechnology in sustainable agriculture [52]. This section explores these challenges in depth.

6.1 *Potential Toxicity and Environmental Risks of Nanomaterials*

Nanomaterials, due to their unique properties, can interact with biological and environmental systems in ways that are not yet fully understood. While these interactions offer advantages for agricultural applications, they also raise concerns about potential toxicity and long-term environmental impacts. One of the most pressing concerns is the potential toxicity of nanomaterials to nontarget organisms [53]. For instance, nanoparticles used in fertilizers, pesticides, or soil conditioners can leach into surrounding ecosystems, affecting beneficial microorganisms, insects, and other wildlife. Studies have shown that certain nanoparticles, such as silver and zinc oxide, can be toxic to soil microbes and aquatic organisms, disrupting natural ecosystems and nutrient cycles.

The small size and high reactivity of nanoparticles allow them to penetrate biological membranes, potentially causing oxidative stress, cellular damage, and inflammation in plant and many other organisms (Fig. 8) [53]. For example, exposure to certain metal-based nanoparticles has been linked to adverse effects in

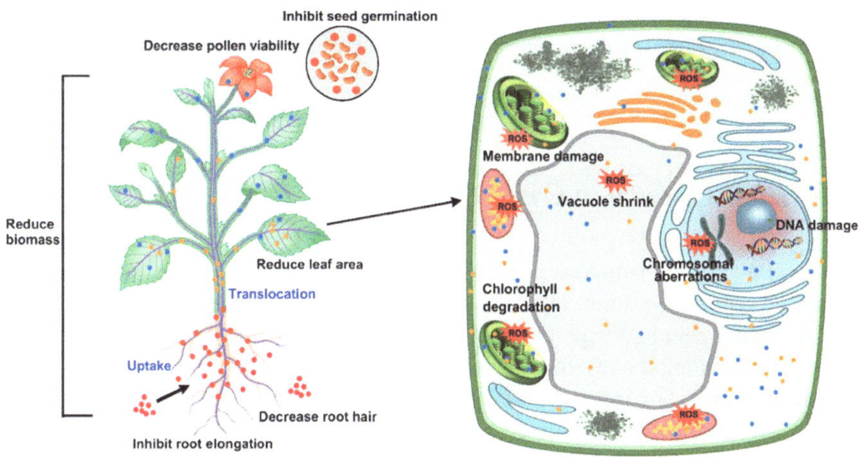

Fig. 8 Uptake and effects of nanomaterials: Nanomaterials absorbed by underground tissues are translocated to aboveground parts, impacting biomass, leaf area, pollen viability, and seed germination. At the cellular level, they induce excess ROS production, causing cytotoxicity and genotoxicity, including membrane damage, chlorophyll degradation, vacuole shrinkage, DNA damage, and chromosomal aberrations. This is a reprinted image of Ref. [89] with permission (Licensed under CC BY)

fish and amphibians, raising concerns about their accumulation in aquatic environments.

6.2 *Bioaccumulation and Persistence*

Another critical issue is the potential for bioaccumulation and persistence of nanomaterials in the environment [54]. Unlike traditional agrochemicals, nanoparticles can accumulate in soil, water, and living organisms over time. This persistence may lead to unintended consequences, such as the transfer of nanoparticles through food chains or the development of resistance in pests and pathogens.

Research has indicated that some nanoparticles, particularly those that are not biodegradable, can persist in soils for extended periods, altering soil chemistry and affecting plant growth [54]. For example, studies on carbon-based nanomaterials, such as graphene oxide, have highlighted their potential to impact soil structure and microbial communities, with long-term implications for soil health.

6.3 *Lack of Comprehensive Risk Assessments*

The novelty of nanotechnology means that comprehensive risk assessments are still lacking. Existing studies often focus on short-term impacts, with a limited understanding of the long-term behavior and interactions of nanomaterials in complex environmental systems [55]. Developing standardized testing protocols and models to evaluate the safety of nanomaterials across diverse agricultural settings is essential to address these gaps.

6.4 *High Production Costs*

The production of nanomaterials often involves advanced techniques, such as chemical vapor deposition, laser ablation, or sol-gel synthesis, which require specialized equipment and expertise. These methods are not only energy-intensive but also costly, making the resulting nanoproducts expensive compared to conventional alternatives [56]. For instance, the cost of nanofertilizers or nanopesticides can be prohibitive for smallholder farmers, who already operate on tight margins.

6.5 *Infrastructure and Resource Limitations*

The adoption of nanotechnology also depends on the availability of infrastructure and resources. Smallholder farmers in remote or resource-constrained areas may lack access to the necessary equipment, training, or distribution networks to implement nano-based solutions effectively [57]. For example, the use of nanosensors for precision agriculture requires reliable power sources, Internet connectivity, and technical expertise, which may not be readily available in rural farming communities.

6.6 *Economic Viability and Market Barriers*

The economic viability of nanotechnology-based agricultural solutions is another concern. While these technologies promise long-term cost savings through improved efficiency, the initial investment required for their adoption can deter smallholder farmers [58]. Additionally, market barriers, such as limited awareness, fragmented supply chains, and competition from cheaper conventional products, further hinder the widespread adoption of nanotechnology in agriculture.

6.7 *Strategies for Addressing Scalability and Affordability*

To overcome these challenges, efforts must focus on developing cost-effective and scalable production methods for nanomaterials. For instance, bioinspired or green synthesis techniques, which use plant extracts or microorganisms to produce nanoparticles, offer a promising alternative to traditional methods. These techniques are not only environmentally friendly but also more affordable, making nano-based solutions accessible to a broader audience [59].

Public–private partnerships and government subsidies can also play a crucial role in promoting the adoption of nanotechnology among smallholder farmers. By providing financial incentives, training programs, and infrastructure support, these initiatives can bridge the gap between advanced technologies and resource-constrained farming communities [59].

6.8 *Regulatory Gaps and Inconsistencies*

One of the major challenges is the lack of harmonized regulatory standards for nanotechnology. Different countries have varying definitions, testing protocols, and approval processes for nanomaterials, creating uncertainty for manufacturers and users. For instance, while some regions classify nanomaterials based on particle size alone, others consider additional factors such as chemical composition and reactivity. This regulatory inconsistency complicates the commercialization and global trade of nano-based agricultural products [60]. It also poses challenges for ensuring product safety, as the absence of standardized testing protocols makes it difficult to evaluate and compare the risks associated with different nanomaterials.

6.9 Ethical Concerns Related to Safety and Equity

The potential risks of nanotechnology raise ethical concerns about its impact on human health and the environment. For example, the aerosolization of nanopesticides during foliar application can expose farmworkers and nearby communities to inhalation risks. Similarly, the presence of nanoparticles in food products may raise questions about consumer safety and transparency.

Ethical considerations also extend to issues of equity and access. The high costs and technical requirements of nanotechnology may exacerbate existing inequalities in agriculture, favoring large-scale commercial farms over smallholder farmers [61]. Ensuring that the benefits of nanotechnology are distributed equitably across all segments of the farming community is a critical ethical challenge.

Developing robust regulatory frameworks that address the unique properties and risks of nanomaterials is a priority. These frameworks should include standardized definitions, safety testing protocols, and labeling requirements to ensure transparency and accountability. International collaboration among regulatory agencies can help harmonize standards and facilitate the safe trade and use of nano-based agricultural products.

Ethical guidelines should focus on minimizing risks to human health and the environment, promoting equitable access, and ensuring transparency in product development and commercialization [62]. Stakeholder engagement, including input from farmers, consumers, scientists, and policymakers, is crucial for shaping these guidelines and addressing diverse perspectives. While nanotechnology holds immense promise for advancing sustainable agriculture, its adoption is not without challenges and limitations. Potential toxicity and environmental risks, scalability and affordability for smallholder farmers, and regulatory and ethical considerations represent significant barriers that must be addressed to maximize the benefits of this transformative technology [62].

By investing in research to understand the long-term impacts of nanomaterials, developing cost-effective and scalable solutions, and establishing robust regulatory and ethical frameworks, stakeholders can ensure the responsible and equitable implementation of nanotechnology in agriculture. Addressing these challenges will not only enhance the sustainability and resilience of agricultural systems but also contribute to global food security and environmental conservation [61]. As the field continues to evolve, collaboration among researchers, policymakers, and industry stakeholders will be essential to overcoming these barriers and unlocking the full potential of nanotechnology in sustainable agriculture.

7 Future Directions and Research Opportunities

As nanotechnology continues to revolutionize agriculture, the potential for further innovation is vast. Future advancements in nano-agriculture will likely stem from integrating nanotechnology with other cutting-edge technologies, such as artificial intelligence (AI), the Internet of Things (IoT), and robotics (Fig. 9) [63]. Additionally, the role of nanotechnology in developing climate-resilient agricultural systems will be pivotal in addressing the growing challenges posed by climate change. This section explores these emerging trends and highlights promising research opportunities.

7.1 Potential for Integrating Nanotechnology with Other Advanced Technologies

The convergence of nanotechnology with AI, IoT, and robotics represents a transformative frontier for agriculture. These integrations enable smarter, more efficient, and highly adaptive farming systems that can address both current and future agricultural challenges [64].

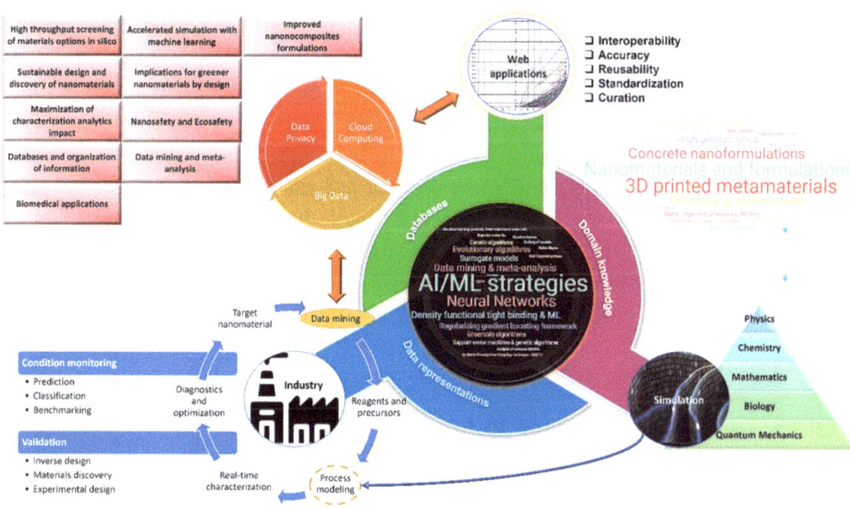

Fig. 9 AI-driven predictive modeling tools enable innovative materials discovery and optimization, facilitating data-driven design of nanomaterials tailored to specific application needs. This is a reprinted image of Ref. [90] with permission (Licensed under CC BY)

7.2 Nanotechnology and Artificial Intelligence

Artificial intelligence has become a cornerstone of precision agriculture, facilitating data-driven decision-making through the analysis of complex datasets. When combined with nanotechnology, AI can unlock new possibilities for optimizing agricultural practices. For instance, nanosensors embedded in soils or crops can generate vast amounts of real-time data on parameters such as moisture, pH, nutrient levels, and pest activity [65].

AI algorithms can analyze this data to provide actionable insights, enabling farmers to make informed decisions about irrigation, fertilization, and pest management. Moreover, AI-powered predictive models can assess the long-term impacts of using nanomaterials, ensuring their safe and sustainable application [65]. By simulating various scenarios, these models can help researchers and policymakers anticipate potential risks and devise strategies to mitigate them. This integration enhances the efficiency of nanotechnology while addressing safety and regulatory concerns.

7.3 Nanotechnology and the Internet of Things (IoT)

The IoT has emerged as a critical enabler of smart farming, connecting devices and systems to create integrated agricultural networks [66]. The incorporation of nanotechnology into IoT frameworks enhances their functionality and precision. For example, nanosensors can be integrated into IoT systems to monitor environmental conditions, crop health, and soil quality in real time. These sensors transmit data to centralized platforms, allowing farmers to remotely monitor and manage their fields [67]. IoT-enabled nanodevices can also be used in automated irrigation systems to optimize water usage. By analyzing data from nanosensors, these systems can deliver water precisely where and when it is needed, minimizing waste and conserving resources. Furthermore, IoT networks can facilitate the remote control of nano-enabled agricultural machinery, such as drones equipped with nanopesticide sprayers or robots applying nanofertilizers [67]. This connectivity streamlines operations, reduces labor costs, and enhances productivity.

7.4 Nanotechnology and Robotics

The integration of nanotechnology with robotics holds significant promise for automating agricultural tasks and improving precision. Robots equipped with nanosensors and nanoactuators can perform a wide range of activities, from planting and harvesting to pest control and soil analysis [68]. These robots can operate with exceptional accuracy, ensuring optimal resource use and minimizing environmental impact. For instance, nano-enabled drones can detect crop stress or pest infestations using nanosensors and apply targeted treatments, such as nanopesticides, with pinpoint precision. Similarly, autonomous robots equipped with nanotechnology can analyze soil samples in real-time, adjusting fertilizer application rates to match crop requirements [68]. These advancements not only increase efficiency but also reduce the environmental footprint of agricultural practices.

7.5 Role in Climate-Resilient Agriculture

Climate change poses a significant threat to global agriculture, with rising temperatures, unpredictable weather patterns, and increased incidences of extreme events jeopardizing food security [69]. Nanotechnology has the potential to play a transformative role in developing climate-resilient agricultural systems that can withstand these challenges.

7.6 Enhancing Crop Tolerance to Abiotic Stresses

Abiotic stresses, such as drought, heat, salinity, and nutrient deficiencies, are major consequences of climate change. Nanotechnology can enhance crop resilience to these stresses through various mechanisms [70]. For example, nanofertilizers can deliver essential nutrients more efficiently, improving plant health and stress tolerance. Nanoparticles that stimulate antioxidant activity in plants can help mitigate the damaging effects of heat and drought stress. Additionally, nanoscale biostimulants can be developed to trigger specific physiological responses in plants, enhancing their ability to adapt to adverse conditions. For instance, silica nanoparticles have been shown to improve water-use efficiency and increase drought tolerance in crops like rice and wheat [71]. Such innovations are critical for maintaining agricultural productivity in the face of climate variability. Healthy soils are the foundation of climate-resilient agriculture, and nanotechnology can play a pivotal role in enhancing soil health. Nanomaterials can improve soil structure, water retention, and nutrient availability, enabling soils to better support crop growth under changing environmental conditions. For example, hydrogel nanoparticles can retain water and release it gradually, ensuring a steady supply to plants during dry periods. Nanotechnology can also aid in the remediation of degraded or contaminated soils, restoring their fertility and productivity [72]. Carbon-based nanomaterials, such as graphene oxide, have been used to remove heavy metals and organic pollutants from soils, reducing their harmful effects on crops and ecosystems [73]. By promoting soil resilience, nanotechnology contributes to the sustainability of agricultural systems in a changing climate.

7.7 Carbon Sequestration and Greenhouse Gas Mitigation

Agriculture is a significant contributor to greenhouse gas emissions, particularly methane and nitrous oxide from livestock and fertilizer use. Nanotechnology offers innovative solutions to mitigate these emissions and reduce the carbon footprint of farming practices. For example, nanocoatings on fertilizers can control the release of nitrogen, minimizing nitrous oxide emissions from soils [74]. Nanomaterials can also be used to enhance carbon sequestration in agricultural soils. Biochar nanoparticles, for instance, can be incorporated into soils to increase their organic carbon content and improve their capacity to store carbon [75]. Such approaches not only mitigate greenhouse gas emissions but also contribute to climate change adaptation by enhancing soil fertility and water retention.

7.8 Enabling Early Warning Systems

Climate-resilient agriculture requires proactive measures to anticipate and respond to climate-related risks. Nanotechnology can facilitate the development of early warning systems that provide timely information about impending threats, such as droughts, floods, or pest outbreaks [76]. Nanosensors capable of detecting subtle changes in environmental conditions can alert farmers to take preventive actions, reducing losses and ensuring food security. For example, nanosensors embedded in IoT networks can monitor soil moisture and weather conditions, providing real-time data to forecast droughts or floods [77]. Similarly, nanosensors that detect pest pheromones or disease markers can enable early intervention, preventing widespread infestations. By enabling precise and timely responses, these systems enhance the resilience of agricultural systems to climate variability.

7.9 Affordable and Scalable Nanotechnology Solutions

Cost-effective production methods and scalable applications of nanotechnology are essential to ensure its accessibility for farmers worldwide, particularly smallholder farmers in developing regions [78].

7.10 Biodegradable and Sustainable Nanomaterials

The development of environmentally friendly nanomaterials that degrade naturally and pose minimal risks to ecosystems is crucial for sustainable agriculture [79].

7.11 Multifunctional Nanosensors

Advancing the design of nanosensors with multifunctional capabilities can provide comprehensive monitoring solutions for soil, crop, and environmental conditions [80].

7.12 Integrated Smart Farming Systems

Research should focus on creating integrated systems that combine nanotechnology with AI, IoT, and robotics, offering seamless solutions for precision agriculture [63].

7.13 Climate Adaptation Strategies

Investigating the mechanisms through which nanotechnology can enhance crop and soil resilience to climate change will be critical for future food security [81]. The future of agriculture lies in the convergence of nanotechnology with other advanced technologies and its application in climate-resilient farming systems. By integrating AI, IoT, and robotics, nanotechnology can enable smarter, more efficient agricultural practices that address the growing challenges of food security and sustainability [82]. Furthermore, its role in enhancing crop and soil resilience to climate change underscores its importance in building adaptive agricultural systems. As research and innovation continue to expand the frontiers of nano-agriculture, the collaboration of scientists, policymakers, and industry stakeholders will be essential to realizing its full potential. With a focus on sustainability, equity, and resilience, nanotechnology can pave the way for a transformative future in global agriculture.

8 Conclusion

Nanotechnology has emerged as a cornerstone for achieving sustainable agriculture, addressing critical challenges such as resource depletion, environmental degradation, and food security. By leveraging the precision of nanofertilizers, farmers can significantly enhance crop yields while minimizing nutrient losses, reducing the environmental burden associated with traditional fertilizers. Similarly, nanopesticides offer targeted pest management, effectively lowering chemical usage and protecting ecosystems from contamination. The integration of nanomaterials into agricultural practices has profound implications for soil health, enhancing fertility and supporting microbial activity. Advanced nanosensors enable real-time monitoring of crucial parameters like soil moisture, nutrient levels, and crop health, allowing farmers to make data-driven decisions that optimize resource use and increase efficiency. Nanotechnology also plays a pivotal role in water conservation through nanofiltration and delivery systems, ensuring efficient irrigation and reduced water wastage. Postharvest innovations, such as nanocoatings and smart packaging, extend the shelf life of agricultural produce, reducing food waste and enhancing marketability. By minimizing dependence on conventional agrochemicals and introducing cost-effective solutions, nanotechnology promotes environmentally friendly practices and ensures accessibility for farmers worldwide. As the agricultural sector grapples with growing challenges, nanotechnology offers a transformative pathway to build resilient, productive, and sustainable systems that will secure global food security for future generations.

References

1. Gupta, G. S. (2019). Land degradation and challenges of food security. *Review of European Studies, 11*, 63.
2. Lal, R. (2016). Feeding 11 billion on 0.5 billion hectare of area under cereal crops. *Food and Energy Security, 5*(4), 239–251.
3. Singh, R., & Singh, G. S. (2017). Traditional agriculture: A climate-smart approach for sustainable food production. *Energy, Ecology and Environment, 2*, 296–316.
4. Dubey, R., Dutta, D., Sarkar, A., & Chattopadhyay, P. (2021). Functionalized carbon nanotubes: Synthesis, properties and applications in water purification, drug delivery, and material and biomedical sciences. *Nanoscale Advances, 3*(20), 5722–5744.
5. Sim, D. H. H., Tan, I. A. W., Lim, L. L. P., & Hameed, B. H. (2021). Encapsulated biochar-based sustained release fertilizer for precision agriculture: A review. *Journal of Cleaner Production, 303*, 127018.
6. Anandhi, S., Saminathan, V. R., Yasotha, P., Saravanan, P. T., & Rajanbabu, V. (2020). Nanopesticides in pest management. *Journal of Entomology and Zoology Studies, 8*(4), 685–690.
7. Hasan, M. K., Shopan, J., & Ahammed, G. J. (2020). Nanomaterials and soil health for agricultural crop production: current status and future prospects. *Nanomaterials for agriculture and forestry applications*, 289–312.
8. Taghipour, L., & Assar, P. (2025). Nano-packaging technology for conserving the quality of fruits and vegetables. In *Sustainable postharvest technologies for fruits and vegetables* (pp. 227–243). CRC Press.
9. Khedr, E. H., & Al-Khayri, J. M. (2024). Biosensors and nanosensors for determination of harvest quality parameters and fruit handling processes monitoring. In *Innovative methods in horticultural crop improvement: Biosensors and nanosensors* (pp. 143–190). Springer Nature Switzerland.
10. Mahesha, K. N., Singh, N. K., Amarshettiwar, S. B., Singh, G., Gulaiya, S., Das, H., & Kumar, J. (2023). Entering a new agricultural era through the impact of Nano-fertilizers on crop development: A review. *International Journal of Plant & Soil Science, 35*(20), 94–102.
11. Hong, J., Wang, C., Wagner, D. C., Gardea-Torresdey, J. L., He, F., & Rico, C. M. (2021). Foliar application of nanoparticles: Mechanisms of absorption, transfer, and multiple impacts. *Environmental Science: Nano, 8*(5), 1196–1210.
12. Abdollahdokht, D., Gao, Y., Faramarz, S., Poustforoosh, A., Abbasi, M., Asadikaram, G., & Nematollahi, M. H. (2022). Conventional agrochemicals towards nano-biopesticides: An overview on recent advances. *Chemical and Biological Technologies in Agriculture, 9*(1), 13.

13. Yuvaraj, M., Priya, R. S., Jagathjothi, N., Saranya, M., Suganthi, N., Sharmila, R., et al. (2023). Silicon nanoparticles (SiNPs): Challenges and perspectives for sustainable agriculture. *Physiological and Molecular Plant Pathology, 102161.*
14. Abd-Elsalam, K. A. (Ed.). (2021). *Silver nanomaterials for agri-food applications*. Elsevier.
15. Mohan, K., Kandasamy, S., Rajarajeswaran, J., Thanigaivel, S., Bjeljac, M., Surendran, R. P., & Ganesan, A. R. (2024). Chitosan-based insecticide formulations for insect pest control management: A review of current trends and challenges. *International Journal of Biological Macromolecules, 135937.*
16. Rajwade, J. M., Oak, M. D., & Paknikar, K. M. (2024). Copper-based nanofungicides: The next generation of novel agrochemicals. In *Nanofungicides* (pp. 141–168). Elsevier.
17. Dhanapal, A. R., Thiruvengadam, M., Vairavanathan, J., Venkidasamy, B., Easwaran, M., & Ghorbanpour, M. (2024). Nanotechnology approaches for the remediation of agricultural polluted soils. *ACS Omega, 9*(12), 13522–13533.
18. Kinoti, I. K., Karanja, E. M., Nthiga, E. W., M'thiruaine, C. M., & Marangu, J. M. (2022). Review of clay-based nanocomposites as adsorbents for the removal of heavy metals. *Journal of Chemistry, 2022*(1), 7504626.
19. Lu, C., Hei, R., Song, X., Fan, Z., Guo, D., Luo, J., & Ma, Y. (2023). Metal oxide nanoparticles inhibit nitrogen fixation and rhizosphere colonization by inducing ROS in associative nitrogen-fixing bacteria Pseudomonas stutzeri A1501. *Chemosphere, 336*, 139223.
20. Roychoudhury, A. (2020). Silicon-nanoparticles in crop improvement and agriculture. *International Journal on Recent Advancement in Biotechnology & Nanotechnology, 3*(1), 2582–1571.
21. Azeez, L., Adejumo, A. L., Lateef, A., Adebisi, S. A., Adetoro, R. O., Adewuyi, S. O., et al. (2019). Zero-valent silver nanoparticles attenuate Cd and Pb toxicities on Moringa oleifera via immobilization and induction of phytochemicals. *Plant Physiology and Biochemistry, 139*, 283–292.
22. Jampílek, J., & Kráľová, K. (2020). Potential of nanoscale carbon-based materials for remediation of pesticide-contaminated environment. In *Carbon nanomaterials for agri-food and environmental applications* (pp. 359–399). Elsevier.
23. Adejumo, A. L., Azeez, L., Kolawole, T. O., Aremu, H. K., Adedotun, I. S., Oladeji, R. D., et al. (2023). Silver nanoparticles strengthen Zea mays against toxic metal-related phytotoxicity via enhanced metal phytostabilization and improved antioxidant responses. *International Journal of Phytoremediation, 25*(12), 1676–1686.
24. Pratama, A., & Rahayu, S. (2024). Biosensors enabled by nanotechnology: Advancing precision agriculture through IoT monitoring. *Journal of Intelligent Connectivity and Emerging Technologies, 9*(1), 25–39.
25. Bharti, A., Jain, U., & Chauhan, N. (2024). From lab to field: Nano-biosensors for real-time plant nutrient tracking. *Plant Nano Biology, 100079.*
26. Flores-Rangel, G., de León-Martínez, L. D., Mizaikoff, B., & Roque-Jiménez, J. A. Nanosensors applied within the food industry: The role of volatile organic compounds sensing. In *Nanobiotechnology for sustainable food management* (pp. 319–339). CRC Press.
27. Khondakar, K. R., & Kaushik, A. K. (Eds.). (2024). Next-generation smart biosensing: Nano-platforms, nano-microfluidics interfaces, and emerging applications of quantum sensing.
28. Manikandan, S., Subbaiya, R., Saravanan, M., Ponraj, M., Selvam, M., & Pugazhendhi, A. (2022). A critical review of advanced nanotechnology and hybrid membrane based water recycling, reuse, and wastewater treatment processes. *Chemosphere, 289*, 132867.
29. Al-Maliki, R. M., Alsalhy, Q. F., Al-Jubouri, S., Salih, I. K., AbdulRazak, A. A., Shehab, M. A., et al. (2022). Classification of nanomaterials and the effect of graphene oxide (GO) and recently developed nanoparticles on the ultrafiltration membrane and their applications: A review. *Membranes, 12*(11), 1043.
30. Khondoker, M., Mandal, S., Gurav, R., & Hwang, S. (2023). Freshwater shortage, salinity increase, and global food production: A need for sustainable irrigation water desalination—A scoping review. *Earth, 4*(2), 223–240.

31. Mustafa, G. (2025). Nanosensor in soil-water agroecosystem. Nanoscience and soil-water interaction in agroecosystem, 1660
32. Buledi, J. A., Mallah, A., Solangi, A. R., & Amin, S. Nanobiotechnology in food science. In *Nanobiotechnology for sustainable food management* (pp. 1–16). CRC Press.
33. Sharma, A., Ranjit, R., Kumar, N., Kumar, M., & Giri, B. S. (2023). Nanoparticles based nanosensors: Principles and their applications in active packaging for food quality and safety detection. *Biochemical Engineering Journal, 193*, 108861.
34. Kraśniewska, K., Galus, S., & Gniewosz, M. (2020). Biopolymers-based materials containing silver nanoparticles as active packaging for food applications–a review. *International Journal of Molecular Sciences, 21*(3), 698.
35. Prasad, R., Bhattacharyya, A., & Nguyen, Q. D. (2017). Nanotechnology in sustainable agriculture: Recent developments, challenges, and perspectives. *Frontiers in Microbiology, 8*, 1014.
36. Yadav, N., Garg, V. K., Chhillar, A. K., & Rana, J. S. (2023). Recent advances in nanotechnology for the improvement of conventional agricultural systems: A review. *Plant Nano Biology, 4*, 100032.
37. Usman, M., Farooq, M., Wakeel, A., Nawaz, A., Cheema, S. A., ur Rehman, H., et al. (2020). Nanotechnology in agriculture: Current status, challenges and future opportunities. *Science of the Total Environment, 721*, 137778.
38. Solanki, P., Bhargava, A., Chhipa, H., Jain, N., & Panwar, J. (2015). Nano-fertilizers and their smart delivery system. *Nanotechnologies in food and agriculture*, 81–101.
39. Al-Juthery, H. W., Lahmod, N. R., & Al-Taee, R. A. (2021). Intelligent, nano-fertilizers: A new technology for improvement nutrient use efficiency (article review). In *IOP conference series: Earth and environmental science (Vol. 735, No. 1, p. 012086)*. IOP Publishing.
40. Vivekanandhan, P., Kannan, S., Pittarate, S., & Krutmuang, P. (2024). Nanoinsecticide: An introduction. In *Nano-insecticide: Today and future perspectives* (pp. 1–23). Springer Nature Switzerland.
41. Salam, M., Zheng, H., Liu, Y., Zaib, A., Rehman, S. A. U., Riaz, N., et al. (2023). Effects of micro (nano) plastics on soil nutrient cycling: State of the knowledge. *Journal of Environmental Management, 344*, 118437.
42. Babatunde, D. E., Denwigwe, I. H., Babatunde, O. M., Gbadamosi, S. L., Babalola, I. P., & Agboola, O. (2019). Environmental and societal impact of nanotechnology. *IEEE Access, 8*, 4640–4667.
43. Sales, H. B., de Carolino, S., de Nunes, A., Macalia, R. Z., Ruzo, C. M., da Pinto, C., et al. (2024). Advances in agricultural technology: A review of slow-release Nanofertilizers and innovative carriers. *Communications in Soil Science and Plant Analysis, 55*(12), 1849–1882.
44. Behl, K., Jaiswal, P., & Pabbi, S. (2024). Recent advances in microbial and nano-formulations for effective delivery and agriculture sustainability. *Biocatalysis and Agricultural Biotechnology, 103180.*
45. Guruprasath, N., Sankarganesh, P., Adeyeye, S. A. O., Babu, A. S., & Parthasarathy, V. (2024). Review on emerging applications of nanobiosensor in food safety. *Journal of Food Science.*
46. Chen, H., & Yada, R. (2011). Nanotechnologies in agriculture: New tools for sustainable development. *Trends in Food Science & Technology, 22*(11), 585–594.
47. Yin, J., Wang, Y., & Gilbertson, L. M. (2018). Opportunities to advance sustainable design of nano-enabled agriculture identified through a literature review. *Environmental Science: Nano, 5*(1), 11–26.
48. Dave, V., Sur, S., & Gupta, N. (2021). Current framework, ethical consideration and future challenges of regulatory approach for nano-based products. *Nanopharmaceutical Advanced Delivery Systems*, 447–472.
49. Gul, M. Z., & Beedu, S. R. (2023). Integrated approaches to Agri-nanotechnology: Applications, challenges, and future perspectives. In *Microbiomes for the Management of Agricultural Sustainability* (pp. 1–28). Springer Nature Switzerland.

50. Rose, D. C., Sutherland, W. J., Parker, C., Lobley, M., Winter, M., Morris, C., et al. (2016). Decision support tools for agriculture: Towards effective design and delivery. *Agricultural Systems, 149*, 165–174.
51. Chandel, N., Kumar, A., & Kumar, R. (2024). Towards sustainable agriculture: Integrating agronomic practices, environmental physiology and plant nutrition. *International Journal of Plant & Soil Science, 36*(6), 492–503.
52. Singh, M., Upadhyay, L., Badekhan, A., Shil, S., Gautam, S. K., & Anand, A. (2024). Leveraging remote sensing and nanotechnology to overcome barriers to agroforestry adoption by small holder farmers. *Journal of Scientific Research and Reports, 30*(5), 113–140.
53. El-Kady, M. M., Ansari, I., Arora, C., Rai, N., Soni, S., Verma, D. K., et al. (2023). Nanomaterials: A comprehensive review of applications, toxicity, impact, and fate to environment. *Journal of Molecular Liquids, 370*, 121046.
54. Mona, C., Salomé, M. M., Judit, K., José-María, N., Eric, B., & María-Luisa, F. C. (2023). Considerations for bioaccumulation studies in fish with nanomaterials. *Chemosphere, 312*, 137299.
55. Wang, Y., Lei, C., & Lin, D. (2021). Environmental behaviors and biological effects of engineered nanomaterials: Important roles of interfacial interactions and dissolved organic matter. *Chinese Journal of Chemistry, 39*(2), 232–242.
56. Jain, K., Takuli, A., Gupta, T. K., & Gupta, D. (2024). Rethinking nanoparticle synthesis: A sustainable approach vs. traditional methods. *Chemistry–An Asian Journal, 19*(21), e202400701.
57. Irewale, A. T., Dimkpa, C. O., Agunbiade, F. O., Oyetunde, O. A., Elemike, E. E., & Oguzie, E. E. (2024). Unlocking sustainable agricultural development in Africa via bio-nanofertilizer application-challenges, opportunities and prospects. *Scientific African*, e02276.
58. Ruzzante, S., Labarta, R., & Bilton, A. (2021). Adoption of agricultural technology in the developing world: A meta-analysis of the empirical literature. *World Development, 146*, 105599.
59. Saleh, H. M., & Hassan, A. I. (2023). Synthesis and characterization of nanomaterials for application in cost-effective electrochemical devices. *Sustainability, 15*(14), 10891.
60. Mishra, H. (2024). Nanobiostimulants and precision agriculture: A data-driven approach to farming and market dynamics. In *Nanobiostimulants: Emerging strategies for agricultural sustainability* (pp. 365–398). Springer Nature Switzerland.
61. Besha, A. T., Liu, Y., Bekele, D. N., Dong, Z., Naidu, R., & Gebremariam, G. N. (2020). Sustainability and environmental ethics for the application of engineered nanoparticles. *Environmental Science & Policy, 103*, 85–98.
62. Amutha, C., Gopan, A., Pushbalatatha, I., Ragavi, M., & Reneese, J. A. (2024). Nanotechnology and governance: Regulatory framework for responsible innovation. In *Nanotechnology in societal development* (pp. 481–503). Springer Nature Singapore.
63. Ashique, S., Raikar, A., Jamil, S., Lakshminarayana, L., Gajbhiye, S. A., De, S., & Kumar, S. (2025). Artificial intelligence integration with nanotechnology: A new frontier for sustainable and precision agriculture. *Current Nanoscience, 21*(2), 242–273.
64. Adir, O., Poley, M., Chen, G., Froim, S., Krinsky, N., Shklover, J., et al. (2020). Integrating artificial intelligence and nanotechnology for precision cancer medicine. *Advanced Materials, 32*(13), 1901989.
65. Das, H., Singh, S. K., Priyadarshini, E., Maurya, C. L., Tiwari, A. K., Gulaiya, S., & Singh, B. V. (2024). Nanotechnology facilitated real time soil monitoring for optimized crop production. *Asian Journal of Soil Science and Plant Nutrition, 10*(2), 582–607.
66. Abdel-Basset, M., Hawash, H., & Abdel-Fatah, L. (2024). *Artificial intelligence and internet of things in smart farming*. CRC Press.
67. Ibrahim, A. A., Abd-Ellatif, S., Abdel Razik, E. S. S., Hamedo, H. A., & Salem, K. F. (2024). Nanosensors for enhancing plant growth and productivity. In *Innovative methods in horticultural crop improvement: Biosensors and nanosensors* (pp. 25–50). Springer Nature Switzerland.

68. Egorov, E., Pieters, C., Korach-Rechtman, H., Shklover, J., & Schroeder, A. (2021). Robotics, microfluidics, nanotechnology and AI in the synthesis and evaluation of liposomes and polymeric drug delivery systems. *Drug Delivery and Translational Research, 11*, 345–352.
69. Behnassi, M., Qureshi, R. H., Al-Shaikh, A. A., Baig, M. B., & Faraj, T. K. A. (2024). Climate-smart and resilient food systems and security: Introduction. In *Climate-smart and resilient food systems and security* (pp. 1–14). Springer Nature Switzerland.
70. Shoukat, A., Pitann, B., Zafar, M. M., Farooq, M. A., Haroon, M., Nawaz, A., et al. (2024). Nanotechnology for climate change mitigation: Enhancing plant resilience under stress environments. *Journal of Plant Nutrition and Soil Science, 187*(5), 604–620.
71. Akhtar, N., Ilyas, N., Hayat, R., Yasmin, H., Noureldeen, A., & Ahmad, P. (2021). Synergistic effects of plant growth promoting rhizobacteria and silicon dioxide nano-particles for amelioration of drought stress in wheat. *Plant Physiology and Biochemistry, 166*, 160–176.
72. Shaghaleh, H., Hamoud, Y. A., & Sun, Q. (2024). Functionalized nanocellulose nanocomposite hydrogels for soil and water pollution prevention, remediation, and monitoring: A critical review on fabrication, application properties, and potential mechanisms. *Journal of Environmental Chemical Engineering, 111892*.
73. Nayak, T., & Pathan, A. (2023). Environmental remediation and application of carbon-based nanomaterials in the treatment of heavy metal-contaminated water: A review. *Materials Today: Proceedings*.
74. Moradi, S., Babapoor, A., Ghanbarlou, S., Kalashgarani, M. Y., Salahshoori, I., & Seyfaee, A. (2024). Toward a new generation of fertilizers with the approach of controlled-release fertilizers: A review. *Journal of Coatings Technology and Research, 21*(1), 31–54.
75. Medina, J., Calabi-Floody, M., Aponte, H., Santander, C., Paneque, M., Meier, S., et al. (2021). Utilization of inorganic nanoparticles and biochar as additives of agricultural waste composting: Effects of end-products on plant growth, C and nutrient stock in soils from a Mediterranean region. *Agronomy, 11*(4), 767.
76. Miley, K. (2020). Global health biosecurity in a vulnerable world–an evaluation of emerging threats and current disaster preparedness strategies for the future. *Global Health Security: Recognizing Vulnerabilities, Creating Opportunities*, 79–102.
77. Ammar, E. E., Sayed, A. A., Rabee, M. M., Keshta, A. E., Daher, M. G., & Ali, G. A. (2024). Environmental and agricultural applications of sensors. In *Handbook of nanosensors: Materials and technological applications* (pp. 1427–1458). Springer Nature Switzerland.
78. Kansotia, K., Naresh, R., Sharma, Y., Sekhar, M., Sachan, P., Baral, K., & Pandey, S. K. (2024). Nanotechnology-driven solutions: Transforming agriculture for a sustainable and productive future. *Journal of Scientific Research and Reports, 30*(3), 32–51.
79. Wahab, A., Muhammad, M., Ullah, S., Abdi, G., Shah, G. M., Zaman, W., & Ayaz, A. (2024). Agriculture and environmental management through nanotechnology: Eco-friendly nanomaterial synthesis for soil-plant systems, food safety, and sustainability. *Science of the Total Environment, 171862*.
80. Sharma, P., Pandey, V., Sharma, M. M. M., Patra, A., Singh, B., Mehta, S., & Husen, A. (2021). A review on biosensors and nanosensors application in agroecosystems. *Nanoscale Research Letters, 16*, 1–24.
81. Vaidya, S., Deng, C., Wang, Y., Zuverza-Mena, N., Dimkpa, C., & White, J. C. (2024). Nanotechnology in agriculture: A solution to global food insecurity in a changing climate? *NanoImpact, 34*, 100502.
82. Rugji, J., Erol, Z., Taşçı, F., Musa, L., Hamadani, A., Gündemir, M. G., et al. (2024). Utilization of AI–reshaping the future of food safety, agriculture and food security–a critical review. *Critical Reviews in Food Science and Nutrition*, 1–45.
83. Shang, Y., Hasan, M. K., Ahammed, G. J., Li, M., Yin, H., & Zhou, J. (2019). Applications of nanotechnology in plant growth and crop protection: A review. *Molecules, 24*(14), 2558.
84. Rajput, V. D., Kumari, A., Upadhyay, S. K., Minkina, T., Mandzhieva, S., Ranjan, A., et al. (2023). Can nanomaterials improve the soil microbiome and crop productivity? *Agriculture, 13*(2), 231.

85. Shalaby, T. A., Bayoumi, Y., Eid, Y., Elbasiouny, H., Elbehiry, F., Prokisch, J., et al. (2022). Can nanofertilizers mitigate multiple environmental stresses for higher crop productivity? *Sustainability, 14*(6), 3480.
86. de Oliveira Filho, J. G., Miranda, M., Ferreira, M. D., & Plotto, A. (2021). Nanoemulsions as edible coatings: A potential strategy for fresh fruits and vegetables preservation. *Food, 10*(10), 2438.
87. Worrall, E. A., Hamid, A., Mody, K. T., Mitter, N., & Pappu, H. R. (2018). Nanotechnology for plant disease management. *Agronomy, 8*(12), 285.
88. Ur Rahim, H., Qaswar, M., Uddin, M., Giannini, C., Herrera, M. L., & Rea, G. (2021). Nano-enable materials promoting sustainability and resilience in modern agriculture. *Nanomaterials, 11*(8), 2068.
89. Yan, A., & Chen, Z. (2019). Impacts of silver nanoparticles on plants: A focus on the phytotoxicity and underlying mechanism. *International Journal of Molecular Sciences, 20*(5), 1003.
90. Konstantopoulos, G., Koumoulos, E. P., & Charitidis, C. A. (2022). Digital innovation enabled nanomaterial manufacturing; machine learning strategies and green perspectives. *Nanomaterials, 12*(15), 2646.

Revealing Molecular Mechanisms behind Nano-Fertilizers, Nano-Pesticides, and Nano-Fungicides

Amisha Rani, Aman Sharma, Neha Soni, Arti Shivraj Nile, and Shivraj Hariram Nile

Contents

A. Rani
Division of Food and Nutritional Biotechnology, National Agri-Food Biotechnology Institute (NABI), Mohali, Punjab, India

Department of Forest Products, Dr. Yashwant Singh Parmar University of Horticulture and Forestry, Nauni Solan, HP, India

A. Sharma · N. Soni · S. H. Nile (✉)
Division of Food and Nutritional Biotechnology, National Agri-Food Biotechnology Institute (NABI), Mohali, Punjab, India
e-mail: shivraj.nile@nabi.res.in

A. S. Nile
Department of Food Science and Technology, School of Health Sciences, Amity University, Sahibzada Ajit Singh Nagar, Punjab, India

R. S. Chaughule, A. V. Shelar (eds.), *Nanotechnology in Agriculture*,
https://doi.org/10.1007/978-3-031-97544-8_2

1 Introduction

Nano-technology offers innovative solutions by utilizing nanoparticles' unique properties, defined as particles with at least one dimension smaller than 100 nm. These nanoparticles exhibit enhanced surface area and chemical properties that facilitate more efficient nutrient delivery and pest control mechanisms. The application of nano-fertilizers in agriculture is of great potential to enhance crop growth and yield while reducing environmental impact. Agriculture is a crucial sector, not only as a primary source of nourishment but also as a significant source of income for many countries, particularly in rural areas, where approximately 86% of the population depends on farming as a livelihood [1]. Nanotechnology presents significant opportunities for optimizing fertilizer production by customizing chemical compositions and enhancing nutrient use efficiency, which can mitigate environmental contamination and increase plant productivity [2]. Additionally, the controlled release and targeted delivery of nanoscale active ingredients (AIs) enable the advancement of sustainable intensification of growth, yield, and biochemical attributes [3]. Nano-fertilizers consist of nanoparticles designed to regulate the release of nutrients, enhancing their efficiency and cost-effectiveness compared to conventional fertilizers, as illustrated in Fig. 1.

These nano-fertilizers contain one or more nutrients encapsulated within nanoparticles, with at least 50% of the particles having diameters smaller than 100 nm [4]. The mechanism of action of nano-fertilizers on primary and secondary metabolism is intricate and involves the regulation and modulation of various biosynthetic pathways (Table 1).

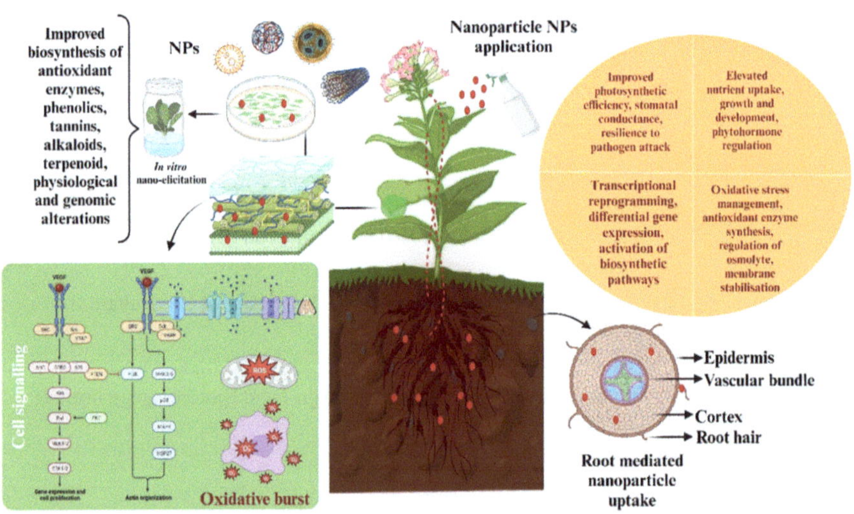

Fig. 1 Generalized mechanism of nano-fertilizer application, uptake within the plant cells, and physiological and molecular alterations upon application (Adopted from [89])

Table 1 Effect of nano-fertilizers on industrially important crops at molecular extent

Sr. no.	Plant	Nanofertilizer type	Effective concentration	Impact assessment on molecular extent	References
1	*Pisum sativum*	Chitosan complex (NPK and polymerizing methacrylic acid)	0.0625 and 0.125 ppm	Upregulation of convicilin, vicilin, and legumin and genotoxic effect	[5]
2	*Triticum aestivum* L.	Composite nano synergists (Carbon and Calcium)	70% compound fertilizer+0.3% composite nano synergist	Upregulation of nitrate transporter gene *TaNRT2.3* and *TaNRT2.2*	[6]
3	*Glycine max* L.	Zinc oxide (ZnO NPs) and their bulk type (BZnO)	5 mg/L	Upregulation of the *EREB, HSF-34, MAPK1*, and *HDA3* genes	[7]
4	*Curcuma longa*	Zinc oxide	10 mg/L	Upregulation of Curcumin synthesis-related genes (*CURS* and *DCS*)	[8]
5	*Ginkgo biloba*	Zinc oxide	25 mg/L	Upregulation of flavonoid synthesis genes (*GbF3H, GbF30H*, and *GbFLS*)	[9]
6	*Zea mays* L.	Zinc oxide	10 mL particle suspension	Overexpression of chlorophyll synthesis (*CHLH, DVR, CHLD, CHLM*, and *POR*), photosystem regulatory (*PetH*), and carbon fixa tion enzyme genes (*pepc, rbcL*, and *rbcS*)	[10]
7	*Punica granatum L.*	Nano-boron and nano-calcium	1.5 g/L	Upregulation of *pgAUX* gene	[11]
8	*Allium cepa* L.	Micronutrient nanofertilizer	4 L/ha	Regulation of aquaporin genes (*PIP1* and *PIP2*)	[12]
9	*Arabidopsis thaliana*	Nano urea	70 μM	Downregulation of chlorophyll catabolism genes	[13]

Agricultural productivity faces considerable challenges, with crop losses attributed to various threats such as animal pests (15–18%), weeds (34%), and microbial diseases (16%). Among these, fungal pathogens alone contribute to 70–80% of yield losses, posing a severe risk to global food security [1]. With around 1.5 million classified species in the fungal kingdom, these pathogens are primarily parasitic and saprophytic, causing extensive damage to a wide range of crops worldwide every year [5, 6]. Currently, disease control in agriculture heavily relies on agrochemicals, including fungicides, due to their rapid action, reliability, and availability.

Nevertheless, conventional fungicides have significant drawbacks despite these benefits. There have been several reports of toxic effects on nontarget organisms that interfere with biosynthetic pathways, disturb the activities of soil microorganisms, and increase pest resistance, which leads to a resurgence in pest populations [7, 8]. Significant environmental pollution results from the estimated 80–90% of sprayed fungicides that are lost to the environment during or after application [7, 9]. This demonstrates the urgent need for high-performing, environmentally friendly fungicides that are both profitable and successful in lessening their negative effects on the environment. As a novel approach to agricultural pest management, nano-fungicides are being utilized to increase the potency and safety of traditional fungicides. These formulations, which use nanoparticles to more efficiently target fungal pathogens while reducing their negative effects on the environment, are essential for inhibiting the growth and activity of various fungi [10]. Ensuring sustainable agricultural productivity and global food security are the two hard-to-do challenges under the scenario of climate change and the explosion of global population [11]. Although global crop productivity has tripled since green revolution, however, excessive use of fertilizers and pesticides have threatened agroecosystem, ultimately hampering crop productivity [12]. Indeed, nearly four million tons of pesticides are used annually with approximately 25% being effective to target pests, leaving a large proportion for environmental damage. Reports suggested that heavy pest infestation can cause up to 50% crop loss, affecting food security. Conventional insecticides have many major drawbacks, like high dosages per unit crop, operational hazards, and drift hazards. They also affect nontarget organisms and vegetation [13]. Thus, sound and experimentally proven technologies should be deployed to overcome the drawbacks of conventional pesticides. The involvement of nanotechnology with the advances in the characteristics of nanoparticles (smaller size, tunable surface chemistry, and targeted delivery system) has become a boon to agriculture with the use of responsive nanoparticles [14].

A nanoparticle is a tiny particle that has at least one dimension smaller than 100 nm. Particles with such smaller sizes possess unique physicochemical properties such as higher surface area and higher volume. Nanoparticles can be of various shapes, i.e., cuboidal, spherical, cylindrical, etc. [15]. Pesticidal active ingredients (AIs) encapsulated and/or formulated at the nano-scale are called nano-pesticides. In comparison to nonnano-scale analogs, nano-pesticides are more efficient with controlled, targeted, and synchronized delivery of AI at the site of infestation through responsive nano-scale delivery platforms (RNDP) with higher efficacy [16]. RNDP helps in increased solubility of the AI with lesser premature loss of AI, increased adherence to the site of infestation, and lesser residual effect on the ecosystem [17]. Reasons for the crop production loss include the loss of vigor and resilience of crops against biotic and abiotic stresses [18]. The loss in resilience occurred particularly due to the loss of soil's ability to suppress pathogens (soil suppressiveness) and the loss of diversity and abundance of the rhizospheric beneficial microbiome [19]. Thus, the scenario compromises crop productivity by increasing opportunistic soil-borne pathogens. By targeting the opportunistic pathogens through responsive nano-pesticides can enhance the diversity, abundance, and structure of beneficial soil microbiome with the resumption of soil suppressiveness. Altogether, these could help in better crop growth with other collateral benefits such as lesser use of resources and negligible damage to the agroecosystem.

2 Molecular Mechanisms Underlying Nano-Fertilizers, Nano-Pesticides, and Nano-Fungicides

2.1 Nano-Fertilizers

In agriculture, these nanoparticles are mostly applied in the form of nano-fertilizers to the leaf surface via foliar application, where they are adsorbed and then deposited, ultimately being absorbed via the cuticle or at the stomata located on the leaf surface. The waxy epidermal cuticle primarily consists of wax, cutin, and pectin, which together function to inhibit water loss during leaf development and provide a natural barrier against the ingress of nanoparticles [20]. Nonetheless, its surface has two distinct channels: one hydrophilic and the other lipophilic [21]. The diameter of the hydrophilic and lipophilic channels varies from 0.6 to 4.8 nm [22]. Hydrophilic channels facilitate the transport of hydrophilic nanoparticles with diameters smaller than 4.8 nm. The lipophilic channels on the cuticle's surface facilitate the diffusion and penetration of lipophilic nanoparticles into the leaves. Carbon dots smaller than 2 nm may infiltrate the cuticular route in cotton leaves, as seen by confocal fluorescence microscopy with great spatial and temporal resolution [23]. Narrow pore channels in cuticles impede the absorption of nanoparticles by plants via their epidermis. Numerous studies indicate that nanoparticles settle in both leaf epidermis and vascular tissues after application to the leaf surface. Conversely, it is widely established that nanoparticles may infiltrate many tissues of a plant. The researchers posited that the nanoparticles infiltrated via another route, referred to as the stomatal pathway. Stomata are apertures on leaves that facilitate the regulation of water and gas exchange inside plants. The typical size range of stomata is between 10 and 100 micrometres. The dimensions and density also differ across plant species. The unusual geometric configuration of stomata and their biological roles have rendered the SEL of stomatal aperture regarding nanoparticle diffusion unclear to this day [24].

2.1.1 Factors Affecting Uptake of NPs

Chemical Composition and Size

The absorption of nanoparticles by roots is significantly influenced by their dimensions. Particles less than 10 nm are often absorbed more readily, except silicon-based and natural polymer-based nanoparticles exceeding 100 nm. Nanoparticles (NPs) of

200 nm and Zein NPs of 135 ± 3 nm have shown absorption by plant roots. Zein, the major storage protein of corn, is widely used in nanoparticle formulations due to its biocompatibility, biodegradability, and natural origin. Moreover, the chemical composition of nanoparticles significantly influences their interaction with plant species, shaping the uptake and response of plant species [25].

Surface Charge

The surface charge of nanoparticles significantly affects their uptake by plant roots. The negatively charged plant root cell walls demonstrate electrostatic attraction to positively charged nanoparticles, leading to their accumulation on the root surface. Nonetheless, the ability to infiltrate root tissues is contingent upon the surface charge of the nanoparticle. Research has shown that plant cells may uptake uncharged and negatively charged nanoparticles (NPs), facilitating their entry into the xylem. Conversely, positively charged nanoparticles tend to accumulate on the root's outer layer without penetrating further [26].

2.1.2 Plant Interaction with Nanoparticles

The transport and deposition of nanoparticles in plants will result in phytotoxicity and interactions that generate increased reactive oxygen species (ROS), leading to oxidative damage and genetic toxicity. When reactive oxygen species (ROS) surpass the capacity of defensive systems, cells enter a condition of "oxidative stress," leading to extensive damage to proteins, nucleic acids, and lipids in cellular membranes. Plants safeguard cells from ROS damage via antioxidant enzymes and antioxidants [27]. The protective role of plants on cellular structures and subcellular systems against the detrimental effects of reactive oxygen species is ascribed to the function of antioxidant enzymes and low-molecular-weight antioxidants [28]. Consequently, the majority of research about oxidative damage induced by nanoparticles emphasizes the function of antioxidant enzymes and the concentration of reactive oxygen species. Extensive investigation and analysis occur when nanoparticles interact with plants at elevated concentrations, resulting in significant nanoparticle accumulation in plants and the activation of the antioxidant system in response to many reactive oxygen species (ROS) [29].

As the concentration of plastic nanoparticles increased, the activity of several antioxidant enzymes in rice roots rose, suggesting that the plant may activate a defensive response to mitigate excessive reactive oxygen species formation [30]. A greater concentration of nanoparticles will influence the levels of plant endogenous hormones. Current research is also available to assess the influence of nanoparticles on plant endogenous hormones. Iron oxide nanoparticles at a concentration of 100 mg/L reduced the production of Bt-transgenic cotton and elevated hormone levels. Nevertheless, the hormone concentration inside the roots of Bt-transgenic cotton decreased [31]. Furthermore, metal-based nanoparticles suffer alterations and release deleterious metal ions in plants, thereby inflicting further harm to DNA and proteins within the plant system and consequently disrupting normal cellular metabolic functions [32].

In addition to promoting the generation of reactive oxygen species (ROS) and altering hormonal equilibrium in plants, nanoparticles have genotoxic properties. Nanoparticles engage with biological macromolecules, including nuclei and lipids, to produce genotoxic effects that may affect the plant cell cycle [33]. For example, silver nanoparticles (AgNPs) internalized in wheat root tips have been shown to disrupt normal cell division, inhibit DNA synthesis, and produce chromosomal abnormalities [34]. Furthermore, the mitotic index of onion cells exposed to PS nanoparticles exhibited a considerable reduction. This was related to chromosomal and nuclear abnormalities. Consequently, genomic stability was undermined by such genotoxic consequences [35].

To decrease nanoparticle toxicity, methods have been created that reduce the release of harmful compounds by the application of surface coatings. Research indicates that the iron coating on zinc oxide nanoparticles (ZnO NPs) greatly reduces the elution of zinc ions into the environment, thereby markedly reducing both elution and toxicity. The Fe-coated zinc oxide nanoparticle does neither inhibit plant germination nor diminish its pigment content [36]. This technique of nanoparticle encapsulation or surface modification to mitigate toxicity is equally applicable to polymer nanoparticles. Understanding the ideal concentration of nanoparticles for plant development, together with the analysis of nanoparticles as agricultural pesticides based on their toxicity, might be beneficial. Natural polymer nanoparticles, being nontoxic, degradable, and biocompatible, may significantly mitigate the detrimental effects associated with metal-based, silicon-based, and organic-based nanoparticles on plants and the environment [37]. Chitosan is the only naturally occurring positively charged polysaccharide. Their characteristics have been identified as beneficial for nanoparticles, exhibiting antibacterial and antiviral effects. The positive charge of chitosan nanoparticles facilitates their interaction with the plant cell membrane, hence augmenting reactivity in plant systems. Research indicates that the use of chitosan-based nanoparticles as growth promoters on seeds and seedlings enhances nutrient absorption, elevates chlorophyll levels, and boosts the photosynthetic rate in plants. Three distinct diameters of chitosan nanoparticles, namely 420, 750, and 970 nm, were applied to the plants. The findings indicated that chlorophyll levels rose by 61%, 81%, and 61%, while photosynthetic rates increased by 29%, 59%, and 72%, respectively [38]. Chitosan nanoparticles have been shown to enhance plant development in wheat and barley agriculture [39].

2.2 *Nano-Pesticides*

Nano-pesticides are mainly categorized into two types, viz., Type-I nano-pesticides, which are silver (Ag), copper (Cu), and titanium (Ti)-based nanomaterials, and Type-II, which are active ingredients (AIs) encapsulated nano-carriers. Type-I nano-pesticides have strong antimicrobial activity due to the adhesive and dissolution activity, which ultimately leads to cytotoxicity, reactive oxygen species (ROS) formation, and genotoxicity-induced cell death. The nano-pesticides can control various plant pathogens, including bacteria and fungi. The strong activity of these nano-pesticides can be quantified by their minimum inhibitory concentration (MIC) value. On the other hand, Type-II nano-pesticides are carrier-based, wherein polymers, clays, and zein nanoparticles are used. The nano-pesticides are response stimuli-based, which work through RNDP to achieve sustainable agricultural goals such as biocompatibility and cost-effectiveness. For polymer-based nanocarriers, common natural polymers, i.e., cellulose, polylactide, and chitosan, are used to encapsulate active ingredients in the form of nano-capsule, nano-sphere, nano-hydrogels, and nano-micelle. On the contrary, mesoporous silica nanoparticles and montmorillonite nanomaterials (NMs) are used for clay-based nano-carriers, and these have higher encapsulation capacity due to the high surface area [16].

Nanoparticles can also be classified based on the formulations, viz., nano-carrier, nano-emulsions, and nano-suspensions. Nano-carriers encapsulate pesticides in nano-shell, which allows slow release of AIs with better penetration to the site of infestation. On the other hand, oil-water emulsion is commonly used for nano-emulsions wherein pesticidal active ingredients are diffused in water in the form of nanoparticles, and surfactants are confined to the pesticide–water interface to form a stable emulsion. Nano-emulsions increase the solubility of pesticides with better absorption [40]. On the contrary, in the case of nano-suspensions, pesticides are dispersed in nanoparticle form, and surfactants are confined to the nanoparticle surface for stable suspension. Nano-suspensions ensure sustained release of AI with higher efficacy [41].

2.2.1 Physicochemical Properties of Nano-Pesticides

Nano-pesticides are deployed to maximize the use of the unique properties of nanoparticles. The unique properties include higher surface area and volume and tunable surface chemistry. Indeed, the physicochemical properties determine the efficiency, efficacy, fate, adherence, and most importantly impact of the agroecosystem. The type-I and type-II nano-pesticides possess characteristics of interest, such as smaller size (in nano-scale), higher surface area, and uniformity in formulations. Besides the abovementioned properties of interest, type-II nano-pesticides have other properties for the efficient loading and release of active ingredients, i.e., Encapsulation efficiency (EE), Loading efficiency (LE), and Release Efficiency (RE). The properties deal with the assembly and release of AI at the infestation site. Nano-carriers have higher EE values due to the higher surface area. For instance, zein nanoparticles have the highest EE value, leading to the highest encapsulation capacity. On the other hand, the RE of nano-pesticides ranges from 50% to 90%, confirming that nano-pesticides cannot release 100% of the active ingredients at the infestation site [42].

The size of nano-pesticides ranges from 1 to 500 nm, although the maximum limit of nanoparticle size is limited to 100 nm. Previous research showed [43] that the upper threshold of nano-pesticides can be of 500 nm without compromising with the unique nano-scale properties, and the findings were in line with the regulations of nano-pesticides given by European Union. They also mentioned that nano-pesticides can also be possible with the particle size above 500 nm; however, such nano-pesticides have not been proven yet. Thus, transmission electron microscope and dynamic light scattering results suggested that type-1 and type-II nano-pesticides can range from 20 to 160 nm and 160 to 360 nm, respectively. The uniformity of particle size in suspension is also a physicochemical property of nano-pesticides, and the property can be quantified by the polydispersity index (PDI). PDI less than 0.2 depicts that the dispersed nanoparticles in suspension have a narrow size range. Type-II nano-pesticides have a lower PDI value than type-I nano-pesticides, indicating that type-II nano-pesticides have more uniformity in suspension. The uniformity of nanoparticles in suspension ensures higher stability, consistent penetrating ability through biological barriers, controlled release, and reduced off-target effects [16].

2.2.2 Mode of Action of Nano-Pesticides

The response of nano-pesticides would depend on their mode of action which is believed to be linked to their nano-scaled features that increase interaction with the target site pest plant or the environment. In Type I, nano-pesticides are either the active ingredient in their nano form or the active ingredient in conventional formulations but are encapsulated in a nanocarrier system. This results in more effective delivery and absorption by target pests or plants; and that is enhanced by their very small size, high surface area, and special physicochemical properties. This increases the efficiency of pesticide use, decreases the amount applied, and reduces any possible adverse influence on the environment [44].

Type-I nano-pesticides cause physical damage or oxidative impairment through their nanometric structures. For example, silver (Ag)- and copper (Cu)-based metal nanoparticles kill cells by producing reactive oxygen species (ROS), which damage the cell membranes and metabolic organelles when introduced to harmful pests. They may also inflict surface or membrane injuries through their interaction with the pests, which leads to pest mortality. The mode of action of nano-pesticides can also be photochemically induced, as in the case of titanium dioxide (TiO_2), which can be activated by light to produce toxic ultraviolet rays to pests but harmless to human beings and the environment [45]. Type-II nano-pesticides are designed to increase delivery efficiency and release control by using the technique of encapsulation. These protective nanocarriers secure the active ingredient from environmental stresses such as UV light, temperature, or degrading microbes. For instance, such pesticides may only release AI when the pests' stomach acid or enzymes that are specifically designed to dissolve the pesticide are triggered. Moreover, due to the targeted specificity, pesticidal AI can be delivered close to the area required, hence minimizing “collateral” damage and increasing the general effectiveness of pest control methods [46].

Another important mode of action of nano-pesticides is enhanced translocation and systemic absorption of AI after translocation in the plants. The use of nano-formulations increases the ability of the active ingredient to penetrate the plant cuticles and cell walls into the systemic circulation. This enhancement guarantees that pests feeding on the treated plants are effectively exposed to the pesticidal AI even if such areas are difficult to reach. In addition, nano-pesticides can enhance the adherence of the pesticide onto the plant surfaces, thus reducing runoff and enhancing pest exposure [47].

2.2.3 Recent Trends of Nano-Pesticides

The domain of nano-pesticides is gaining momentum globally with growing emphasis on sustainability, precision, and efficiency in solving agricultural problems. One such trend is the increase in the development of nano-formulations, which are readily biodegradable and safe. These formulations can use natural carriers like chitosan, starch, and cellulose to coat the active substances, making it possible to slowly release these substances while minimizing environmental pollution. There is also an interest in plant essential oils nano-emulsions of neem and eucalyptus as potent yet environmentally friendly sources of energy. These nano-emulsions not only increase the stability and bioavailability of these oils but can also be used in a targeted manner to control pests with minimal effects on beneficial organisms [48].

The delivery systems’ further development in targeted delivery and controlled release is modifying the approach to pest control. Nanoparticles with stimuli-responsive properties, such as sensitivity to pH, temperature, or moisture, are being used for precision applications, where they release active ingredients in response to specific environmental triggers. This reduces pesticidal impact and applies them very efficiently. In addition, silver (Ag) and copper (Cu) nanoparticles, which are multifunctional and possess both pesticidal and antimicrobial properties, can also be considered effective and versatile tools in crop protection. Drones and IoT devices can precisely deliver and manage these nano-pesticides, thus reducing waste and improving pest monitoring through the integration of these nano-pesticides with agriculture technologies [49].

The creation of novel nano-herbicides and bio-nano-pesticides is prompted by sustainability and resistance management. The implementation of encapsulated biopesticides such as *Bacillus thuringiensis* or *Beauveria bassiana*, as well as polymeric-directed nano-formulations, aims to reduce pesticide residues and avoid pest resistance [49]. There is ongoing research on metal-organic framework (MOF), which enables the development of high-loading and controlled-release conjugation of pesticides. At the same time, coupling nano-pesticides with sustainable agricultural methods strives to achieve efficient pest control without harming the environment, and advanced pest management strategies of the future encompass these nano-pesticides [50]. These developments illustrate the ability of nano-pesticides to optimize the crop protection process without risk to the ecology or people (Table 2).

Table 2 List of nano-agrochemicals with product name, nanoparticle used, active ingredients, and pests targeted

Sr. no.	Product name	Nanoparticle used	Active Ingredient	Target pest
Nano-based insecticide				
1.	Nano-thiamethoxam	Silica nanoparticle	Thiamethoxam	Agricultural pests
2.	Vayego® Nano	Polymeric nanoparticle	Tetraniliprole	Caterpillars, beetles
3.	Nano-pyrethroid	Lipid nanoparticle	Pyrethroid	Flies, crop pests
4.	Nano-imidacloprid	Polymer-based nano pesticide	Imidacloprid	Aphids, leafhopper
5.	EcoNano	Nanoemulsion	Neem oil	White flies, aphids, mites
Nano-based fungicide				
6.	Nano-chitosan	Nanoparticle	Chitosan	*Botrytis* sp. *Fusarium* sp.
7.	Nano-fungicide silica	Silica nanoparticle	Triazole	Blight and fungal leaf spot
8.	Nanocide	Nano emulsion	Sulfur	Rust and powdery mildew
9.	Nano-copper	Metal nanoparticle	Copper oxide	Blight and mildew
Nano-based biopesticide				
10.	Nano-Trichoderma	Biodegradable nano-carrier	*Trichoderma* sp.	Fungal pathogen
11.	Nano-Bacillus	Encapsulated bacteria	*Bacillus thuringiensis* (Bt)	Caterpillar and beetles
12.	Nano-Beauveria	Polymer nanoparticle	*Beauveria bassiana*	White flies and aphids
13.	BioNanoCide	Nano emulsion	Plant essential oil	Insects, Fungi

2.3 *Nano-Fungicides*

Nano-fungicides can function through two primary mechanisms: Direct Mode of Action and use as nanocarriers. Nanoparticles have demonstrated promising anti-fungal properties, directly inhibiting the growth and development of various fungi. According to [51], copper sulfate ($CuSO_4$) and sodium tetraborate ($Na_2B_4O_7$) were 2 out of 15 micronutrients that were notably effective in controlling rust disease in peas. Postharvest fungal diseases caused by species like *Botrytis cinerea* and *Penicillium expansum* were successfully treated with zinc oxide nanoparticles. These nanoparticles distorted the hyphae formation of *B. cinerea* and prevented conidiophore and conidia formation in *Penicillium expansum* [10, 52]. Similarly, it was observed that *Aspergillus flavus* colonization was significantly decreased by zinc nanoparticles at 15 mg/mL. Other metal nanoparticles, such as selenium (Se), nickel (Ni), magnesium (Mg), palladium (Pd), and iron (Fe), have demonstrated promising antifungal properties, although silver (Ag) and copper (Cu) nanoparticles are the most researched for controlling phytopathogenic fungi [53–59]. In vivo, the study of selenium nanoparticles (SeNPs) against *Sclerospora graminicola* at concentrations up to 1000 ppm demonstrated a significant antifungal activity. These Se nanoparticles were synthesized using six strains of *Trichoderma* spp.—*T. asperellum*, *T. harzianum*, *T. atroviride*, *T. virens*, *T. longibrachiatum*, and *T. brevicompactum*. Out of the given strains, *T. asperellum* in the culture filtrate yielded the most effective results, highlighting its antifungal properties and potential as a biological agent for Se nanoparticle synthesis [57]. The mechanics and possible use of nanoparticles as nano-fungicides, detailing their mechanisms and application potential, is thoroughly discussed in an article by Ayaseelan et al. [60]. Silver nanoparticles stabilized with polyvinylpyrrolidone (PVP) have demonstrated significant antifungal effects against multiple yeast and mold strains, including *Candida glabrata*, *C. tropicalis*, *Aspergillus niger*, *C. albicans*, and *C. krusei* [61].

2.3.1 Use as Nanocarriers

Nanoparticles are increasingly employed as nanocarriers in the agroindustry, where they encapsulate, entrap, and attach active compounds to create highly effective formulations. Silica nanoparticles, for instance, are designed to achieve specific shapes, structures, and sizes to maximize delivery efficiency [62]. These particles are typically spherical with pore-like structures, such as mesoporous silica nanoparticles (MSNs) or porous hollow silica nanoparticles (PHSNs). MSNs and PHSNs contain fungicides within an internal cavity, which prevents the active compounds from volatilizing, allowing precise and controlled administration. In order to preserve the stability and long-term functionality, the protective shell of PHSNs also shields the active compounds from UV degradation. Several studies have shown that silicon improves plant tolerance against both biotic and abiotic stressors, making silica nanoparticles a suitable platform for agro-industrial formulations focused on fungal management [6]. Solid lipid nanoparticles (SLNs) and chitosan-based nanoparticles are two of the most common nanocarrier systems used in plant disease control, demonstrating potential in targeted and sustainable fungicide delivery. Carbendazim (MBC), known as methyl-2-benzimidazole carbamate, and Tebuconazole (TBZ), chemically denoted as (RS)-1-(4-chlorophenyl)-4,4-dimethyl-3-(1H-1,2,4-triazol-1-ylmethyl) pentane-3-ol, are two widely utilized fungicides in agriculture for the treatment of various fungal pathogens [6]. As a delivery system, solid lipid nanoparticles offer numerous advantages, such as reduced toxicity to humans and the environment, adaptive release profiles for active compounds, and fewer leaching losses. These nano-formulated fungicides provide an alternative solution for controlling fungal diseases among multiple crop species [6].

Chitosan nanoparticles are used in the agricultural sector as both growth enhancers and robust antifungal agents for a range of fungal infections. Furthermore, chitosan acts as a nanocarrier for agrochemicals [63], allowing the development of a potent nano-delivery system for agrochemical formulations by encapsulating active chemicals in a polymer matrix of chitosan or by covalent or ionic bonding [64]. It has also demonstrated potential as a sustainable pesticide against harmful bacteria, viruses, and fungi, regardless of whether it is combined with or without macronutrient combinations. Chitosan, whether used independently or in conjunction with other active agents, has shown significant effectiveness as an alternative to traditional fungicides, successfully controlling diseases such as wilt and Fusarium head blight in chickpeas and wheat, rice blast, post-flowering stalk rot in maize, maize leaf spot, and finger millet blast [63].

Chitosan nanoparticles are of great interest due to their biocompatibility, biodegradability, and low toxicity, making them suitable for various biomedical and pharmaceutical applications [64]. Additionally, chitosan displays a wide range of beneficial properties, such as antifungal, antimicrobial, antioxidant, antiviral, bioadhesive, adsorption-enhancing, and anti-inflammatory effects, highlighting its usefulness in medical and engineering fields [64]. Its mechanism of action against microbial pathogens can be classified based on its extracellular, intracellular, or combined impacts, depending on the antimicrobial target [65–68]. Low-molecular-weight (low-MW) chitosan, besides its extracellular antimicrobial activity, can also penetrate cells, thereby impacting protein synthesis, mitochondrial functionality, and RNA synthesis [66, 69], whereas high-molecular-weight (high-MW) chitosan, extracellular antimicrobial action is predominant, as it generally cannot penetrate microbial cell membranes and walls. It is thus effective in inhibiting nutrient uptake extracellularly, altering cell permeability, and chelating essential metals [66, 70]. Studies indicate that chitosan exerts an inhibitory effect on various microbial organisms [69], though its antimicrobial efficiency depends on factors such as the degree of acetylation, molecular mass, and specific microbial characteristics. Due to its hydrophobic nature, chitosan is less soluble in water. Therefore, to increase its solubility, it is mixed with different organic and inorganic copolymers [71]. The presence of hydroxyl and amine functional groups in chitosan allows for various functionalizing processes, including the promotion of ionic bonding, and undergoing grafting reactions, thereby enhancing the properties. The strong adhesion of chitosan to the stem and leaf surface protein assists the uptake of bioactive molecules and extends the contact time of chitosan with the target surfaces [71].

Synthesis of biodegradable chitosan-lactide copolymer (CPLA) is used as a hydrophobic carrier for fungicide pyraclostrobin, creating a nanoparticle formulation having efficacy higher than the conventional methods for the treatment against *Colletotrichum gossypii* [72]. Similarly, nanoparticles containing the fungicides tebuconazole or chlorothalonil have been developed using polyvinylpyridine (PVP), copolymers of PVP and styrene (PVP-co-St), and blends of PVP with hyperbranched polyesters (HBPs) as matrices. These nanoparticles were tested against *Gloeophyllum trabeum* on southern yellow pine, exhibiting promising antifungal effects [73]. Table 3 provides further information on the nanoformulations. Besides these formulations, other examples of alternatives to traditional fungicides include essential oils and chitosan-dazomet-hexaconazole, chitosan-dazomet, and chitosan-hexaconazole.

The nanocarrier delivery system-related research has been categorized into nine categories by the Fungicide Resistance Action Committee (FRAC), among which

Table 3 Summary of investigations using nanoparticles as actives targeting fungi and carriers of different fungicides (Adopted from [88])

Fungicide (FRAC Code)	Nanoparticles type	Plant Species	Target Fungi	Overall impact	References
Chitosan–Dazomet	Chitosan and Dazomet	Oil palm	*G. boninense*	It has high biocompatibility and biodegradability of formulation constituents	[71]
Chitosan–hexaconozole	Chitosan and hexaconozole	Oil palm	*G. boninense*	Effective nanodelivery system	[71]
Carbendazim (1)	Chitosan/Pectin	Cucumber Maize Tomato	*A. Parasiticus* and *F. oxysporum*	Control release	[72]
Prochloraz (3)	PHSN	Cucumber	*B. cinerea*	It facilitates the controlled nutrient transfer and increases crop protection	[82]
Cymbopogon martini essential oil	Chitosan	Maize grains	*F. graminearum*	Reduce volatility and enhance fungal disease control	[83, 84]
Tebuconazole (3)	Bacterial ghosts	Wheat, cucumber, and barley	*L. nodorum*, *P. teres*, *S. fuliginea*, and *E. graminis*,	It possesses a multifunctional delivery platform	[85]
Tebuconazole (3) Chlorothalonil (M 05)	PVP and PVP copolymer	Southern yellow pine	*G. trabeum*	It has excellent solubility in solvents of different polarities, good binding properties, and a stabilizing effect	[86, 87]
Eugenol oil	Eugenol oil nanoemulsion	Seed cotton	*F. oxysporum*	Displayed better antifungal activity compared to the conventional form	[81]

silica, chitosan, and polymer-based carriers are researched extensively. These nano-fungicides have been tested on several species of fungi to determine their efficacy, highlighting the potential of nanoparticle formulations in targeted and efficient fungicide delivery. A comprehensive review of the advancements and potential of nanoparticle-based fungicides, including their efficacy and mechanisms, has been presented by [74].

2.3.2 Molecular Mechanism of Working of Nano-Fungicides

Nano-fungicides use a variety of molecular pathways to achieve their antifungal activity. For example, hyphae of *Sclerotium sclerotiorum* treated with silver nanoparticles (AgNPs) showed extreme structural damage, including hyphal wall separation and hyphal collapse, as demonstrated by [75]. AgNPs also induced significant alterations in *P. chrysogenum*, such as hyphal shortening and condensation, cell plasmolysis, increased vacuolization, accumulation of membranous structures, collapsed cytoplasm, lipid material build-up, condensed mitochondria, organelle disintegration, nuclear deformation, and chromatin condensation and fragmentation [76]. Besides this, chitosan nanoparticles have also exhibited potential antifungal activity. According to a study by [77], it has been reported that such nanoparticles further inhibit the synthesis of ATP and energy metabolism, inhibiting H+-ATPase activity and disrupting energy metabolism. Furthermore, conidial growth, conidia spore generation, and germ tube development were all reported to be restricted by these nanoparticles as per [78, 79]. Metal nanoparticles interact with fungal cell membranes, leading to changes in membrane integrity and increase the permeability of the membrane [80, 81]. They further inhibit protein oxidation, interfere with the electron transport chain, alter the potential of the mitochondrial membrane, and contribute to fungal cell damage by promoting oxidative stress through an increase in the formation of reactive oxygen species (ROS) [82]. In addition, metal nanoparticles may induce ROS-induced genotoxicity. The recent research by [83] showed that AgNPs inhibited fungal growth, cellular structure, and energy metabolism pathways. In *F. graminearum*, an RNA-Seq study indicated an increased number of azole-related ATP-binding cassette (ABC) transporters without affecting azole control efficacy, upon the treatment with AgNP. The nanoparticles also generates ROS, which in turn activated the genes involved in DON biosynthesis, peroxisome formation, and mycotoxin production, as illustrated in Fig. 2.

3 Conclusion

The application of nano-agrochemicals such as nano-fertilizers, nano-pesticides, and nano-fungicides has revolutionized modern agriculture by enhancing efficiency and reducing environmental impacts, thereby addressing the limitations of conventional agrochemicals. These nanomaterials alter molecular mechanisms within the plant,

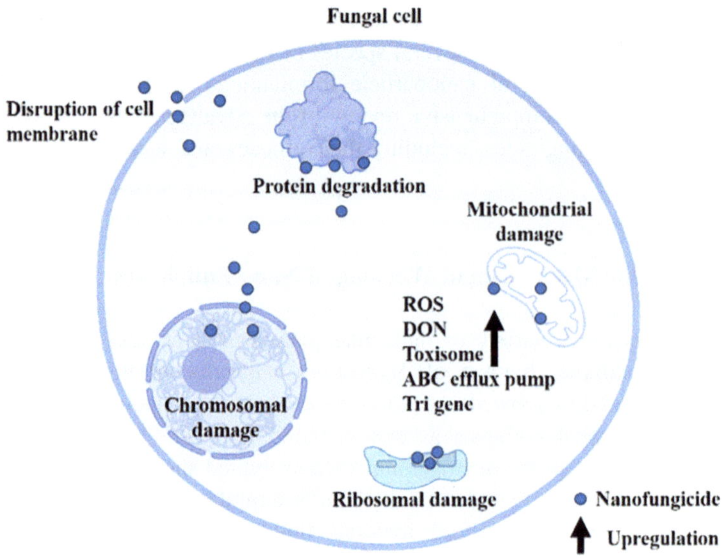

Fig. 2 Mechanism of working of nano-fungicides

which improve nutrient uptake, targeted delivery of nutrients, and regulation of cellular and subcellular processes to yield optimized metabolism and plant protection against pests. Unique properties of nanomaterials such as high surface area, and controlled release as compared to conventional agrochemicals add up to greater specificity, reduced dosing requirements, and longer duration of activity. Despite such advantages, significant challenges have yet to be overcome before nanomaterials become a part of the environment with sustainability. The primary challenge includes understanding the long-term impacts of nanomaterials on the environment and its ecosystem by determining the accumulation, toxicity, and interactions with nontargeted organisms in both water and soil systems. Furthermore, increasing production while maintaining the cost-effectiveness and stability of nano-agrochemicals remains challenging, especially for widespread use in limited areas. Safety regulations regarding use, registration, and disposal are not yet established properly, creating uncertainty for producers and consumers. In addition, the lack of knowledge about the molecular mechanisms of nano-agrochemical interactions with plants and pests highlights the need for further research using omics and imaging technologies for crop-specific formulation and application guides. Public awareness and acceptance are also important issues, as farmers and consumers are still not convinced of the safety of nanotechnology in agriculture. Integrating nanotechnology with permaculture practices creates additional challenges that require new methods to balance technological advancement with ecological protection. In conclusion, effective risk assessment tools and predictive models need to be developed to assess potential risks and support accountability. In summary, while nano-agrochemicals provide a transformative opportunity for agriculture, addressing these challenges through integrated research, integrated management, and stakeholder engagement is essential to ensure they are safe, effective, and durable.

References

1. Atiq, M., Naeem, I., Sahi, S. T., Rajput, N. A., Haider, E., Usman, M., Shahbaz, H., Fatima, K., Arif, E., & Qayyum, A. (2020). Nanoparticles: A safe way towards fungal diseases. *Phytopathol Plant Prot*, 781–792. https://doi.org/10.1080/03235408.2020.1792599
2. Babu, S., Singh, R., Yadav, D., Rathore, S. S., Raj, R., Avasthe, R., Yadav, S. K., Das, A., Yadav, V., Yadav, B., Shekhawat, K., Upadhya, P. K., Yadav, D. K., & Singh, D. K. (2022). Nanofertilizers for agricultural and environmental sustainability. *Chemosphere, 292*, 133451. https://doi.org/10.1016/j.chemosphere.2021.133451
3. Ijaz, M., Khan, F., Ahmed, T., Noman, M., Zulfiqar, F., Rizwan, M., Chen, J., Siddique, K. H. M., & Li, B. (2023). Nanobiotechnology to advance stress resilience in plants: Current opportunities and challenges. *Materials Today Bio, 22*, 100759. https://doi.org/10.1016/j.mtbio.2023.100759
4. Yadav, A., Yadav, K., & Abd-Elsalam, K. A. (2023). Nanofertilizers: Types, delivery, and advantages in agricultural sustainability. *Agrochemicals, 2*(2), 296–336. https://doi.org/10.3390/agrochemicals2020029
5. Khalifa, N. S., & Hasaneen, M. N. (2018). The effect of chitosan–PMAA–NPK nanofertilizer on *Pisum sativum* plants. 3. *Biotech, 8*(4), 1–12. https://doi.org/10.1007/s13205-018-1221-3
6. Yang, M., Dong, C., & Shi, Y. (2023). Nano fertilizer synergist effects on nitrogen utilization and related gene expression in wheat. *BMC Plant Biology, 23*(1), 1–13. https://doi.org/10.1186/s12870-023-04046-9
7. Mirakhorli, T., Ardebili, Z. O., Ladan-Moghadam, A., & Danaee, E. (2021). Bulk and nanoparticles of zinc oxide exert beneficial effects by modifying transcription factors, histone deacetylase, carbon and nitrogen assimilation, antioxidant biomarkers, and secondary metabolism in soybean. *PLoS One, 16*(9), e0256905. https://doi.org/10.1371/journal.pone.0256905
8. Khattab, S., Alkuwayti, M. A., Yap, Y. K., Meligy, A. M. A., Bani Ismail, M., & El Sherif, F. (2023). Foliar spraying of ZnO nanoparticles on *Curcuma longa* increased growth, yield, expression of curcuminoid synthesis genes, and curcuminoid accumulation. *Horticulturae, 9*(3), 355. https://doi.org/10.3390/horticulturae9030355
9. Wang, Q., Xu, S., Zhong, L., Zhao, X., & Wang, L. (2023). Effects of zinc oxide nanoparticles on growth, development, and flavonoid synthesis in *Ginkgo biloba*. *International Journal of Molecular Sciences, 24*(21), 15775. https://doi.org/10.3390/ijms242115775
10. Guo, S., Zhang, X., & Sun, H. (2023). Transcriptomic mechanism for foliar-applied nano-ZnO alleviating phytotoxicity of nanoplastics in corn (*Zea mays* L.) plants. *Sci Total Environ, 905*, 166818. https://doi.org/10.1016/j.scitotenv.2023.166818
11. Al-Janabi, A., Al-Zurfi, M. A., & Al-Presem, W. F. (2021). Influence of spraying nano-fertilizers with boron and calcium on vegetative growth characteristics and *PGAUX* gene expression of pomegranate transplants. *Int J Agric Stat Sci*. 1–6. https://faculty.uobasrah.edu.iq/uploads/publications/1654085057.pdf
12. Berna-Sicilia, J. A., Quizhpe-Romero, M., Hurtado-Navarro, M., Pascual, J. A., Carvajal, M., & Bárzana, G. (2024). Combined soil microorganism amendments and foliar micronutrient nanofertilization increased the production of *Allium cepa* L. through aquaporin gene regulation. *Life, 14*(1), 4. https://doi.org/10.3390/life14010004
13. Jangir, N., Marik, D., Verma, D., Dey, A., Shekhawat, R. S., Patel, D., Yadav, P., Sankhala, K., & Sadhukhan, A. (2024). Effects of nano urea on growth and gene expression of *Arabidopsis*

thaliana in hydroponics. *bioRxiv*;2024.03.20.585664. https://doi.org/10.1101/2024.03.20.585664

14. Flood, J. (2010). The importance of plant health to food security. *Food Security, 2*(3), 215–231. https://doi.org/10.1007/s12571-010-0072-5
15. Worrall, E. A., Hamid, A., Mody, K. T., Mitter, N., & Pappu, H. R. (2018). Nanotechnology for plant disease management. *Agronomy, 8*(12), 285. https://doi.org/10.3390/agronomy8120285
16. Stephenson, G., Coats, J., Yamamoto, H. (2001). Pesticide use and world food production: risks and benefits. Expert Committee on Weeds Proceedings;1–16. https://citeseerx.ist.psu.edu/document?repid=rep1&type=pdf&doi=00cb930df32d36febb3267ff1e9eee8b6d5c74bd.
17. Zaller, J. G., & Brühl, C. A. (2019). Editorial: Non-target effects of pesticides on organisms inhabiting agroecosystems. *Frontiers in Environmental Science, 7*, 449950. https://doi.org/10.3389/fenvs.2019.00499
18. Ghormade, V., Deshpande, M. V., & Paknikar, K. M. (2011). Perspectives for nano-biotechnology enabled protection and nutrition of plants. *Biotechnology Advances, 29*(6), 792–803. https://doi.org/10.1016/j.biotechadv.2011.06.007
19. Khan, M., & TRPP J. (2014). Nanotechnology: Scope and application in plant disease management. *ResearchGate*. https://doi.org/10.3923/ppj.2014.214.231
20. Gitz, V., Meybeck, A., Lipper, L., Young, C. D., & Braatz, S. (2016). Climate change and food security: Risks and responses. *FAO Report*. https://ciheam.org/uploads/attachments/250/06_Meybeck_WL_37.pdf
21. Tilman, D., Cassman, K. G., Matson, P. A., Naylor, R., & Polasky, S. (2002). Agricultural sustainability and intensive production practices. *Nature, 418*(6898), 671–677. https://doi.org/10.1038/nature01014
22. Quandahor, P., Kim, L., Kim, M., Lee, K., Kusi, F., & Jeong, I. H. (2024). Effects of agricultural pesticides on decline in insect species and individual numbers. *Environments, 11*(8), 182. https://doi.org/10.3390/environments11080182
23. An, C., Sun, C., Li, N., Huang, B., Jiang, J., Shen, Y., Wang, C., Zhao, X., Bui, B., Wang, C., Li, X., Zhan, S., Gao, F., Zeng, Z., Cui, H., & Wang, Y. (2022). Nanomaterials and nanotechnology for the delivery of agrochemicals: Strategies towards sustainable agriculture. *Journal of Nanbiotechnology, 20*(1), 1–19. https://doi.org/10.1186/s12951-021-01214-7
24. Joudeh, N., & Linke, D. (2022). Nanoparticle classification, physicochemical properties, characterization, and applications: A comprehensive review for biologists. *Journal of Nanbiotechnology, 20*(1), 1–29. https://doi.org/10.1186/s12951-022-01477-8
25. Wang, D., Saleh, N. B., Byro, A., Zepp, R., Sahle-Demessie, E., Luxton, T. P., Ho, K. T., Burgess, R. M., Flury, M., White, J. C., & Su, C. (2022). Nano-enabled pesticides for sustainable agriculture and global food security. *Nature Nanotechnology, 17*(4), 347–360. https://doi.org/10.1038/s41565-022-01082-8
26. Li, M., Sun, X., Yin, M., Shen, J., & Yan, S. (2023). Recent advances in nanoparticle-mediated co-delivery system: A promising strategy in medical and agricultural fields. *International Journal of Molecular Sciences, 24*(6), 5121. https://doi.org/10.3390/ijms24065121
27. Kissoudis, C., van de Wiel, C., Visser, R. G. F., & van der Linden, G. (2014). Enhancing crop resilience to combined abiotic and biotic stress through the dissection of physiological and molecular crosstalk. *Frontiers in Plant Science, 5*(MAY), 90871. https://doi.org/10.3389/fpls.2014.00871
28. Steinberg, C., Edel-Hermann, V., Alabouvette, C., & Lemanceau, P. (2019). Soil suppressiveness to plant diseases. In J. D. Van Elsas, J. K. Jansson, & J. T. Trevors (Eds.), *Modern soil microbiology* (3rd ed., pp. 343–359). CRC Press. https://doi.org/10.1201/9780429059186-21
29. Yang, C., Powell, C. A., Duan, Y., Shatters, R., & Zhang, M. (2015). Antimicrobial nanoemulsion formulation with improved penetration of foliar spray through citrus leaf cuticles to control citrus Huanglongbing. *PLoS One, 10*(7), e0133826. https://doi.org/10.1371/journal.pone.0133826
30. Avellan, A., Yun, J., Zhang, Y., Spielman-Sun, E., Unrine, J. M., Thieme, J., Li, J., Lombi, E., Bland, G., & Lowry, G. V. (2019). Nanoparticle size and coating chemistry control foliar uptake

pathways, translocation, and leaf-to-rhizosphere transport in wheat. *ACS Nano, 13*(5), 5291–5305. https://doi.org/10.1021/acsnano.8b09781

31. Eichert, T., Kurtz, A., Steiner, U., & Goldbach, H. E. (2008). Size exclusion limits and lateral heterogeneity of the stomatal foliar uptake pathway for aqueous solutes and water-suspended nanoparticles. *Physiologia Plantarum, 134*(1), 151–160. https://doi.org/10.1111/j.1399-3054.2008.01135.x
32. Hu, P., An, J., Faulkner, M. M., Wu, H., Li, Z., Tian, X., & Giraldo, J. P. (2020). Nanoparticle charge and size control foliar delivery efficiency to plant cells and organelles. *ACS Nano, 14*(7), 7970–7986. https://doi.org/10.1021/acsnano.9b09178
33. Ha, N., Seo, E., Kim, S., & Lee, S. J. (2021). Adsorption of nanoparticles suspended in a drop on a leaf surface of *Perilla frutescens* and their infiltration through the stomatal pathway. *Scientific Reports, 11*(1), 1–13. https://doi.org/10.1038/s41598-021-91073-x
34. Khan, I., Awan, S. A., Rizwan, M., Hassan, Z. U., Akram, M. A., Tariq, R., Brestic, M., & Xie, W. (2022). Nanoparticle uptake and translocation mechanisms in plants via seed priming, foliar treatment, and root exposure: A review. *Environmental Science and Pollution Research, 29*(60), 89823–89833. https://doi.org/10.1007/s11356-022-23945-2
35. Spielman-Sun, E., Avellan, A., Bland, G. D., Tappero, R. V., Acerbo, A. S., Unrine, J. M., Giraldo, J. P., & Lowry, G. V. (2019). Nanoparticle surface charge influences translocation and leaf distribution in vascular plants with contrasting anatomy. *Environmental Science. Nano, 6*(8), 2508–2519. https://doi.org/10.1039/C9EN00626E
36. Biczak, R., Telesiński, A., & Pawłowska, B. (2016). Oxidative stress in spring barley and common radish exposed to quaternary ammonium salts with hexafluorophosphate anion. *Plant Physiology and Biochemistry, 107*, 248–256. https://doi.org/10.1016/j.plaphy.2016.05.030
37. Qamer, Z., Chaudhary, M. T., Du, X., Hinze, L., & Azhar, M. T. (2021). Review of oxidative stress and antioxidative defense mechanisms in *Gossypium hirsutum* L. in response to extreme abiotic conditions. *Journal of Cotton Research, 4*(1), 1–9. https://doi.org/10.1186/s42397-021-00086-4
38. Jurkow, R., Pokluda, R., Sękara, A., & Kalisz, A. (2020). Impact of foliar application of some metal nanoparticles on antioxidant system in oakleaf lettuce seedlings. *BMC Plant Biology, 20*(1), 1–12. https://doi.org/10.1186/s12870-020-02490-5
39. Zhou, C. Q., Lu, C. H., Mai, L., Bao, L. J., Liu, L. Y., & Zeng, E. Y. (2021). Response of rice (*Oryza sativa* L.) roots to nanoplastic treatment at seedling stage. *Journal of Hazardous Materials, 401*, 123412. https://doi.org/10.1016/j.jhazmat.2020.123412
40. Van Nhan, L., Ma, C., Rui, Y., Cao, W., Deng, Y., Liu, L., & Xing, B. (2016). The effects of Fe2O3 nanoparticles on physiology and insecticide activity in non-transgenic and Bt-transgenic cotton. *Frontiers in Plant Science, 6*, 151860. https://doi.org/10.3389/fpls.2015.00912
41. Moazzami Farida, S. H., Karamian, R., & Albrectsen, B. R. (2020). Silver nanoparticle pollutants activate oxidative stress responses and rosmarinic acid accumulation in sage. *Physiologia Plantarum, 170*(3), 415–432. https://doi.org/10.1111/ppl.13172
42. Youssef, M. S., & Elamawi, R. M. (2020). Evaluation of phytotoxicity, cytotoxicity, and genotoxicity of ZnO nanoparticles in *Vicia faba*. *Environmental Science and Pollution Research, 27*(16), 18972–18984. https://doi.org/10.1007/s11356-018-3250-1
43. Abdelsalam, N. R., Abdel-Megeed, A., Ali, H. M., Salem, M. Z. M., Al-Hayali, M. F. A., & Elshikh, M. S. (2018). Genotoxicity effects of silver nanoparticles on wheat (*Triticum aestivum* L.) root tip cells. *Ecotoxicology and Environmental Safety, 155*, 76–85. https://doi.org/10.1016/j.ecoenv.2018.02.050
44. Maity, S., Chatterjee, A., Guchhait, R., De, S., & Pramanick, K. (2020). Cytogenotoxic potential of a hazardous material, polystyrene microparticles on *Allium cepa* L. *Journal of Hazardous Materials, 385*, 121560. https://doi.org/10.1016/j.jhazmat.2019.121560
45. Sturikova, H., Krystofova, O., Huska, D., & Adam, V. (2018). Zinc, zinc nanoparticles and plants. *Journal of Hazardous Materials, 349*, 101–110. https://doi.org/10.1016/j.jhazmat.2018.01.040

46. Aranaz, I., Harris, R., & Heras, A. (2010). Chitosan amphiphilic derivatives: Chemistry and applications. *Current Organic Chemistry, 14*(3), 308–330. https://doi.org/10.2174/138527210790820580
47. Nguyen Van, S., Dinh Minh, H., & Nguyen Anh, D. (2013). Study on chitosan nanoparticles on biophysical characteristics and growth of Robusta coffee in greenhouse. *Biocatalysis and Agricultural Biotechnology, 2*(4), 289–294. https://doi.org/10.1016/j.bcab.2013.05.005
48. Behboudi, F., Sarvestani, Z. T., Kassaee, M. Z., Sanavi, S. A. M. M., & Sorooshzadeh, A. (2017). Phytotoxicity of chitosan and SiO2 nanoparticles to seed germination of wheat (*Triticum aestivum* L.) and barley (*Hordeum vulgare* L.) plants. *Notulae Scientia Biologicae, 9*(2), 242–249. https://doi.org/10.15835/nsb9210075
49. Nuruzzaman, M., Rahman, M. M., Liu, Y., & Naidu, R. (2016). Nanoencapsulation, nano-guard for pesticides: A new window for safe application. *Journal of Agricultural and Food Chemistry, 64*(7), 1447–1483. https://doi.org/10.1021/acs.jafc.5b05214
50. Bocca, B., Barone, F., Petrucci, F., Benetti, F., Picardo, V., Prota, V., & Amendola, G. (2020). Nanopesticides: Physicochemical characterization by a combination of advanced analytical techniques. *Food and Chemical Toxicology, 146*, 111816. https://doi.org/10.1016/j.fct.2020.111816
51. Rajna, S., Paschapur, A. U., & Raghavendra, K. V. (2019). Nanopesticides: Its scope and utility in pest management. *Indian Farmer, 6*(1), 17–21. https://indianfarmer.net/assets/archieves/2019/JANUARY%202019.pdf#page=19
52. Min, Y., Suminda, G. G. D., Heo, Y., Kim, M., Ghosh, M., & Son, Y. O. (2023). Metal-based nanoparticles and their relevant consequences on cytotoxicity cascade and induced oxidative stress. *Antioxidants, 12*(3), 703. https://doi.org/10.3390/antiox12030703
53. Zainab, R., Hasnain, M., Ali, F., Abideen, Z., Siddiqui, Z. S., Jamil, F., Hussain, M., & Park, Y. K. (2024). Prospects and challenges of nanopesticides in advancing pest management for sustainable agricultural and environmental service. *Environmental Research, 261*, 119722. https://doi.org/10.1016/j.envres.2023.119722
54. Oliveira, H. C., Stolf-Moreira, R., Martinez, C. B. R., Grillo, R., De Jesus, M. B., & Fraceto, L. F. (2015). Nanoencapsulation enhances the post-emergence herbicidal activity of atrazine against mustard plants. *PLoS One, 10*(7), e0132971. https://doi.org/10.1371/journal.pone.0132971
55. Machado, S., Pereira, R., & Sousa, R. M. O. F. (2023). Nanobiopesticides: Are they the future of phytosanitary treatments in modern agriculture? *Science of The Total Environment, 896*, 166401. https://doi.org/10.1016/j.scitotenv.2023.166401
56. Camara, M. C., Campos, E. V. R., Monteiro, R. A., Do Espirito Santo Pereira, A., De Freitas Proença, P. L., & Fraceto, L. F. (2019). Development of stimuli-responsive nano-based pesticides: Emerging opportunities for agriculture. *Journal of Nanobiotechnology, 17*(1), 1–19. https://doi.org/10.1186/s12951-019-0533-8
57. Vishnu, M., Kannan, M., Soundararajan, R. P., Suganthi, A., Subramanian, A., Senthilkumar, M., Rameash, K., Madesh, K., & Govindaraju, K. (2024). Nano-bioformulations: Emerging trends and potential applications in next generation crop protection. *Environmental Science. Nano, 11*(7), 2831–2860. https://doi.org/10.1039/d4en00263f
58. Shangguan, W., Chen, H., Zhao, P., Cao, C., Yu, M., Huang, Q., & Cao, L. (2024). Scenario-oriented nanopesticides: Shaping nanopesticides for future agriculture. *Advanced Agrochem, 3*(4), 265–278. https://doi.org/10.1016/j.advag.2024.11.002
59. Singh, D., Kumar, A., Singh, A. K., & Tripathi, H. S. (2013). Induction of resistance in field pea against rust disease through various chemicals/micronutrients and their impact on growth and yield. *Plant Pathology Journal (Faisalabad), 12*(2), 36–49. https://doi.org/10.3923/ppj.2013.36.49
60. Ayaseelan, C., Rahuman, A. A., Kirthi, A. V., Marimuthu, S., Santhoshkumar, T., Bagavan, A., Gaurav, K., Karthik, L., & Rao, K. V. B. (2012). Novel microbial route to synthesize ZnO nanoparticles using Aeromonas hydrophila and their activity against pathogenic bacteria and

fungi. *Spectrochimica Acta. Part A, Molecular and Biomolecular Spectroscopy, 90*, 78–84. https://doi.org/10.1016/J.SAA.2012.01.006
61. Ahmed, A., Yadav, D., & Lee, Y. S. (2016). Applications of nickel nanoparticles for control of Fusarium wilt on lettuce and tomato. *International Journal of Innovative Research in Science, Engineering and Technology, 3*(5), 7378–7385. https://www.researchgate.net/profile/Afra-Roughani/post/Can-we-coat-seeds-with-Nano-particles-to-control-soil-born-plant-pathogen/attachment/5a6c65234cde266d5886ba64/AS%3A587377928056837%401517053219184/download/Ahmedetal2016.pdf
62. Asghar, M. A., Zahir, E., Shahid, S. M., Khan, M. N., Iqbal, J., & Walker, G. (2018). Iron, copper and silver nanoparticles: Green synthesis using green and black tea leaves extracts and evaluation of antibacterial, antifungal and aflatoxin B1 adsorption activity. *LWT., 90*, 98–107. https://doi.org/10.1016/j.lwt.2017.12.009
63. Ismail, A. W., Sidkey, N., Arafa, R., Fathy, R., & El-Batal, A. (2016). Evaluation of in vitro antifungal activity of silver and selenium nanoparticles against Alternaria solani caused early blight disease on potato. *British Biotechnology Journal, 12*(3), 1–11. https://doi.org/10.9734/bbj/2016/24155
64. Jagana, D., Hegde, Y. R., & Lella, R. (2017). Green nanoparticles—a novel approach for the management of banana anthracnose caused by Colletotrichum musae. *International Journal of Current Microbiology and Applied Sciences, 6*(10), 1749–1756. https://doi.org/10.20546/ijcmas.2017.610.211
65. Nandini, B., Hariprasad, P., Prakash, H. S., Shetty, H. S., & Geetha, N. (2017). Trichogenic-selenium nanoparticles enhance disease suppressive ability of Trichoderma against downy mildew disease caused by Sclerospora graminicola in pearl millet. *Scientific Reports, 7*(1), 1–11. https://doi.org/10.1038/s41598-017-02737-6
66. Osonga, F. J., Kalra, S., Miller, R. M., Isika, D., & Sadik, O. A. (2020). Synthesis, characterization and antifungal activities of eco-friendly palladium nanoparticles. *RSC Advances, 10*(10), 5894–5904. https://onlinelibrary.wiley.com/doi/full/10.1002/ps.5735
67. Vrandečić, K., Ćosić, J., Ilić, J., Ravnjak, B., Selmani, A., Galić, E., Pem, B., Barbir, R., Vrček, I. V., & Vinković, T. (2020). Antifungal activities of silver and selenium nanoparticles stabilized with different surface coating agents. *Pest Management Science, 76*(6), 2021–2029. https://doi.org/10.1002/ps.5735
68. Cruz-Luna, A. R., Cruz-Martínez, H., Vásquez-López, A., & Medina, D. I. (2021). Metal nanoparticles as novel antifungal agents for sustainable agriculture: Current advances and future directions. *Journal of Fungi, 7*(12), 1033. https://doi.org/10.3390/jof7121033
69. Bryaskova, R., Pencheva, D., Nikolov, S., & Kantardjiev, T. (2011). Synthesis and comparative study on the antimicrobial activity of hybrid materials based on silver nanoparticles (AgNps) stabilized by polyvinylpyrrolidone (PVP). *Journal of Chemical Biology, 4*(4), 185–191. https://doi.org/10.1007/s12154-011-0063-9/metrics
70. Mody, V. V., Cox, A., Shah, S., Singh, A., Bevins, W., & Parihar, H. (2014). Magnetic nanoparticle drug delivery systems for targeting tumor. *Applied Nanoscience (Switzerland), 4*(4), 385–392. https://doi.org/10.1007/s13204-013-0216-y
71. Maluin, F. N., Hussein, M. Z., Yusof, N. A., Fakurazi, S., Abu Seman, I., Zainol Hilmi, N. H., & Jeffery Daim, L. D. (2019). Enhanced fungicidal efficacy on *Ganoderma boninense* by simultaneous co-delivery of hexaconazole and dazomet from their chitosan nanoparticles. *RSC Advances, 9*(46), 27083–27095. https://doi.org/10.1039/c9ra05417k
72. Maluin, F. N., & Hussein, M. Z. (2020). Chitosan-based agronanochemicals as a sustainable alternative in crop protection. *Molecules, 25*(7), 1611. https://doi.org/10.3390/molecules25071611
73. Varlamov, V. P., & Mysyakina, I. S. (2018). Chitosan in biology, microbiology, medicine, and agriculture. *Microbiology (Russian Federation), 87*(5), 712–715. https://doi.org/10.1134/S0026261718050168

74. Kravanja, G., Primožič, M., Knez, Ž., & Leitgeb, M. (2019). Chitosan-based (nano)materials for novel biomedical applications. *Molecules, 24*(10), 1960. https://doi.org/10.3390/molecules24101960
75. Ke, C. L., Deng, F. S., Chuang, C. Y., & Lin, C. H. (2021). Antimicrobial actions and applications of chitosan. *Polymers, 13*(6), 904. https://doi.org/10.3390/polym13060904
76. Hosseinnejad, M., & Jafari, S. M. (2016). Evaluation of different factors affecting antimicrobial properties of chitosan. *International Journal of Biological Macromolecules, 85*, 467–475. https://doi.org/10.1016/j.ijbiomac.2016.01.061
77. Seankamsorn, A., Cherdthong, A., & Wanapat, M. (2020). Combining crude glycerin with chitosan can manipulate in vitro ruminal efficiency and inhibit methane synthesis. *Animals, 10*(1), 37. https://doi.org/10.3390/animals10010037
78. Ke, C. L., Liao, Y. T., & Lin, C. H. (2021). MSS2 maintains mitochondrial function and is required for chitosan resistance, invasive growth, biofilm formation and virulence in *Candida albicans*. *Virulence, 12*(1), 281–297. https://doi.org/10.1080/21505594.2020.1870082
79. Li, M., Huang, Q., & Wu, Y. (2011). A novel chitosan-poly(lactide) copolymer and its submicron particles as imidacloprid carriers. *Pest Management Science, 67*(7), 831–836. https://doi.org/10.1002/ps.2120
80. Xu, L., Cao, L. D., Li, F. M., Wang, X. J., & Huang, Q. L. (2014). Utilization of chitosan-lactide copolymer nanoparticles as controlled release pesticide carrier for Pyraclostrobin against *Colletotrichum gossypii* Southw. *Journal of Dispersion Science and Technology, 35*(4), 544–550. https://doi.org/10.1080/01932691.2013.800455
81. Liu, Y., Laks, P., & Heiden, P. (2002). Controlled release of biocides in solid wood. I. Efficacy against brown rot wood decay fungus (*Gloeophyllum trabeum*). *Journal of Applied Polymer Science, 86*(3), 596–607. https://doi.org/10.1002/app.10896
82. Kumar, S., Kumar, D., & Dilbaghi, N. (2017). Preparation, characterization, and bio-efficacy evaluation of controlled release carbendazim-loaded polymeric nanoparticles. *Environmental Science and Pollution Research, 24*(1), 926–937. https://doi.org/10.1007/s11356-016-7774-y
83. Khoo, K. S., Chia, W. Y., Tang, D. Y. Y., Show, P. L., Chew, K. W., & Chen, W. H. (2020). Nanomaterials utilization in biomass for biofuel and bioenergy production. *Energies (Basel), 13*(4), 1–23. https://doi.org/10.3390/en13040908
84. Zhao, P., Cao, L., Ma, D., Zhou, Z., Huang, Q., & Pan, C. (2018). Translocation, distribution and degradation of prochloraz-loaded mesoporous silica nanoparticles in cucumber plants. *Nanoscale, 10*(4), 1798–1806. https://doi.org/10.1039/C7NR08107C
85. Kalagatur, N. K., Nirmal Ghosh, O. S., Sundararaj, N., & Mudili, V. (2018). Antifungal activity of chitosan nanoparticles encapsulated with *Cymbopogon martinii* essential oil on plant pathogenic fungi *Fusarium graminearum*. *Frontiers in Pharmacology, 9*(JUN), 1–10. https://doi.org/10.3389/fphar.2018.00534
86. Hatfaludi, T., Liska, M., Zellinger, D., Ousman, J. P., Szostak, M., Ambrus, Á., Jalava, K., & Lubtiz, W. (2004). Bacterial ghost technology for pesticide delivery. *Journal of Agricultural and Food Chemistry, 52*(18), 5627–5634. https://doi.org/10.1021/jf040084f
87. Li, R., & Liu, Q. (2020). Engineered bacterial outer membrane vesicles as multifunctional delivery platforms. *Frontiers in Materials, 7*, 1–17. https://doi.org/10.3389/fmats.2020.00019
88. Kutawa, A. B., Ahmad, K., Ali, A., Hussein, M. Z., Wahab, M. A. A., Adamu, A., Ismaila, A. A., Gunasena, M. T., Rahman, M. Z., & Hossain, M. I. (2021). Trends in nanotechnology and its potentialities to control plant pathogenic fungi: A review. *Biology (Basel), 10*(9), 881. https://doi.org/10.3390/biology10090881
89. Rani, A., Soni, K., Sharma, Y. P., Singh, A., Chaudhary, A., Kumar, P., & Sharma, R. (2022). *Nano-elicitation in medicinal and aromatic plants: Applications and challenges* (pp. 225–253). Propag Pharmacopeia.

Nanopesticides for Effective Management of Pest Infections in Agriculture

Messenbet Geremew Kassa, Neela Satheesh, Shivraj Hariram Nile, and Desye Alemu Teferi

Contents

M. G. Kassa · D. A. Teferi
College of Agriculture, Food, Climate Science, Injibara University, Injibara, Ethiopia

N. Satheesh (✉)
Department of Food Science and Technology, School of Health Sciences, Amity University, Mohali, Punjab, India

S. Hariram Nile
Division of Food and Nutritional Biotechnology, National Agri-food Biotechnology Institute (NABI), Sahibzada Ajit Singh Nagar, Punjab, India

R. S. Chaughule, A. V. Shelar (eds.), *Nanotechnology in Agriculture*,
https://doi.org/10.1007/978-3-031-97544-8_3

1 Introduction

Agriculture has been challenged by pests since its inception about 10,000 years ago, with growers contending with various threats including animals, insects, and microorganisms [1]. Today, these pests pose significant challenges in crop production worldwide, leading to crop losses ranging from 16% to 80% [2, 3]. Approximately 4.6 million tons of synthetic chemical pesticides are utilized annually, much of which contributes to environmental pollution and residues in food products [4]. The reliance on conventional pesticides not only raises concerns about pest resistance but also leads to significant ecological consequences, including water and soil contamination [5]. Moreover, pesticide application losses are exceedingly high, with estimates suggesting that more than 99% can be lost due to factors such as runoff and spray drift [6].

To address these pressing challenges, nanoparticle-based pesticides, commonly referred to as nanopesticides, have emerged as a promising solution. These nanoscale materials (1–100 nm) offer advantages such as increased surface area and higher penetration capacity, which enhance their effectiveness while potentially minimizing environmental impacts [7, 8]. Unlike conventional pesticides, nanopesticides can be formulated to release active ingredients in a controlled manner, leading to more efficient pest management with reduced quantities of active ingredients [9]. This targeted approach improves pest control efficacy and helps mitigate chemical pesticides' adverse effects on ecosystems.

The global challenge of food security is further exacerbated by a rapidly growing population, projected to reach 9.7 billion by 2050 [10]. As arable land per capita declines, the demand for efficient food production systems becomes increasingly urgent. The Food and Agriculture Organization (FAO) reported that approximately 795 million people suffer from undernourishment, which represents about 14.3% of the world's population [11]. This is particularly critical in developing regions such as Sub-Saharan Africa, where food insecurity is prevalent and exacerbated by high population growth rates [12]. Sustainable pest management strategies are essential to improve crop yields and address the looming food crisis [13, 14]. Integrated Pest Management (IPM) combines various control methods including biological, mechanical, cultural, and chemical approaches to maintain pest populations below damaging levels while protecting ecological balance [15]. The adoption of nanopesticides within IPM frameworks can enhance pest control effectiveness while reducing environmental risks [16].

Nanotechnology's applications in agriculture are broad, with potential uses in seed management, targeted delivery of nutrients, and the detection of pathogens. For instance, nanoparticles can enhance the effectiveness of agrochemicals, ensuring their controlled release and minimizing environmental contamination [17, 18]. The unique properties of nanomaterials allow for innovative applications, such as the development of nanopesticides that can act as "magic bullets," delivering active ingredients directly to specific plant tissues [19]. These nanoparticles can also improve the solubility of poorly soluble active ingredients, thereby enhancing their effectiveness and stability [20, 21].

Overall, the integration of nanotechnology into agricultural practices, particularly through nanopesticides, offers significant opportunities to enhance crop protection, improve food security, and create sustainable agricultural systems. As the global population continues to rise, the innovative application of nanotechnology may provide critical solutions to some of the most pressing challenges facing the agricultural sector today.

2 Role of Pesticides in Agriculture and Problems with Traditional Pesticides

Pesticides are crucial in modern agriculture, serving as essential tools for managing pests, weeds, and diseases that threaten crop production [22]. Their use has significantly boosted agricultural productivity, ensuring food security for a growing global population [23]. By protecting crops from various harmful organisms, pesticides enable farmers to achieve higher yields, maintain produce quality, and minimize economic losses [24]. Despite concerns regarding their environmental and health impacts, pesticides remain integral to agricultural practices, particularly in large-scale farming [25].

A primary function of pesticides is controlling insect pests, which can cause extensive crop damage [26]. Insects reduce yields by feeding on plants, transmitting diseases, or competing for nutrients [27]. Without effective pest control measures, many crops would suffer substantial losses, leading to decreased food availability and increased prices [3]. Pesticides, especially insecticides, play a vital role in managing these pests, ensuring crops reach maturity.

In addition to insect control, pesticides are essential for managing plant diseases caused by fungi, bacteria, and viruses [28]. These diseases can rapidly spread, resulting in significant reductions in crop yields and quality [29]. Fungicides specifically target fungal pathogens and are commonly used to prevent and treat diseases like powdery mildew, rust, and blight [30]. By controlling these diseases, fungicides help maintain plant health, improve crop quality, and extend the shelf life of harvested produce [31], thereby contributing to reduced food waste and enhanced availability of fresh produce in markets.

Furthermore, pesticides support the sustainability of agricultural systems through IPM strategies. This IPM combines biological, cultural, physical, and chemical methods to manage pests sustainably. In IPM programs, pesticides are used judiciously alongside natural predators, crop rotation, and resistant crop varieties [32]. This approach reduces reliance on chemical inputs, minimizes the risk of pest resistance, and lessens the environmental impact of pesticide use [33].

However, the use of pesticides is not without challenges. Overreliance on chemical pesticides has led to several issues, including the development of resistance, the decline of beneficial organisms like pollinators, and the contamination of soil and water resources [34]. Pesticide residues on food products also raise concerns about food safety and human health [35]. These challenges highlight the need for responsible pesticide use, adherence to recommended application rates, and the adoption of safer, more targeted formulations [36].

Chemical pesticides such as organophosphates, carbamates, and synthetic pyrethroids became standard tools for managing a broad spectrum of pests [37]. While effective, their use has created a cycle of resistance, as pests evolve and adapt to these chemicals, necessitating higher doses or stronger alternatives [38]. Additionally, the environmental impact of these pesticides is concerning; they are often nonselective, harming beneficial species such as bees and natural pest predators [39]. The decline in pollinator populations poses a serious threat to global food production, as many crops depend on pollination for fruit and seed development [40].

Human health risks also raise critical concerns. Farmers and agricultural workers are particularly vulnerable to pesticide exposure through direct contact [41]. Acute poisoning can lead to various symptoms, while chronic exposure has been linked to serious health issues, including cancer and reproductive disorders [42, 43]. Moreover, the overreliance on chemical pesticides has contributed to soil health degradation, disrupting the balance of soil microorganisms and impairing fertility [44]. This degradation can reduce crop yields over time and make agricultural systems more susceptible to pests and diseases [45].

3 Nanotechnology of the Pesticides

Nanotechnology is a transformative field that operates at the scale of atoms and molecules, typically between 1 and 100 nm [46, 47]. Working at this scale allows scientists and engineers to manipulate the fundamental building blocks of matter, leading to new materials, devices, and systems with unique properties and functionalities [48]. The applications of nanotechnology are diverse and rapidly expanding [49]. The core principles of nanotechnology stem from the fact that materials at the nanoscale exhibit different physical, chemical, and biological properties compared to their bulk counterparts [50]. These differences arise mainly due to the high surface area-to-volume ratio of nanomaterials and quantum effects that become significant at such small dimensions. For instance, nanoparticles can exhibit enhanced reactivity, superior strength, unique optical properties, and different electrical behavior [51]. This enables the creation of new materials with capabilities that are not possible at larger scales, such as ultra-strong materials, highly efficient catalysts, and advanced electronic devices.

Nanomaterials are a key component of nanotechnology [52]. These materials can be structured in various forms, including nanoparticles, nanotubes, nanoemulsions, and thin films [53]. Each form has distinct characteristics that make it suitable for specific applications [54]. The versatility of nanomaterials lies in their tunability; by controlling their size, shape, and surface chemistry, researchers can tailor these materials to meet specific needs [54].

Fabrication techniques in nanotechnology are broadly classified into two approaches: top-down and bottom-up [55] (Fig. 1). The top-down approach involves scaling down larger structures to the nanoscale, such as through lithography or etching processes used in semiconductor manufacturing [56]. Conversely, the bottom-up approach (Fig. 1) assembles nanostructures atom by atom or molecule

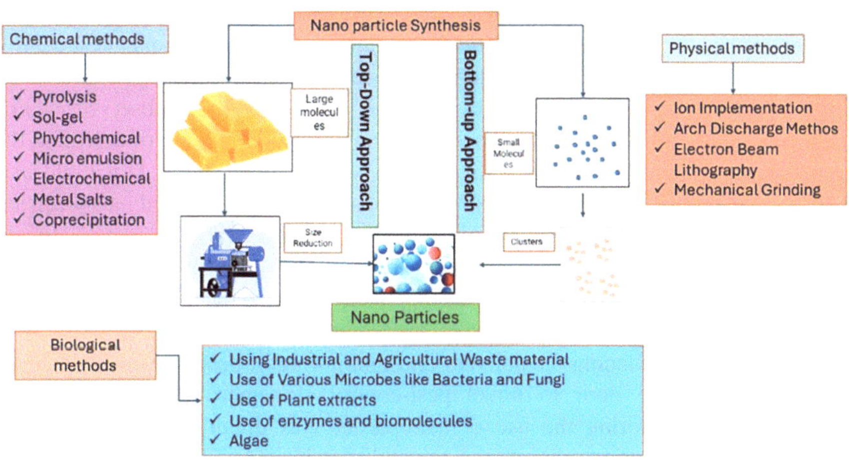

Fig. 1 Schematic diagram of different methods used to synthesize nanoparticles

by molecule, typically using self-assembly or chemical synthesis methods [57]. These approaches allows for precise control over the composition, structure, and functionality of nanomaterials, enabling the production of highly specialized nanoscale devices and systems [58].

4 Nanotechnology in Pest Management

The advantage of nanotechnology in pest management is the ability to develop nanoformulations of pesticides [59]. These nanoformulations involve the encapsulation of active ingredients within nanoparticles or other nanocarriers, which can improve the pesticide's stability, solubility, and controlled release [60]. Traditional pesticides often degrade quickly under environmental conditions, requiring frequent applications that increase costs and environmental exposure [35]. In contrast, nanoformulated pesticides can be designed to release their active ingredients slowly and in a targeted manner, reducing the frequency of applications and minimizing the amount of pesticide needed [61]. This targeted delivery not only enhances the effectiveness of pest control but also reduces the impact on nontarget species, such as beneficial insects and pollinators.

Moreover, nanotechnology enables the preparation of pesticide carriers that can be engineered to respond to specific environmental triggers [62]. For instance, smart nanocarriers can release their payload only in response to certain pH levels, temperatures, or the presence of specific enzymes associated with pests [63]. This level of precision ensures that the pesticide is released only where it is needed, further reducing environmental contamination and the risk of off-target effects [64]. Additionally, such smart delivery systems can help combat the problem of pesticide resistance by ensuring that pests are exposed to the correct dose of the active ingredient at the right time, preventing them from developing resistance [65].

Another promising application of nanotechnology in pest management is the use of nanosensors for early pest detection and monitoring [66]. Traditional pest control often relies on reactive measures, where pesticides are applied after a pest infestation has been identified [67]. However, by the time visible damage is observed, significant crop losses may have already occurred. Nanosensors, capable of detecting the presence of pests, their metabolites, or the early signs of plant stress at the molecular level, offer a proactive approach [68]. These sensors can be integrated into precision agriculture systems, providing real-time data that allows farmers to take timely action before pests cause significant damage [66]. This early intervention reduces the need for broad-spectrum pesticide applications, thereby lowering costs and minimizing environmental impact [68].

In addition to enhancing the delivery and monitoring of pesticides, nanotechnology also opens the door to novel pest-control strategies [66]. For example, researchers are exploring the use of nanoparticles as direct pest control agents [13]. Certain nanoparticles have shown the ability to interfere with the physiological processes of pests, such as by disrupting their digestive systems, altering their reproductive cycles, or damaging their protective exoskeletons [69]. These nanoparticles can be used alone or in combination with traditional pesticides to enhance their efficacy and reduce the likelihood of resistance development [70].

5 Formulations of Nanopesticides

The formulation of nanopesticides involves incorporating active pesticide ingredients into nanoscale carriers or converting them into nano-sized particles [71]. These formulations are designed to overcome many of the limitations associated with conventional pesticides, such as poor stability, rapid degradation, off-target effects, and the development of pest resistance [70]. By engineering the physical and chemical properties of pesticide formulations at the nanoscale, nanopesticides offer enhanced performance in terms of efficacy, safety, and sustainability [14].

One of the key components in the formulation of nanopesticides is the use of nanocarriers, which are materials designed to encapsulate, adsorb, or bind the active ingredients [72]. These carriers can be made from a variety of materials, including polymers, lipids, metals, or inorganic substances like silica and clay [73]. Nanocarriers provide several benefits, including protecting the active ingredient from environmental degradation, controlling its release over time, and improving its solubility and bioavailability [74]. For instance, encapsulating a pesticide in polymer-based nanocarriers can protect it from sunlight, moisture, and microbial degradation, extending its shelf life and reducing the need for repeated applications [75]. Moreover, the controlled release of the pesticide ensures that it is delivered in a sustained manner, enhancing its effectiveness and minimizing the amount required [76]. Another important aspect of nanopesticide formulation is the development of nanoemulsions, which are stable mixtures of oil and water that contain nano-sized droplets of the active ingredient [77]. Nanoemulsions are particularly advantageous for delivering hydrophobic (water-insoluble) pesticides, as they enhance the solubility and dispersion of these compounds in aqueous environments [77]. This improves the uniformity of pesticide application and ensures better coverage on plant surfaces. Additionally, nanoemulsions can penetrate plant tissues more effectively, allowing for systemic protection against pests that feed on internal plant structures [78]. This deep penetration capability reduces the likelihood of pests developing resistance, as they are exposed to the active ingredient at multiple stages of their life cycle [79].

Nanopesticide formulations can also include solid nanoparticles, which are small, solid particles that contain or coated with the active ingredient [80]. These nanoparticles can be engineered to have specific surface properties, such as positive or negative charges, which influence their interaction with pests and plants [81]. For example, positively charged nanoparticles are more likely to adhere to the negatively charged surfaces of plant leaves, enhancing the retention and uptake of the pesticide [82]. This increased adhesion not only improves the effectiveness of the pesticide but also reduces runoff and environmental contamination, addressing a major concern associated with conventional pesticide usage [76]. Furthermore, nanopesticides can be designed for targeted delivery, ensuring that the active ingredient is released specifically at the site of pest infestation [83]. This can be achieved through the use of stimuli-responsive nanocarriers, which release their payload in response to environmental triggers such as pH changes, temperature fluctuations, or the presence of specific enzymes produced by pests [84]. For example, pH-sensitive nanocarriers might release their pesticide load only in the acidic gut of an insect, ensuring that nontarget organisms and the environment are not exposed to the active ingredient [60]. This precision targeting reduces the overall environmental impact of pesticide use and enhances the safety profile of the product.

The formulation of nanopesticides also allows for the incorporation of multiple active ingredients within a single nanocarrier, enabling synergistic effects that can improve pest control outcomes [85]. By combining different modes of action, nanopesticides can target multiple pest species or life stages simultaneously, reducing the likelihood of resistance development [86]. For instance, a nanopesticide formulation might include both a contact insecticide and a systemic pesticide, providing immediate knockdown of pests as well as long-term protection [71].

The classification of nanopesticides is divided into two main categories: metals and their oxides used directly as nanopesticides, and composite systems that utilize carrier materials to deliver conventional pesticide active compounds. Metal and metal oxide nanopesticides possess potent antibacterial properties and can be easily fabricated into specific nanoscale sizes and geometries [87]. Another category of nanopesticides primarily consists of composite materials, such as polymers and inorganic nonmetals. These carriers load active compounds through adsorption or encapsulation and can provide controlled release in response to external stimuli [88]. These nanopesticides not only function as nano-delivery systems but also offer great potential for applications in crop nutrition, soil restoration, and stress mitigation.

5.1 Silver Nanoparticles

Silver nanoparticles (AgNPs) have garnered significant interest in agriculture due to their antimicrobial properties and potential for improving crop health. Synthesized from sources like pomegranate peel extract, AgNPs have been tested on pests such as *Spodoptera litura* and store grain pests like *Sitophilus oryzae*, demonstrating dose-dependent growth reduction and mortality rates ranging from 16.4% to 86.4% for different larval stages [89, 90]. When combined with other nanoparticles like zinc oxide (ZnO) and titanium oxide (TiO_2), similar effects have been observed, including reduced larval and pupal periods, adult emergence, and longevity [91].

AgNPs also show potential for use in pest control by inhibiting gut protease activity, as seen in *Helicoverpa armigera*, where a decrease in larval weight and longevity was noted [92]. The biological synthesis of AgNPs, using plant extracts like *Ficus religiosa*, is favored over chemical or physical methods due to its eco-friendly, cost-effective nature [93]. This bio-route avoids the need for harsh chemicals and extreme conditions, making it a sustainable approach for agricultural applications. The development of AgNP-based pest control strategies holds promise for sustainable agriculture, providing an innovative method to manage pests and enhance plant health [94].

5.2 Silica Nanoparticles

Silica nanoparticles (SNPs) have proven effective in pest control and agriculture applications due to their biocompatibility and efficient delivery system. Ayoub et al. demonstrated that amorphous silica nanoparticles exhibit high entomotoxicity against *Spodoptera littoralis*, causing up to 96% mortality within 3 days [95]. This effectiveness is primarily due to the nanoparticles' ability to impair the digestive tract and cause dehydration. Furthermore, the nanoparticles reduced both carbohydrate and protein levels, resulting in malformed pupae. In another study, Wibowo et al. [96] developed silica nanocapsules for the sustained release of fipronil, showing that thicker capsule walls reduced diffusion, leading to a slower release and 100% termite mortality in 6 days. Nasif et al. [97] used silica nanoparticles as a herbicide delivery system, creating a hexagonal mesoporous structure that reduced herbicide leaching and improved effectiveness. Similarly, Qin et al. [98] immobilized a biopesticide enzyme on functionalized silica nanoparticles, achieving enhanced nematicidal activity against *Caenorhabditis elegans*. These studies highlight the potential of silica nanoparticles in both pest control and agricultural sustainability, offering promising alternatives to traditional pesticides and delivery systems.

5.3 Titanium Dioxide Nanoparticles

Titanium dioxide nanoparticles (TiO_2NPs) possess unique properties that make them valuable in agriculture and environmental applications. TiO_2 is known for its hardness and anticorrosive properties, but it also enhances plant growth and photosynthesis, especially through its role as a photocatalyst [99]. This photocatalytic activity enables TiO_2 to degrade pesticides effectively, a feature that can be further enhanced by using thin films or dye-doping techniques [100]. The synthesis of TiO_2 NPs is often done using plant extracts, such as *Psidium guajava*, producing particles around 32.58 nm in size. These nanoparticles have shown strong antimicrobial activity, particularly against *Staphylococcus aureus* [101]. TiO_2NPs also positively impact plant growth. For instance, they induce root elongation in crops like rapeseed and wheat by increasing hydromineral flux [102]. However, their effects vary depending on the plant species and nanoparticle characteristics [103]. TiO_2NPs have been tested on a variety of crops, including spinach, wheat, and watermelon, with studies showing improvements in plant biomass, chlorophyll content, and germination rates [104].

5.4 Zinc Oxide Nanoparticles

Zinc oxide nanoparticles (ZnO NPs) are increasingly recognized for their potential in agriculture and antimicrobial applications. Zinc deficiency is a common issue in arid regions due to high calcium carbonate concentrations that raise soil pH and reduce zinc solubility [105]. Traditional zinc fertilizers like zinc sulfate and zinc oxide often show limited uptake by plants [106]. Nanostructured ZnO, with its high surface energy and dissolution rate, enhances zinc bioavailability in soil, improving crop nutrition [107]. ZnO NPs, particularly those under 100 nm, exhibit strong antimicrobial activity through the production of reactive oxygen species (ROS), which leads to cell death in pathogens [108]. Furthermore, ZnO NPs synthesized from biological extracts, such as *Moringa oleifera*, *Catharanthus roseus*, and *Parthenium hysterophorus*, have demonstrated effectiveness against pathogens like *Staphylococcus aureus* and *Aspergillus niger* [109]. ZnO NPs also show promise in postharvest pathogen control, inhibiting *Botrytis cinerea* and *Penicillium expansum*, two major contributors to fruit loss [110]. The use of ZnO in remediation, combined with fungicides like thiram, enhances pathogen inhibition while reducing environmental burdens [111].

5.5 Copper Nanoparticles

Copper nanoparticles (CuNPs) offer a cost-effective alternative to traditional copper-based fungicides, which face issues like resistance development and adverse effects on beneficial organisms [112]. CuNPs have shown significant efficacy in managing plant diseases at much lower concentrations than conventional copper compounds. For example, CuNPs effectively suppressed *Xanthomonas axonopodis* pv. *punicae* at 0.2 ppm, which is 10,000 times lower than commercial copper-oxychloride [112]. Additionally, CuNPs have demonstrated superior antifungal activity compared to traditional fungicides against pathogens like *Phoma destructiva* and *Fusarium oxysporum* [112]. The biological synthesis of CuNPs offers a sustainable approach to pest control, exhibiting strong insecticidal properties against pests like *Tribolium castaneum* and *Spodoptera littoralis* [113]. Despite promising results, further research is needed to fully understand the efficacy and formulation of CuNPs for broader agricultural applications [114].

5.6 Sulfur Nanoparticles

Sulfur, historically renowned for its fungicidal, miticidal, and insecticidal properties, has been used extensively both alone and in combination with lime. However, lime sulfur's high phytotoxicity and detrimental effects on leaf and fruit quality limit its effectiveness, particularly against apple scabs [115]. Additionally, its use as an acaricide can harm beneficial predatory species [116]. Nano-sulfur formulations, such as IARI nano-S (50–90 nm), have shown superior performance by inhibiting conidial germination of *Erysiphe cichoracearum* and causing significant structural damage to fungal spores, outperforming conventional sulfur products. Nano-sulfur also demonstrated enhanced efficacy against *Tetranychus urticae* [117].

5.7 Aluminum Nanoparticles

Nanostructured aluminum oxide (nanoalumina) has been explored as a potential solution for protecting stored grains, exhibiting insecticidal properties comparable to commercial insecticides. Research comparing nanoalumina with diatomaceous earth demonstrated superior or equal effectiveness across different insect species and humidity conditions [118]. Nanoalumina offers a promising alternative to traditional grain protection products, although its mode of action remains unclear. Further research is necessary to optimize its formulation and ensure its efficacy under various environmental conditions. Metal-based nanoparticles, including nanoalumina, have also demonstrated significant insecticidal activity against pests like *Sitophilus oryzae* [90].

5.8 Chitosan Nanoparticles

Chitosan, a biopolymer, has been recognized for its ability to induce systemic resistance, enhance yields, and regulate cell death and stomatal closure in plants [119]. Although bulk chitosan has limited antifungal activity due to its insolubility in water, chitosan nanoparticles (CNPs) have shown promising results as plant protectants. Copper-based CNPs (196.4 $\pm$ 2.2 nm) at 0.1% concentration effectively inhibited the growth of *Alternaria alternata*, *Macrophomina phaseolina*, and *Rhizoctonia solani* by 89.5%, 63.0%, and 60.1%, respectively [120]. Further studies demonstrated that Copper-CNPs (374.3 $\pm$ 8.2 nm) significantly improved tomato seed germination, seedling growth, and reduced disease severity in comparison to traditional treatments [121]. Recent research also highlights the efficacy of CNPs (90 $\pm$ 5 nm) in activating plant defense mechanisms, including upregulation of antioxidant genes and increased levels of phenolics and nitric oxide, at markedly lower doses than bulk chitosan [122].

5.9 Nano Encapsulation

Nanoencapsulation involves enclosing substances within nanometer-sized shells to create novel materials with enhanced properties. The core, or internal phase, is encased by a shell, or external phase, ranging from 1 to 100 nm [123]. This method is increasingly applied to pesticides, leveraging various nanomaterials such as solid lipid nanoparticles, nanoclays, and polymer-based nanomaterials [6, 124]. Polymer-based nanoencapsulation, particularly using biodegradable polymers, has gained attention due to its eco-friendly nature and ability to release pesticides in a controlled manner [125, 126]. Techniques like emulsion diffusion and solvent evaporation are commonly used to create nanocapsules, with polymers such as polyethylene glycol (PEG) and poly-ε-caprolactone (PCL) showing promise [106]. For instance, PEG-coated nanocapsules demonstrated effective pesticide delivery and stability [107], while PCL nanocapsules have shown time-dependent aqueous dispersion [109].

In addition to polymeric methods, nanoemulsions with nanoscale droplets are utilized for encapsulating hydrophobic pesticides. High-energy and low-energy methods, such as phase inversion temperature (PIT) and emulsion inversion point (EIP) methods, are employed to achieve nanoemulsions. For instance, PIT methods have been used to prepare kinetically stable nanoemulsions, resulting in significant improvements in bioefficacy [47]. Similarly, the EIP method has successfully produced stable O/W nanoemulsions from natural pyrethrins [97].

5.10 Nanospheres

Nanospheres, which are spherical nanostructures, are increasingly utilized for the uniform distribution of active compounds in various applications, including agro-based solutions. These nanospheres are characterized by their solid, spherical nature and a size range typically between 10 and 200 nm [96]. They are particularly effective in encapsulating active ingredients within a polymer matrix, offering consistent dispersion and stability.

The fabrication of nanospheres involves a process where nanostructures bind to polymer units, followed by polymerization and solidification, resulting in a spherical nanostructure [125]. For example, polycaprolactone (PCL) is often used due to its effectiveness in agro-applications. PCL-based nanospheres loaded with azadirachtin, an active compound, have shown an average diameter of 150–200 nm and can be stabilized for up to 2 months through surfactant treatment [107].

The emulsion-solvent diffusion method has been used to prepare azadirachtin-loaded PCL nanospheres, which are then freeze-dried to preserve their UV stability [98]. Similarly, the flash nanoprecipitation technique has been employed to create polymer-stabilized bifenthrin nanoparticles, which are utilized for their pesticidal properties [99]. Nanospheres can be made from various materials, including inorganic compounds and polymers, and are known for their high drug-loading capacities and effectiveness in protecting active ingredients from environmental degradation [100]. They also offer benefits in pest control and plant protection, as evidenced by their use in controlling nematodes and enhancing pesticide delivery in agricultural settings [101].

5.11 Micelles

Micelles can encapsulate both water-insoluble and water-soluble agents. Water-insoluble agents are encapsulated using polymers or surfactants that self-assemble above the critical micelle concentration (CMC), whereas water-soluble agents can be incorporated through direct dissolution or film casting methods [112]. For instance, PEG-based amphiphilic block copolymers are commonly used for this purpose due to their stability and ability to form micelles with controlled size and properties [17].

Research indicates that micelles made from PEG and chitosan derivatives can encapsulate various pesticides, such as azadirachtin and imidacloprid, effectively. These micelles typically range in size from 26 to 500 nm and provide controlled release properties that are responsive to environmental conditions [48]. For example, imidacloprid-loaded micelles demonstrated a slower release profile compared to conventional formulations, offering prolonged pesticide efficacy and stability [17].

Micelles also exhibit responsiveness to external stimuli, such as temperature and light, which enhances their utility in controlled release systems. For instance, nanomicelles with cross-linked hydrogen bonds can swell or shrink in response to environmental changes, facilitating the regulated release of encapsulated pesticides [14]. This characteristic is particularly valuable for developing sustainable and efficient pest management solutions in agriculture.

5.12 Nanogels

Nanogels are nanoscale hydrogel particles formed through the physical or chemical cross-linking of polymer networks in aqueous environments. These versatile materials are utilized primarily as carriers for active compounds in drug delivery systems. Key methods for synthesizing nanogels include polymerization in homogeneous or heterogeneous phases, cross-linking of preformed polymers, template-assisted nanofabrication, and physical assembly of interactive polymers [15]. Nanogels are characterized by their ability to swell in water, which aids in the encapsulation of active materials via van der Waals, electrostatic, or hydrophobic interactions.

In agro-based applications, nanogels have been employed for pesticide delivery. For instance, researchers have used chitosan and cashew gum to encapsulate Lippia sidoides oil, enhancing its performance and extending its fumigant toxicity. Similarly, encapsulation of *Carum copticum* oil using myristic acid was reported, to achieve prolonged efficacy. Researchers developed chitosan (CHI) nanogels loaded with copper ions for combating *Fusarium graminearum* in cereal crops. This method, involving physical gelation in reverse emulsion, demonstrated significant efficacy due to the synergistic effect between CHI and copper ions [106].

Nanogels have also shown promise in pest management. Researchers prepared a pheromone-loaded nanogel using a low-molecular-mass gelator, which effectively controlled *Bactrocera dorsalis* populations. This simple and environmentally friendly method improved the stability and effectiveness of the pheromone bait [105].

5.13 Lipid-Modified Nanopesticides

Lipid-based nanomaterials are increasingly recognized for their effectiveness in delivering bioactive substances due to their ability to encapsulate both hydrophilic and hydrophobic materials [15]. Solid-lipid conjugated nanoparticles, first introduced in 1991, are notable for their potential in controlled-release systems and are considered superior to other colloidal carriers like liposomes and nanoemulsions. These conjugates require specific synthesis conditions, including the use of solid lipids, surfactants, and emulsifiers [112]. The primary synthesis methods are hot and cold homogenization, with the latter addressing the issue of hydrophilic compound loss seen in hot homogenization. The loading capacity of these conjugates is influenced by the lipid's chemical and physical properties and the solubility of the active compounds. Various techniques and emulsifiers, such as phospholipids, are employed to stabilize the solid lipid systems and enhance their effectiveness [123].

Nanoliposomes, nanoscale versions of traditional liposomes, maintain fundamental properties similar to their bulk counterparts [115]. They are widely used for encapsulating and delivering bioactive substances in agricultural and biological fields [119]. These structures offer increased surface area, enhanced solubility, and improved bioavailability of active compounds, making them suitable for controlled, precise release at target sites [11]. Nanoliposomes are vesicles with a bilayer lipid membrane and aqueous interior, forming phospholipids when dissolved in water. Inadequate energy distribution during formation leads to flat membrane-like structures. Synthesis methods for nanoliposomes include both mechanical techniques (e.g., ultrasonication and high-pressure homogenization) and nonmechanical techniques (e.g., reversed-phase evaporation). Cholesterol is often added to stabilize these structures by modulating fluidity and preventing crystallization. Modern methods such as freeze-drying and super-critical fluid technology further enhance stability. For example, etofenprox encapsulated in lecithin and cholesterol was successfully reduced to 150 nm with improved performance through ultrasonic homogenization and chitosan coating. Despite their advantages, nanoliposomes face issues like low payload and rapid release, which can be improved through better synthesis techniques [127].

5.14 Nanopesticides from Botanical Sources

The emergence of resistant fungal pathogens has been attributed to the excessive use of traditional synthetic fungicides. Green synthesis for fabricating metal nanoparticles (NPs) has gained prominence as an eco-friendly and nontoxic method for producing nanopesticides. AgNPs synthesized from plant extracts have demonstrated remarkable bio-efficacy in controlling *Alternaria zinniae*, the causative agent of leaf spot in marigolds, by effectively reducing mycelial growth [16]. AgNPs produced from *Saccharum officinarum* leaf biomass exhibited significant inhibition against fungal pathogens such as *Cladosporium fulvum*, *Colletotrichum acutatum*, and *Phytophthora capsica* [19]. Similarly, green-synthesized AgNPs from *Amaranthus retroflexus* proved effective in controlling *A. alternata*, *Macrophomina phaseolina*, and *Fusarium oxysporum* [17]. AgNPs synthesized using *Tagetes patula* leaf extracts displayed high toxicity against *Colletotrichum chlorophyti*. Further studies show *Artemisia absinthium*-derived AgNPs inhibiting *Phytophthora* infections. Additionally, *Aspergillus versicolor*-mediated AgNPs exhibited antifungal properties against strawberry pathogens *Sclerotinia sclerotiorum* and *Botrytis cinerea*. Moreover, *Prunus yedoensis*-derived platinum nanoparticles (PtNPs) inhibited the growth of *C. acutatum* and *C. fulvum* [41].

Microbial nanotechnology offers an efficient biosynthetic approach for large-scale nanoparticle production. The fungus *Trichoderma longibrachiatum* facilitates extracellular biosynthesis of silver nanoparticles (AgNPs), which significantly inhibit fungal pathogens like *Fusarium verticillioides*, *Pyricularia grisea*, and *Helminthosporium oryzae*, achieving up to 90% efficiency [18]. Similarly, *Trichoderma harzianum*-derived AgNPs suppress sclerotia production in *Sclerotinia sclerotiorum* [39]. *Taxus baccata*-mediated AgNPs exhibit potent antibacterial activity, particularly against *Bacillus subtilis*. Moreover, *Cryptococcus laurentii* and *Rhodotorula glutinis* synthesized nanoparticles demonstrate comparable effectiveness to traditional fungicides like iprodione in controlling postharvest fungal pathogens [117]. Additionally, *Bipolaris maydis* biosynthesis of AgNPs proves effective against *Exserohilum turcicum*, which causes northern corn leaf blight. These mycosynthesized nanoparticles present eco-friendly alternatives for combating diverse plant pathogens, showcasing their potential in sustainable agriculture (Table 1).

6 Nanopesticides: Mode of Action

Nanopesticides represent a cutting-edge advancement in pest management, utilizing nanotechnology to enhance the efficacy and precision of pest control [21]. Understanding the mode of action of nanopesticides is crucial for optimizing their use, ensuring their safety, and maximizing their effectiveness [7]. Nanopesticides operate through several unique mechanisms (Fig. 2), leveraging their nanoscale properties to overcome many limitations associated with conventional pesticides [73].

Table 1 The key differences between conventional pesticides and nano pesticides

Aspect	Conventional pesticides	Nanopesticides	Reference
Particle Size	Typically in the micron range (larger particles)	Nanoscale particles, typically 1–100 nm	[8]
Efficacy	Requires higher concentrations for effectiveness	Higher efficacy at lower concentrations due to increased surface area and reactivity	[72]
Targeting	Less specific, affects a broader range of organisms, including nontarget species	Can be designed for targeted delivery to specific pests or sites	[83]
Environmental Impact	Often results in environmental contamination and persistence	Reduced environmental impact due to lower usage rates and potential for biodegradability	[13]
Degradation	Can persist in the environment, leading to bioaccumulation and pollution	Controlled degradation, with some formulations designed to break down after use	[19]
Resistance Development	High risk of pests developing resistance due to broad application	Lower risk of resistance due to novel modes of action and targeted delivery	[14]
Application Frequency	Requires frequent reapplication due to degradation and lower efficiency	The less frequent application is needed due to controlled release and higher efficacy	[74]
Toxicity to Nontarget Species	Often toxic to beneficial organisms, such as pollinators and soil microbes	Can be designed to minimize impact on nontarget species, including beneficial organisms	[27]
Regulatory Challenges	Well-established regulatory frameworks exist	Emerging regulatory frameworks, with challenges due to novel properties	[18]
Cost	Generally lower production costs but higher usage rates	Potentially higher initial production costs but lower overall usage	[22]
Stability	Can degrade quickly under environmental conditions	Enhanced stability through encapsulation and controlled release mechanisms	[91]
Mode of Action	Often chemical, with broad-spectrum effects	Can include physical, chemical, and biological modes of action, offering greater flexibility	[14]

Several nano-based insecticides, such as imidacloprid [12], thiamethoxam, and thiram [96], are currently available. Nanopesticides reduce the toxic effects of conventional pesticides by minimizing the pesticide usage rate [6]. Researchers demonstrated the bioefficacy of nano-deltamethrin, finding that its strength could be reduced by 5–10 times compared to commercial formulations. Nano-calcium carbonate improved pest control against California red scale and oriental fruit flies [27]. Nanoinsecticidal alumina dust outperformed conventional powders for controlling stored pests such as *Sitophilus oryzae* and *Rhyzopertha dominica* [90]. Hydrophobic silica nanoparticles effectively treated powdery mildew in

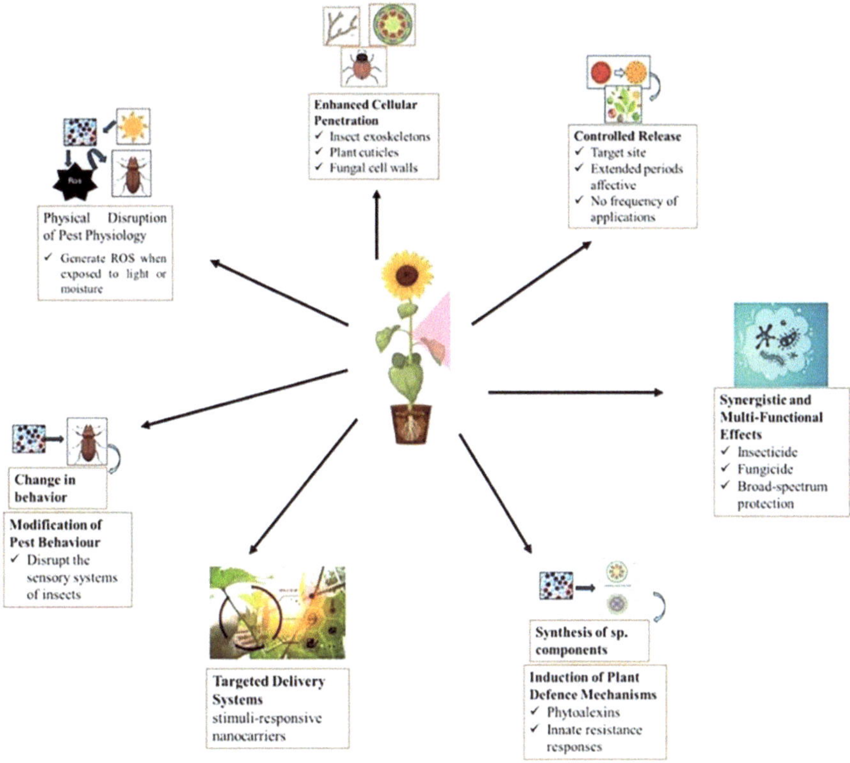

Fig. 2 Nanopesticides and their mode of action against pests

crops and beetle infestations in storage. Moreover, nano-silica applied to squash plants at 500 ppm significantly controlled *Spodoptera littoralis* [28].

Plant-derived nanoparticles (NPs) show promising insecticidal properties. For example, citrus peel essential oil nanoparticles (EONPs) have proven effective against the tomato pest *Tuta absoluta*, offering increased pesticidal activity with reduced toxicity. Camphor essential oil nanoemulsions are active against pests like *Sitophilus granarius* [87]. Nanospheres containing oils from *Zanthoxylum rhoifolium* exhibit antimicrobial effects against *Bemisia tabaci* [109]. NeemAzal, impregnated with polyvinyl alcohol/alginate-montmorillonite (PVA/Alg-MMT), showed enhanced pest control [82]. Nano-diatomaceous earth (nano-DE) was found effective against pests like *Tribolium confusum* and *Tribolium castaneum* [43]. Geranium essential oil-loaded lipid nanoparticles showed high effectiveness against *Phthorimaea operculella* [66].

Green silver nanoparticles (GSNs) have emerged as an eco-friendly alternative to synthetic nematicides for controlling root-knot nematodes, a major pest affecting eggplants (*Solanum melongena*). The GSNs were synthesized using extracts from two algal species, *Ulva lactuca* and *Turbinaria turbinata*. Under controlled conditions, GSNs significantly reduced the second juvenile (J2) stage of *Meloidogyne javanica* in soil, demonstrating effective nematode control comparable to chemical methods. Importantly, GSNs did not cause phytotoxic effects on the treated eggplants, making them a safer, environmentally sustainable solution for nematode management [78].

Copper-based bactericides are commonly used in agriculture to protect crops, but they raise concerns about environmental impact and bacterial resistance. Nanotechnology offers solutions by enhancing copper's bioavailability. Nano-copper-loaded silica gel (ZnO-nCuSi) exhibits strong antimicrobial properties and effectively controls citrus canker at lower copper doses than traditional formulations [103]. The AgNPs synthesized using plant extracts like neem and tulsi, inhibit the growth of *Xanthomonas axonopodis*, the bacterium causing bacterial blight in pomegranates [68]. The AgNPs also prevent seedling infection in melons caused by *Acidovorax citrulli*. Furthermore, magnesium oxide nanoparticles (MgONPs) reduce bacterial wilt in tomatoes by inducing systemic resistance. Copper oxide nanoparticles (CuONPs), derived from papaya leaves, effectively manage bacterial wilt in tobacco caused by *Ralstonia solanacearum* [103].

Nanomaterials have emerged as promising antiviral agents, leveraging their unique properties to combat viral infections in plants. The intricate ability of viruses to infect host cells, deliver their nucleic acid genomes, self-replicate, and exit the host systemically makes them suitable targets for nanotechnology [110]. Silver nanoparticles (AgNPs), synthesized from newly isolated bacterial strains, such as *Bacillus pumilus*, *Bacillus persicus*, and *Bacillus licheniformis*, demonstrate efficacy in controlling bean yellow mosaic virus infections in faba beans [93]. Research into antiviral nanomaterials holds significant promise for managing detrimental viral diseases in crop plants.

The land snail (*Eobania vermiculata*) poses significant threats to agricultural productivity. Traditionally, metaldehyde has been employed to mitigate damage from snails, slugs, and mollusks; however, this commercially available molluscicide has drawbacks, including high cost, nonspecificity, and harmful effects on nontarget organisms. The AgNPs, biologically synthesized from *Raphanus sativus* var. *aegyptiacus*, have proven effective in controlling land snail populations [127]. Additionally, researchers developed a molluscicidal polymeric nanoparticle made from polylactic acid, encapsulating curcumin and nisin, which demonstrated substantial efficacy against *Biomphalaria pfeifferi*. The following modes of action are reported by the nanopesticides [42].

6.1 Enhanced Cellular Penetration

One of the most notable advantages of nanopesticides is their ability to penetrate biological barriers more effectively than conventional pesticides [14]. Due to their extremely small size, nanoparticles can traverse insect exoskeletons, plant cuticles, and fungal cell walls more easily. This enhanced penetration allows nanoparticles to deliver active ingredients directly to their target sites within pests or plants. For instance, nanoparticles can enter the digestive systems of insects or penetrate plant tissues, ensuring that the pesticide reaches critical areas where it can disrupt vital physiological processes [85]. This improved delivery enhances the efficacy of the pesticide and reduces the need for high application rates.

6.2 *Targeted Delivery Systems*

Nanopesticides can release their active ingredients in a controlled and targeted manner [88]. This targeted delivery is achieved through various mechanisms, including using stimuli-responsive nanocarriers [70]. These carriers can be designed to release their payload only in response to specific environmental conditions, such as changes in pH, temperature, or the presence of certain enzymes [62]. For example, pH-sensitive nanoparticles might release their pesticide load only in the acidic environment of an insect's gut, thereby minimizing exposure to nontarget organisms and reducing environmental impact [14]. This level of precision ensures that the pesticide is applied only where it is needed, improving its effectiveness and reducing waste [66].

6.3 *Controlled Release Mechanisms*

Another significant feature of nanopesticides is their ability to provide a controlled release of active ingredients [39]. Conventional pesticides often degrade rapidly in the environment, leading to the need for frequent applications and potential environmental contamination. Nanopesticides, however, can be formulated to release their active components gradually over time. This controlled release maintains effective pesticide concentrations at the target site for extended periods, reducing the frequency of applications and the total amount of pesticide required [76]. The result is a more efficient pest control strategy that minimizes environmental residues and the risk of pesticide runoff [77].

6.4 *Synergistic and Multifunctional Effects*

Nanopesticides often incorporate multiple active ingredients or utilize synergistic effects to enhance their pest control capabilities [71]. By combining different modes of action within a single formulation, nanopesticides can address multiple pest species or life stages simultaneously. For instance, nanopesticides may include both an insecticide and a fungicide, providing broad-spectrum protection against various pests and diseases [117]. This synergistic approach not only improves the overall effectiveness of the pesticide but also helps mitigate the development of resistance in target organisms, a common problem with single-mode-of-action pesticides.

6.5 Physical Disruption of Pest Physiology

In addition to their chemical modes of action, some nanopesticides exert their effects through physical disruption of pest physiology [124]. Certain nanoparticles, such as metal oxides or carbon-based materials, can generate reactive oxygen species (ROS) when exposed to light or moisture. These ROS can damage cellular structures, disrupt metabolic processes, and ultimately lead to the death of pests. This physical mechanism of action offers an alternative to chemical-based pesticides and can reduce the likelihood of resistance development, as it does not rely on biochemical interactions [17].

6.6 Induction of Plant Defense Mechanisms

Nanopesticides can also enhance plant defense mechanisms, offering an indirect mode of action [129]. By interacting with plant tissues, nanoparticles can trigger the production of defensive compounds such as phytoalexins or enhance the plant's innate resistance responses. This induction of natural defense mechanisms provides additional protection against pests, reducing the need for external pesticide applications [37]. Nanoparticles can act as elicitors, promoting the plant's defense systems and contributing to overall pest management strategies.

6.7 Modification of Pest Behavior

Some nanopesticides can alter pest behavior, further contributing to their effectiveness. For example, nanoparticles might disrupt the sensory systems of insects, making them less likely to locate or feed on plants [57]. This behavioral modification reduces the overall impact of pests on crops and complements the direct killing action of the pesticide. By influencing pest behavior, nanopesticides can contribute to a more integrated approach to pest management [89].

7 Applications of Nanopesticides

Nanopesticides, utilizing the principles of nanotechnology, are increasingly being recognized for their potential to revolutionize pest management in agriculture. By harnessing the unique properties of nanoscale materials, nanopesticides offer a range of applications that improve the efficacy, sustainability, and safety of pest control methods [73]. These applications address many of the limitations associated with traditional pesticides, such as environmental contamination, pest resistance, and nontarget effects, thereby offering innovative solutions for modern agriculture [98].

7.1 Enhanced Crop Protection

Nanopesticides are designed to provide superior protection against a wide array of pests, including insects, fungi, bacteria, and weeds [130]. Their ability to penetrate plant tissues more effectively allows for targeted pest control and improved coverage. For example, nanopesticides can be formulated as nanoemulsions or nanoencapsulated formulations that enhance the solubility and dispersion of active ingredients. This results in better adhesion to plant surfaces and more efficient uptake by the plant. Enhanced crop protection reduces the overall incidence of pest damage and can lead to higher yields and better-quality produce [79].

7.2 Precision Agriculture

In precision agriculture, the goal is to optimize resource use and improve crop management through data-driven techniques. Nanopesticides fit seamlessly into this approach by providing targeted delivery and controlled release of active ingredients [88]. For instance, nanosensors can be integrated into precision agriculture systems to monitor pest populations and environmental conditions in real time [66]. Based on this data, nanopesticides can be applied in a precisely targeted manner, reducing the need for broad-spectrum pesticide applications. This precision minimizes waste, lowers costs, and reduces the environmental impact of pesticide use [77].

7.3 Controlled Release Systems

One of the significant advantages of nanopesticides is their ability to provide a controlled and sustained release of active ingredients [24]. Traditional pesticides often require multiple applications due to their rapid degradation or loss of efficacy over time [35]. Nanopesticides, on the other hand, can be engineered to release their active components gradually, maintaining effective concentrations at the target site for extended periods [119]. This controlled release not only improves pest control efficiency but also reduces the frequency of applications, thereby minimizing environmental residues and the risk of pesticide runoff.

7.4 Integrated Pest Management (IPM)

Nanopesticides are increasingly being integrated into IPM programs, which combine biological, cultural, mechanical, and chemical control methods to manage pests sustainably. In IPM, nanopesticides can complement biological control agents, such as natural predators or parasitoids, by providing targeted chemical control when needed [59]. Their ability to be precisely applied and to reduce the likelihood of resistance development makes them a valuable tool in IPM strategies [98]. By integrating nanopesticides, farmers can achieve a more balanced approach to pest management, enhancing overall effectiveness while minimizing negative environmental impacts.

7.5 Disease Vector Control

Nanopesticides also show promise in controlling disease vectors. Nanoencapsulated insecticides can be applied to breeding sites or environments where these vectors are prevalent, providing targeted and long-lasting control [131]. The controlled release properties of nanopesticides ensure that the insecticides remain effective over time, reducing the overall frequency of applications and minimizing the risk of resistance development in vector populations [63].

7.6 Specialty Crop Protection

Nanopesticides offer specific advantages for specialty crops, such as fruits, vegetables, and ornamental plants, which often require more delicate pest management approaches [90]. The ability of nanoparticles to deliver active ingredients precisely and with minimal impact on the surrounding environment makes them ideal for protecting high-value crops [85]. For instance, nanopesticides can be used to manage pests that are difficult to control with conventional methods, such as those that affect the appearance or quality of specialty crops [69]. This targeted approach ensures that the crop receives the necessary protection while minimizing the risk of damaging nontarget plants or beneficial organisms.

8 Regulatory Considerations and Safety Assessment

As nanotechnology continues to advance and integrate into agricultural practices, the development and use of nanopesticides have garnered significant attention [88]. While these innovations offer numerous advantages over traditional pesticides, they also introduce unique challenges regarding regulation and safety. Effective regulatory frameworks and thorough safety assessments are essential to ensure that nanopesticides are safe for human health, the environment, and nontarget organisms, while also providing the benefits promised by this technology [77].

8.1 Regulatory Frameworks

The regulation of nanopesticides is complex due to their novel characteristics, which differ significantly from those of conventional pesticides. Regulatory bodies such as the U.S. Environmental Protection Agency (EPA), the European Food Safety Authority (EFSA), and other international agencies are actively working to establish guidelines and standards for the registration and approval of nanopesticides [116]. These frameworks must account for the unique properties of nanoparticles, such as their size, shape, surface area, and reactivity, which can influence their behavior, efficacy, and safety [121]. One of the primary challenges in regulating nanopesticides is determining whether existing pesticide regulations are sufficient or whether new, nanoparticle-specific guidelines are needed [107]. In some regions, nanopesticides may be regulated under the same frameworks as conventional pesticides, with additional requirements for nanomaterial characterization. However, the potential for nanoparticles to interact differently with biological systems and the environment has led to calls for more stringent and tailored regulatory approaches.

8.2 Risk Assessment and Safety Evaluation

The safety assessment of nanopesticides involves a comprehensive evaluation of their potential risks to human health, nontarget organisms, and the environment [77]. This process includes the identification of hazards, exposure assessment, dose-response analysis, and risk characterization [22]. Given the unique properties of nanoparticles, traditional toxicological assessment methods may not always be applicable. Therefore, new methodologies and approaches are being developed to assess the specific risks associated with nanopesticides [87]. The potential risks to human health from nanopesticides are a major concern, particularly for farmworkers, consumers, and others exposed to these products. The small size of nanoparticles allows them to penetrate biological barriers, such as the skin, lungs, and digestive tract, more easily than larger particles [29]. As a result, there is a need for rigorous testing to determine the potential for toxicity, bioaccumulation, and other adverse effects. Animal-based inhalation studies, dermal exposure assessments, and ingestion tests are commonly used to evaluate human health risks, with a focus on determining safe levels of exposure and identifying any long-term health implications [58]. Nanopesticides can interact with the environment in different ways from conventional pesticides [31]. Their small size, high surface area, and potential for increased mobility raise concerns about their persistence, bioaccumulation, and potential to harm nontarget species. For example, nanoparticles may penetrate soil and water systems more readily, leading to unknown effects on soil microbiota, aquatic organisms, and other components of ecosystems [116]. Environmental risk assessments for nanopesticides must consider factors such as degradation, transport, and the potential for bioaccumulation in various environmental compartments [31]. One of the key considerations in the safety assessment of nanopesticides is their impact on nontarget organisms, including beneficial insects (e.g., pollinators like bees), birds, aquatic life, and soil organisms [100]. The selective toxicity of nanopesticides is one of their main advantages, but unintended effects on nontarget species remain a concern [108]. Regulatory bodies require comprehensive testing to assess the potential risks to these organisms, including acute and chronic toxicity studies, reproductive and developmental assessments, and behavioral studies [117].

8.3 Labeling and Transparency

Transparency in the labeling of nanopesticides is crucial for ensuring that end-users, such as farmers and consumers, are fully informed about the products they are using. Regulatory agencies may require specific labeling information related to the nano-particle content, safety precautions, and disposal instructions. Clear labeling helps mitigate risks by providing users with the knowledge needed to handle nanopesticides safely and responsibly.

8.4 Post-Market Surveillance and Monitoring

Even after a nano pesticide is approved and released onto the market, ongoing surveillance and monitoring are necessary to ensure that it does not pose unforeseen risks. Post-market surveillance programs track the long-term effects of nanopesticides on human health and the environment. This data helps regulatory agencies update safety guidelines and take corrective actions if necessary, such as revising usage recommendations or, in extreme cases, removing a product from the market.

8.5 International Harmonization

Given the global nature of agricultural trade, international harmonization of nanopesticide regulations is essential. Variations in regulatory approaches between countries can create trade barriers and complicate the approval process for new products. Efforts are being made to align regulatory frameworks and safety standards across different regions, promoting the safe and responsible use of nanopesticides worldwide.

9 Future Perspectives

Despite these challenges, the future of nanopesticides is bright, with several emerging trends and opportunities that could drive their development and adoption. Continued advancements in nanotechnology are likely to lead to the development of more sophisticated and effective nanopesticides. Innovations such as smart delivery systems, which release active ingredients in response to specific environmental triggers, could further enhance the precision and efficacy of pest control. The use of biodegradable nanoparticles and green synthesis methods could address some of the environmental and cost challenges associated with nanopesticides. Nanopesticides are well-suited for integration into precision agriculture systems, which aim to optimize inputs such as water, fertilizers, and pesticides. By combining nanopesticides with data-driven tools, such as remote sensing, GPS mapping, and predictive analytics, farmers can apply these products with pinpoint accuracy, reducing waste and maximizing crop protection. This integration could lead to significant improvements in agricultural sustainability and productivity. Addressing the challenges of nanopesticides requires collaboration between researchers, industry, regulators, and farmers. Multidisciplinary research efforts that combine expertise in nanotechnology, toxicology, agronomy, and environmental science are essential to developing safe and effective nanopesticides. Public–private partnerships and international collaborations can also play a key role in advancing the commercialization of these products while ensuring that they meet rigorous safety and efficacy standards.

10 Conclusion

In conclusion, while the advent of nanopesticides marks a significant leap forward in agricultural pest management, their integration into mainstream farming practices requires careful consideration. The enhanced precision, efficacy, and environmental safety offered by nanopesticides present a promising alternative to conventional chemical pesticides, potentially mitigating many of the issues associated with traditional pest control methods. However, the successful adoption of these novel technologies hinges on addressing the regulatory, economic, and environmental challenges that accompany their use. Developing robust regulatory frameworks to ensure the safe and effective deployment of nanopesticides is crucial, as is investing in research to better understand their long-term impacts on ecosystems and human health. Additionally, efforts must be made to make these technologies accessible and affordable to farmers worldwide, particularly in developing regions where resources may be limited. Collaboration between scientists, policymakers, and industry stakeholders will be essential in overcoming these hurdles and realizing the full potential of nanopesticides in sustainable agriculture. If these challenges are met, nanopesticides could play a pivotal role in ensuring food security, reducing environmental contamination, and promoting the health of agricultural ecosystems for future generations.

References

1. Oerke, E. C. (2006). Crop losses to pests. *Journal of Agricultural Science, 144*(1), 31–43. https://doi.org/10.1017/S0021859605005708
2. Padmavathi, S., & Anuradha, C. S. (2022). Nanotechnology in plant disease management-an overview. *Journal of Advanced Scientific Research, 13*(10), 01–06.
3. Popp, J., Pető, K., & Nagy, J. (2013). Pesticide productivity and food security. A review. *Agronomy for Sustainable Development, 33*, 243–255.
4. Donkor, A., Osei-Fosu, P., Dubey, B., Kingsford-Adaboh, R., Ziwu, C., & Asante, I. (2016). Pesticide residues in fruits and vegetables in Ghana: A review. *Environmental Science and Pollution Research, 23*, 18966–18987.
5. Gill, H. K., & Garg, H. (2014). Pesticide: environmental impacts and management strategies. *Pesticides-toxic Aspects, 8*(187), 10–5772.
6. Nuruzzaman, M. D., Rahman, M. M., Liu, Y., & Naidu, R. (2016). Nanoencapsulation, nanoguard for pesticides: A new window for safe application. *Journal of Agricultural and Food Chemistry, 64*(7), 1447–1483. https://doi.org/10.1021/acs.jafc.5b05214
7. Li, X., Chen, Y., Xu, J., Lynch, I., Guo, Z., Xie, C., & Zhang, P. (2023). Advanced nanopesticides: Advantage and action mechanisms. *Plant Physiology and Biochemistry, 108051*. https://doi.org/10.1016/j.plaphy.2023.108051
8. Bocca, B., Barone, F., Petrucci, F., Benetti, F., Picardo, V., Prota, V., & Amendola, G. (2020). Nanopesticides: Physico-chemical characterization by a combination of advanced analytical techniques. *Food and Chemical Toxicology, 146*, 111816. https://doi.org/10.1016/j.fct.2020.111816
9. Kumar, S., Chauhan, N., Gopal, M., Kumar, R., & Dilbaghi, N. (2015). Development and evaluation of alginate–chitosan nanocapsules for controlled release of acetamiprid. *International Journal of Biological Macromolecules, 81*, 631–637. https://doi.org/10.1016/j.ijbiomac.2015.08.062
10. Lal, R. (2013). Food security in a changing climate. *Ecohydrology & Hydrobiology, 13*(1), 8–21. https://doi.org/10.1016/j.ecohyd.2013.03.006
11. FAO, IFAD, & WFP. (2015). *The State of Food Insecurity in the World 2015. Meeting the 2015 international hunger targets: Taking stock of uneven progress*. FAO.
12. Cohen, J. E. (2005). Human population: The next half century. *Science, 309*(5731), 569–570. https://doi.org/10.1126/science.1088665
13. Deka, B., Babu, A., Baruah, C., & Barthakur, M. (2021). Nanopesticides: A systematic review of their prospects with special reference to tea pest management. *Frontiers in Nutrition, 8*, 686131. https://doi.org/10.3389/fnut.2021.686131
14. Chaud, M., Souto, E. B., Zielinska, A., Severino, P., Batain, F., Oliveira-Junior, J., & Alves, T. (2021). Nanopesticides in agriculture: Benefits and challenge in agricultural productivity, toxicological risks to human health and environment. *Toxics, 9*(6), 131. https://doi.org/10.3390/toxics9060131
15. Meijerink, G., & Roza, P. (2007). The role of agriculture in development. In *Markets, chains and sustainable development strategy and policy paper, no. 5*. Stichting DLO.

16. Singh, A., Shraogi, N., Verma, R., Saji, J., Kar, A. K., Tehlan, S., Ghosh, D., & Patnaik, S. (2024). Challenges in current pest management practices: Navigating problems and a way forward by integrating controlled release system approach. *Chemical Engineering Journal, 154989*. https://doi.org/10.1016/j.cej.2024.154989
17. Khan, A. S. (2017). Nanotechnology applications in food and agriculture. In *2017 IEEE 17th international conference on nanotechnology (IEEE-NANO)* (pp. 877–878). IEEE.
18. Gupta, A., Rayeen, F., Mishra, R., Tripathi, M., & Pathak, N. (2023). Nanotechnology applications in sustainable agriculture: An emerging eco-friendly approach. *Plant Nano Biology, 4*, 100033.
19. Saritha, G. N. G., Anju, T., & Kumar, A. (2022). Nanotechnology-big impact: How nanotechnology is changing the future of agriculture? *Journal of Agriculture and Food Research, 10*, 100457. https://doi.org/10.1016/j.jafr.2022.100457
20. Usman, M., Farooq, M., Wakeel, A., Nawaz, A., Cheema, S. A., ur Rehman, H., et al. (2020). Nanotechnology in agriculture: Current status, challenges and future opportunities. *Science of the Total Environment, 721*, 137778. https://doi.org/10.1016/j.scitotenv.2020.137778
21. Prasad, R., Bhattacharyya, A., & Nguyen, Q. D. (2017). Nanotechnology in sustainable agriculture: Recent developments, challenges, and perspectives. *Frontiers in Microbiology, 8*, 1014. https://doi.org/10.3389/fmicb.2017.01014
22. Lykogianni, M., Bempelou, E., Karamaouna, F., & Aliferis, K. A. (2021). Do pesticides promote or hinder sustainability in agriculture? The challenge of sustainable use of pesticides in modern agriculture. *Science of the Total Environment, 795*, 148625. https://doi.org/10.1016/j.scitotenv.2021.148625
23. Carvalho, F. P. (2006). Agriculture, pesticides, food security and food safety. *Environmental Science & Policy, 9*(7–8), 685–692. https://doi.org/10.1016/j.envsci.2006.08.002
24. Meunier, E., Smith, P., Griessinger, T., & Robert, C. (2024). Understanding changes in reducing pesticide use by farmers: Contribution of the behavioural sciences. *Agricultural Systems, 214*, 103818. https://doi.org/10.1016/j.agsy.2023.103818
25. Chagnon, M., Kreutzweiser, D., Mitchell, E. A., Morrissey, C. A., Noome, D. A., & Van der Sluijs, J. P. (2015). Risks of large-scale use of systemic insecticides to ecosystem functioning and services. *Environmental Science and Pollution Research, 22*, 119–134. https://doi.org/10.1007/s11356-014-3277-x
26. Pimentel, D. (2009). Pesticides and pest control. *Integrated pest management: innovation-development process: volume 1*, 83–87. https://doi.org/10.1007/978-1-4020-8992-3_3.
27. Huber, D., Römheld, V., & Weinmann, M. (2012). Relationship between nutrition, plant diseases and pests. In *Marschner's mineral nutrition of higher plants* (pp. 283–298). Academic Press. https://doi.org/10.1016/B978-0-12-384905-2.00010-8
28. Ayaz, M., Li, C. H., Ali, Q., Zhao, W., Chi, Y. K., Shafiq, M., et al. (2023). Bacterial and fungal biocontrol agents for plant disease protection: Journey from lab to field, current status, challenges, and global perspectives. *Molecules, 28*(18), 6735. https://doi.org/10.3390/molecules28186735
29. Yadav, A. N., Kour, D., Kaur, T., Devi, R., & Yadav, N. (2020). Functional annotation of agriculturally important fungi for crop protection: Current research and future challenges. *Agriculturally important fungi for sustainable agriculture: volume 2: functional annotation for crop protection*, 347–356. https://doi.org/10.1007/978-3-030-48474-3_12.
30. Kim, B. S., & Hwang, B. K. (2007). Microbial fungicides in the control of plant diseases. *Journal of Phytopathology, 155*(11–12), 641–653. https://doi.org/10.1111/j.1439-0434.2007.01314.x
31. de la Cruz Quiroz, R., Cruz Maldonado, J. J., Rostro Alanis, M. D. J., Torres, J. A., & Parra Saldívar, R. (2019). Fungi-based biopesticides: Shelf-life preservation technologies used in commercial products. *Journal of Pest Science, 92*, 1003–1015. https://doi.org/10.1007/s10340-019-01117-5

32. Rossi, V., Caffi, T., Salotti, I., & Fedele, G. (2023). Sharing decision-making tools for pest management may foster the implementation of integrated pest management. *Food Security, 15*(6), 1459–1474. https://doi.org/10.1007/s12571-023-01402-3
33. Bueno, A. D. F., Sutil, W. P., Jahnke, S. M., Carvalho, G. A., Cingolani, M. F., Colmenarez, Y. C., & Corniani, N. (2023). Biological control as part of the soybean integrated pest management (IPM): Potential and challenges. *Agronomy, 13*(10), 2532. https://doi.org/10.3390/agronomy13102532
34. Wakhungu, N. C. (2023). Loss of soil biodiversity through judicious use of synthetic pesticides, a case study of trans-Nzoia County Kenya-review. *Scientific Reports in Life Sciences (SRLS), 4*, 4.
35. Leskovac, A., & Petrović, S. (2023). Pesticide use and degradation strategies: Food safety, challenges and perspectives. *Food, 12*(14), 2709. https://doi.org/10.3390/foods12142709
36. Boonupara, T., Udomkun, P., Khan, E., & Kajitvichyanukul, P. (2023). Airborne pesticides from agricultural practices: A critical review of pathways, influencing factors, and human health implications. *Toxics, 11*(10), 858. https://doi.org/10.3390/toxics11100858
37. Ayilara, M. S., Adeleke, B. S., Akinola, S. A., Fayose, C. A., Adeyemi, U. T., Gbadegesin, L. A., et al. (2023). Biopesticides as a promising alternative to synthetic pesticides: A case for microbial pesticides, phytopesticides, and nanobiopesticides. *Frontiers in Microbiology, 14*, 1040901. https://doi.org/10.3389/fmicb.2023.1040901
38. Onstad, D. W., & Knolhoff, L. M. (2023). IPM and insect resistance management. In *Insect resistance management* (pp. 527–549). Academic Press. https://doi.org/10.1016/B978-0-12-823787-8.00010-6
39. Pinheiro, R. A., Duque, T. S., Barroso, G. M., Soares, M. A., Cabral, C. M., Zanuncio, J. C., & Dos Santos, J. B. (2023). Herbicides may threaten advances in biological control of diseases and pests. *Environmental Science and Pollution Research, 30*(52), 111850–111870. https://doi.org/10.1007/s11356-023-30198-0
40. Nath, R., Singh, H., & Mukherjee, S. (2023). Insect pollinators decline: An emerging concern of Anthropocene epoch. *Journal of Apicultural Research, 62*(1), 23–38. https://doi.org/10.1080/00218839.2022.2088931
41. Zanchi, M. M., Marins, K., & Zamoner, A. (2023). Could pesticide exposure be implicated in the high incidence rates of depression, anxiety and suicide in farmers? A systematic review. *Environmental Pollution, 331*, 121888. https://doi.org/10.1016/j.envpol.2023.121888
42. Sabarwal, A., Kumar, K., & Singh, R. P. (2018). Hazardous effects of chemical pesticides on human health–cancer and other associated disorders. *Environmental Toxicology and Pharmacology, 63*, 103–114. https://doi.org/10.1016/j.etap.2018.08.018
43. Faria, N. M. X., Rosa, J. A. R. D., & Facchini, L. A. (2009). Poisoning by pesticides among family fruit farmers, Bento Gonçalves, southern Brazil. *Revista de Saúde Pública, 43*, 335–344. https://doi.org/10.1590/S0034-89102009005000014
44. Rashid, A., Schutte, B. J., Ulery, A., Deyholos, M. K., Sanogo, S., Lehnhoff, E. A., & Beck, L. (2023). Heavy metal contamination in agricultural soil: Environmental pollutants affecting crop health. *Agronomy, 13*(6), 1521. https://doi.org/10.3390/agronomy13061521
45. Ratnadass, A., Fernandes, P., Avelino, J., & Habib, R. (2012). Plant species diversity for sustainable management of crop pests and diseases in agroecosystems: A review. *Agronomy for Sustainable Development, 32*, 273–303. https://doi.org/10.1007/s13593-011-0022-4
46. Ratner, M. A., & Ratner, D. (2003). *Nanotechnology: A gentle introduction to the next big idea*. Prentice Hall Professional.
47. Ochigbo, S. S., & Sadiku, R. E. (2024). Introduction to the synthesis and characterization of nanostructured materials. In *Nanostructured materials for biomedical applications* (pp. 1–34). Elsevier. https://doi.org/10.1016/B978-0-323-90838-2.00004-7
48. Ariga, K. (2024). Materials nanoarchitectonics at dynamic interfaces: Structure formation and functional manipulation. *Materials, 17*(1), 271. https://doi.org/10.3390/ma17010271
49. Belay, T., Worku, L. A., Bachheti, R. K., Bachheti, A., & Husen, A. (2023). Nanomaterials: Introduction, synthesis, characterization, and applications. In *Advances in smart*

nanomaterials and their applications (pp. 1–21). Elsevier. https://doi.org/10.1016/B978-0-323-99546-7.00027-6
50. Ulusoy, U. (2023). A review of particle shape effects on material properties for various engineering applications: From macro to nanoscale. *Minerals, 2023*(13), 91. https://doi.org/10.3390/min13010091
51. Bhar, M., Bhunia, N., Debnath, G. H., Waldeck, D. H., & Mukherjee, P. (2024). Optical properties of semiconductor nanoparticles doped with 3d and 4f block elements. *Chemical. Physics Reviews, 5*, 1. https://doi.org/10.1063/5.0172408
52. Suhag, D., Kaushik, S., & Taxak, V. B. (2024). Nanomaterials in biomedicine. In *Handbook of biomaterials for medical applications, Volume 1: Fundamentals* (pp. 91–120). Springer Nature Singapore. https://doi.org/10.1007/978-981-97-4818-1_4
53. Mekuye, B., & Abera, B. (2023). Nanomaterials: An overview of synthesis, classification, characterization, and applications. *Nano Select, 4*(8), 486–501. https://doi.org/10.1002/nano.202300038
54. Jagadeesh, P., Rangappa, S. M., & Siengchin, S. (2024). Advanced characterization techniques for nanostructured materials in biomedical applications. *Advanced Industrial and Engineering Polymer Research, 7*(1), 122–143. https://doi.org/10.1016/j.aiepr.2023.03.002
55. Boddula, R., Borane, N., Odedara, N., & Singh, J. (2024). Fabricating and designing graphene-based nanomaterials using different Current 'Top-down' and 'bottom-up' techniques. In *Graphene-based nanomaterials* (pp. 33–46). CRC Press.
56. Pakzad, S. Z., Akinci, S., Karimzadehkhouei, M., & Alaca, B. E. (2023). Simplified top-down fabrication of sub-micron silicon nanowires. *Semiconductor Science and Technology, 38*(12), 125005. https://doi.org/10.1088/1361-6641/ad0791
57. Vinnacombe-Willson, G. A., Conti, Y., Stefancu, A., Weiss, P. S., Cortés, E., & Scarabelli, L. (2023). Direct bottom-up in situ growth: A paradigm shift for studies in wet-chemical synthesis of gold nanoparticles. *Chemical Reviews, 123*(13), 8488–8529. https://doi.org/10.1021/acs.chemrev.2c00914
58. Saleh, T. A. (2024). Trends in nanomaterial types, synthesis methods, properties and uses: Toxicity, environmental concerns and economic viability. *Nano-Structures & Nano-Objects, 37*, 101109. https://doi.org/10.1016/j.nanoso.2024.101109
59. Yousef, H. A., Fahmy, H. M., Arafa, F. N., Abd Allah, M. Y., Tawfik, Y. M., El Halwany, K. K., et al. (2023). Nanotechnology in pest management: Advantages, applications, and challenges. *International Journal of Tropical Insect Science, 43*(5), 1387–1399. https://doi.org/10.1007/s42690-023-01053-z
60. Tao, R., You, C., Qu, Q., Zhang, X., Deng, Y., Ma, W., & Huang, C. (2023). Recent advances in the design of controlled-and sustained-release micro/nanocarriers of pesticide. *Environmental Science Nano, 10*(2), 351–371. https://doi.org/10.1039/D2EN00446A
61. Osman, Y. A., Eltanahy, E., Hegazy, A. S., & Razak, A. A. (2023). Nanoformulations of agrochemicals used to apply pesticides and fertilizers for crop improvement. In *Nanoformulations for sustainable agriculture and environmental risk mitigation* (pp. 103–117). CABI.
62. Zhou, J., Liu, G., Guo, Z., Wang, M., Qi, C., Chen, G., et al. (2023). Stimuli-responsive pesticide carriers based on porous nanomaterials: A review. *Chemical Engineering Journal, 455*, 140167. https://doi.org/10.1016/j.cej.2022.140167
63. Fincheira, P., Hoffmann, N., Tortella, G., Ruiz, A., Cornejo, P., Diez, M. C., et al. (2023). Eco-efficient systems based on nanocarriers for the controlled release of fertilizers and pesticides: Toward smart agriculture. *Nanomaterials, 13*(13), 1978. https://doi.org/10.3390/nano13131978
64. Rathore, A., Hasan, W. N. M. R., Pujar, K., Singh, R., Panotra, N., & Satapathy, S. N. (2024). Nanotech for crop protection: Utilizing nanoparticles for targeted pesticide delivery. *Uttar Pradesh Journal of Zoology, 45*(6), 46–71. https://doi.org/10.56557/upjoz/2024/v45i63950

65. Kumar, S., Nehra, M., Dilbaghi, N., Marrazza, G., Hassan, A. A., & Kim, K. H. (2019). Nano-based smart pesticide formulations: Emerging opportunities for agriculture. *Journal of Controlled Release, 294*, 131–153. https://doi.org/10.1016/j.jconrel.2018.12.012
66. Zain, M., Ma, H., Chaudhary, S., Nuruzaman, M., Azeem, I., Mehmood, F., Rahman, S. U., Aiwang, D., & Sun, C. (2023). Nanotechnology in precision agriculture: Advancing towards sustainable crop production. In *Plant physiology and biochemistry* (Vol. 206, p. 108244). Elsevier. https://doi.org/10.1016/j.plaphy.2023.108244
67. Van den Bosch, R. (2023). *The pesticide conspiracy*. Univ of California Press.
68. Sharma, P., Pandey, V., Sharma, M. M. M., Patra, A., Singh, B., Mehta, S., & Husen, A. (2021). A review on biosensors and nanosensors application in agroecosystems. *Nanoscale Research Letters, 16*, 1–24. https://doi.org/10.1186/s11671-021-03593-0
69. Manna, S., Roy, S., Dolai, A., Ravula, A. R., Perumal, V., & Das, A. (2023). Current and future prospects of "all-organic" nanoinsecticides for agricultural insect pest management. *Frontiers in Nanotechnology, 4*, 1082128. https://doi.org/10.3389/fnano.2022.1082128
70. Wang, D., Saleh, N. B., Byro, A., Zepp, R., Sahle-Demessie, E., Luxton, T. P., Ho, K. T., Burgess, R. M., Flury, M., & White, J. C. (2022). Nano-enabled pesticides for sustainable agriculture and global food security. *Nature Nanotechnology, 17*(4), 347–360. https://doi.org/10.1038/s41565-022-01082-8
71. Gahukar, R. T., & Das, R. K. (2020). Plant-derived nanopesticides for agricultural pest control: Challenges and prospects. *Nanotechnology For Environmental Engineering, 5*, 1–9. https://doi.org/10.1007/s41204-020-0066-2
72. Okey-Onyesolu, C. F., Hassanisaadi, M., Bilal, M., Barani, M., Rahdar, A., Iqbal, J., & Kyzas, G. Z. (2021). Nanomaterials as nanofertilizers and nanopesticides: An overview. *ChemistrySelect, 6*(33), 8645–8663. https://doi.org/10.1002/slct.202102379
73. Zainab, R., Hasnain, M., Ali, F., Abideen, Z., Siddiqui, Z. S., Jamil, F., Hussain, M., & Park, Y.-K. (2024). Prospects and challenges of nanopesticides in advancing pest management for sustainable agricultural and environmental service. *Environmental Research*. Elsevier, *261*, 119722.
74. Ayala-Fuentes, J. C., & Chavez-Santoscoy, R. A. (2021). Nanotechnology as a key to enhance the benefits and improve the bioavailability of flavonoids in the food industry. *Food, 10*(11), 2701. https://doi.org/10.3390/foods10112701
75. Kalia, A., Sharma, S. P., Kaur, H., & Kaur, H. (2020). Novel nanocomposite-based controlled-release fertilizer and pesticide formulations: Prospects and challenges. In *Multifunctional hybrid nanomaterials for sustainable Agri-food and ecosystems* (pp. 99–134). Elsevier. https://doi.org/10.1016/B978-0-12-821354-4.00005-4
76. Singh, A., Dhiman, N., Kar, A. K., Singh, D., Purohit, M. P., Ghosh, D., & Patnaik, S. (2020). Advances in controlled release pesticide formulations: Prospects to safer integrated pest management and sustainable agriculture. *Journal of Hazardous Materials, 385*, 121525. https://doi.org/10.1016/j.jhazmat.2019.121525
77. Mustafa, I. F., & Hussein, M. Z. (2020). Synthesis and technology of nanoemulsion-based pesticide formulation. *Nanomaterials, 10*(8), 1608. https://doi.org/10.3390/nano10081608
78. Subramanya, S., Sumanth, K., Gupta, P. K., Chayapathy, V., Keshamma, E., & Murugan, K. (2022). Formulation of green nanoemulsions for controlling agriculture insects. In *Bio-based Nanoemulsions for Agri-food applications* (pp. 165–176). Elsevier. https://doi.org/10.1016/B978-0-323-89846-1.00002-4
79. Chaudhary, M., Sharma, S., Gupta, S., & Mohan, L. (2024). Phytogenic microemulsions: Boosting solubility, bioavailability and safety of insecticidal formulations against insect pests and vectors–a comprehensive review. *Environmental Nanotechnology, Monitoring & Management, 22*, 100980. https://doi.org/10.1016/j.enmm.2024.100980
80. Raj, S. N., Anooj, E. S., Rajendran, K., & Vallinayagam, S. (2021). A comprehensive review on regulatory invention of nano pesticides in agricultural nano formulation and food system. *Journal of Molecular Structure, 1239*, 130517. https://doi.org/10.1016/j.molstruc.2021.130517

81. Bian, T., Gardin, A., Gemen, J., Houben, L., Perego, C., Lee, B., Elad, N., Chu, Z., Pavan, G. M., & Klajn, R. (2021). Electrostatic co-assembly of nanoparticles with oppositely charged small molecules into static and dynamic superstructures. *Nature Chemistry, 13*(10), 940–949. https://doi.org/10.1038/s41557-021-00752-9
82. Avellan, A., Yun, J., Morais, B. P., Clement, E. T., Rodrigues, S. M., & Lowry, G. V. (2021). Critical review: Role of inorganic nanoparticle properties on their foliar uptake and in planta translocation. *Environmental Science & Technology, 55*(20), 13417–13431. https://doi.org/10.1021/acs.est.1c00178
83. Scott-Fordsmand, J. J., Fraceto, L. F., & Amorim, M. J. B. (2022). Nano-pesticides: The lunch-box principle—Deadly goodies (semio-chemical functionalised nanoparticles that deliver pesticide only to target species). *Journal of Nanobiotechnology, 20*(1), 13. https://doi.org/10.1186/s12951-021-01216-5
84. Xiao, D., Wu, H., Zhang, Y., Kang, J., Dong, A., & Liang, W. (2022). Advances in stimuli-responsive systems for pesticides delivery: Recent efforts and future outlook. *Journal of Controlled Release, 352*, 288–312. https://doi.org/10.1016/j.jconrel.2022.10.028
85. de Oliveira, J. L. (2021). Nano-biopesticides: Present concepts and future perspectives in integrated pest management. In *Advances in nano-fertilizers and nano-pesticides in agriculture* (pp. 1–27). Elsevier. https://doi.org/10.1016/B978-0-12-820092-6.00001-X
86. Santiago, E. F., Pontes, M. S., Arruda, G. J., Caires, A. R. L., Colbeck, I., Maldonado-Rodriguez, R., & Grillo, R. (2020). Understanding the interaction of nanopesticides with plants. In *Nanopesticides: From research and development to mechanisms of action and sustainable use in agriculture* (pp. 69–109). Springer. https://doi.org/10.1007/978-3-030-44873-8_4
87. Nuruzzaman, M., Liu, Y., Rahman, M. M., Dharmarajan, R., Duan, L., Uddin, A. F. M. J., & Naidu, R. (2019). Nanobiopesticides: Composition and preparation methods. In *Nanobiopesticides today and future perspectives* (pp. 69–131). Academic Press. https://doi.org/10.1016/B978-0-12-815829-6.00004-8
88. Jampílek, J., & Kráľová, K. (2017). Nanopesticides: Preparation, targeting, and controlled release. In *New pesticides and soil sensors* (pp. 81–127). Academic Press. https://doi.org/10.1016/B978-0-12-804299-1.00004-7
89. Bharani, R. A., & Namasivayam, S. K. R. (2017). Biogenic silver nanoparticles mediated stress on developmental period and gut physiology of major lepidopteran pest Spodoptera litura (Fab.)(Lepidoptera: Noctuidae)—An eco-friendly approach of insect pest control. *Journal of Environmental Chemical Engineering, 5*(1), 453–467. https://doi.org/10.1016/j.jece.2016.12.023
90. Buteler, M., Sofie, S. W., Weaver, D. K., Driscoll, D., Muretta, J., & Stadler, T. (2015). Development of nanoalumina dust as insecticide against Sitophilus oryzae and Rhyzopertha Dominica. *International Journal of Pest Management, 61*(1), 80–89. https://doi.org/10.1080/09670874.2014.1001008
91. Gutiérrez-Ramírez, J. A., Betancourt-Galindo, R., Aguirre-Uribe, L. A., Cerna-Chávez, E., Sandoval-Rangel, A., Ángel, E. D., et al. (2021). Insecticidal Effect of Zinc Oxide and Titanium Dioxide Nanoparticles against Bactericera cockerelli Sulc.(Hemiptera: Triozidae) on Tomato Solanum lycopersicum. *Agronomy, 11*, 1460. https://doi.org/10.3390/agronomy11081460
92. Pathipati, U. R., & Kanuparthi, P. L. (2021). Silver nanoparticles for insect control: Bioassays and mechanisms. In *Silver nanomaterials for agri-food applications* (pp. 471–494). Elsevier. https://doi.org/10.1016/B978-0-12-823528-7.00007-X
93. Saware, K., & Venkataraman, A. (2014). Biosynthesis and characterization of stable silver nanoparticles using Ficus religiosa leaf extract: A mechanism perspective. *Journal of Cluster Science, 25*, 1157–1171. https://doi.org/10.1007/s10876-014-0697-1
94. Kim, D. Y., Patel, S. K., Rasool, K., Lone, N., Bhatia, S. K., Seth, C. S., & Ghodake, G. S. (2023). Bioinspired silver nanoparticle-based nanocomposites for effective control of plant

pathogens: A review. *Science of the Total Environment, 168318*. https://doi.org/10.1016/j.scitotenv.2023.168318
95. Ayoub, H. A., Khairy, M., Rashwan, F. A., & Abdel-Hafez, H. F. (2017). Synthesis and characterization of silica nanostructures for cotton leaf worm control. *Journal of Nanostructure in Chemistry, 7*, 91–100. https://doi.org/10.1007/s40097-017-0229-2
96. Wibowo, D., Zhao, C. X., Peters, B. C., & Middelberg, A. P. (2014). Sustained release of fipronil insecticide in vitro and in vivo from biocompatible silica nanocapsules. *Journal of Agricultural and Food Chemistry, 62*(52), 12504–12511. https://doi.org/10.1021/jf504455x
97. Nasif, S. O., Nuruzzaman, M., & Naidu, R. (2024). Porous silica nanocarriers: Advances in structural orientation and modification to develop sustainable pesticide delivery systems. *ACS Agricultural Science & Technology, 4*(2), 144–172. https://doi.org/10.1021/acsagscitech.3c00436
98. Qin, X., Xiang, X., Sun, X., Ni, H., & Li, L. (2016). Preparation of nanoscale Bacillus thuringiensis chitinases using silica nanoparticles for nematicide delivery. *International Journal of Biological Macromolecules, 82*, 13–21. https://doi.org/10.1016/j.ijbiomac.2015.10.030
99. Sang, X., Fei, M., Sheng, L., Zhao, X., Yu, X., Hong, J., et al. (2014). Immunomodulatory effects in the spleen-injured mice following exposure to titanium dioxide nanoparticles. *Journal of Biomedical Materials Research Part A, 102*(10), 3562–3572. https://doi.org/10.1002/jbm.a.35034
100. Pelaez, M., Nolan, N. T., Pillai, S. C., Seery, M. K., Falaras, P., Kontos, A. G., et al. (2012). A review on the visible light active titanium dioxide photocatalysts for environmental applications. *Applied Catalysis B: Environmental, 125*, 331–349. https://doi.org/10.1016/j.apcatb.2012.05.036
101. Santhoshkumar, T., Rahuman, A. A., Jayaseelan, C., Rajakumar, G., Marimuthu, S., Kirthi, A. V., et al. (2014). Green synthesis of titanium dioxide nanoparticles using Psidium guajava extract and its antibacterial and antioxidant properties. *Asian Pacific Journal of Tropical Medicine, 7*(12), 968–976. https://doi.org/10.1016/S1995-7645(14)60171-1
102. Larue, C., Veronesi, G., Flank, A. M., Surble, S., Herlin-Boime, N., & Carrière, M. (2012). Comparative uptake and impact of TiO2 nanoparticles in wheat and rapeseed. *Journal of Toxicology and Environmental Health, Part A, 75*(13–15), 722–734. https://doi.org/10.1080/15287394.2012.689800
103. Margenot, A. J., Rippner, D. A., Dumlao, M. R., Nezami, S., Green, P. G., Parikh, S. J., & McElrone, A. J. (2018). Copper oxide nanoparticle effects on root growth and hydraulic conductivity of two vegetable crops. *Plant and Soil, 431*, 333–345. https://doi.org/10.1007/s11104-018-3741-3
104. Raliya, R., Biswas, P., & Tarafdar, J. C. (2015). TiO2 nanoparticle biosynthesis and its physiological effect on mung bean (Vigna radiata L.). *Biotechnology Reports, 5*, 22–26. https://doi.org/10.1016/j.btre.2014.10.009
105. Takkar, P. N., & Walker, C. D. (1993). The distribution and correction of zinc deficiency. In A. D. Robson (Ed.), *Zinc in soils and plants*. Academic.
106. Mortvedt, J. J., & Gilkes, R. J. (1993). Zinc fertilizers. In *Zinc in soils and plants: Proceedings of the international symposium on 'zinc in soils and plants' held at the University of Western Australia, 27–28 September, 1993* (pp. 33–44). Springer Netherlands. https://doi.org/10.1007/978-94-011-0878-2_3
107. Gangloff, W. J., Westfall, D. G., Peterson, G. A., & Mortvedt, J. J. (2006). Mobility of organic and inorganic zinc fertilizers in soils. *Communications in Soil Science and Plant Analysis, 37*(1–2), 199–209. https://doi.org/10.1080/00103620500403200
108. Xie, S., Zhu, J., Yang, D., Xu, Y., Zhu, J., & He, D. (2021). Low concentrations of zinc oxide nanoparticles cause severe cytotoxicity through increased intracellular reactive oxygen species. *Journal of Biomedical Nanotechnology, 17*(12), 2420–2432. https://doi.org/10.1166/jbn.2021.3209

109. Chikkanna, M. M., Neelagund, S. E., & Rajashekarappa, K. K. (2019). Green synthesis of zinc oxide nanoparticles (ZnO NPs) and their biological activity. *SN Applied Sciences, 1*(1), 117. https://doi.org/10.1007/s42452-018-0095-7
110. Singh, A., Singh, N. Á., Afzal, S., Singh, T., & Hussain, I. (2018). Zinc oxide nanoparticles: A review of their biological synthesis, antimicrobial activity, uptake, translocation and biotransformation in plants. *Journal of Materials Science, 53*(1), 185–201. https://doi.org/10.1007/s10853-017-1544-1
111. Gupta, M., Tomar, R. S., Kaushik, S., Mishra, R. K., & Sharma, D. (2018). Effective antimicrobial activity of green ZnO nano particles of Catharanthus roseus. *Frontiers in Microbiology, 9*, 2030. https://doi.org/10.3389/fmicb.2018.02030
112. Mondal, K. K., & Mani, C. (2012). Investigation of the antibacterial properties of nanocopper against Xanthomonas axonopodis pv. Punicae, the incitant of pomegranate bacterial blight. *Annals of Microbiology, 62*, 889–893. https://doi.org/10.1007/s13213-011-0382-7
113. Selvan, K. M., Sridharan, N., & John, S. (2012). Roadkill animals on national highways of Karnataka, India. *Journal of Ecology and the Natural Environment, 4*(14), 363–365. https://doi.org/10.5897/JENE11.068
114. Kah, M., & Hofmann, T. (2014). Nanopesticide research: Current trends and future priorities. *Environment International, 63*, 224–235. https://doi.org/10.1016/j.envint.2013.11.015
115. Holb, I. J., De Jong, P. F., & Heijne, B. (2003). Efficacy and phytotoxicity of lime Sulphur in organic apple production. *Annals of Applied Biology, 142*(2), 225–233. https://doi.org/10.1111/j.1744-7348.2003.tb00245.x
116. Jain, P., Satapathy, T., & Pandey, R. K. (2021). Acaricidal activity and biochemical analysis of Citrus limetta seed oil for controlling Ixodid tick Rhipicephalus microplus infesting cattle. *Systematic and Applied Acarology, 26*(7), 1350–1360. https://doi.org/10.11158/saa.26.7.13
117. Chhipa, H., & Joshi, P. (2016). Nanofertilisers, nanopesticides and nanosensors in agriculture. *Nanoscience in Food and Agriculture, 1*, 247–282. https://doi.org/10.1007/978-3-319-39303-2_9
118. Stadler, T., López García, G. P., Gitto, J. G., & Buteler, M. (2017). Nanostructured alumina: Biocidal properties and mechanism of action of a novel insecticide powder. *Bulletin of Insectology 70*(1): 17–25
119. Jiang, Y., Huang, J., Wu, X., Ren, Y., Li, Z., & Ren, J. (2020). Controlled release of silver ions from AgNPs using a hydrogel based on konjac glucomannan and chitosan for infected wounds. *International Journal of Biological Macromolecules, 149*, 148–157. https://doi.org/10.1016/j.ijbiomac.2020.01.221
120. Kora, A. J. (2022). Copper-based nanopesticides. Copper Nanostructures: Next-Generation of Agrochemicals for Sustainable Agroecosystems, 133–153. https://doi.org/10.1016/B978-0-12-823833-2.00019-2
121. Mosa, M. A., & El-Abeid, S. E. (2023). Chitosan-loaded copper oxide nanoparticles: A promising antifungal nanocomposite against fusarium wilt disease of tomato plants. *Sustainability, 15*(19), 14295. https://doi.org/10.3390/su151914295
122. Karamchandani, B. M., Dalvi, S. G., Bagayatkar, M., Banat, I. M., & Satpute, S. K. (2024). Prospective applications of chitosan and chitosan-based nanoparticles formulations in sustainable agricultural practices. *Biocatalysis and Agricultural Biotechnology, 103210*. https://doi.org/10.1016/j.bcab.2024.103210
123. Pires-Oliveira, R., Kfouri, M. S., Mendonça, B., & Cardoso-Gustavson, P. (2020). Nanopesticides: from the bench to the market. *In Nanopesticides: From research and development to mechanisms of action and sustainable use in agriculture*, pp. 317–348. Cham: Springer International Publishing
124. Ghormade, V., Deshpande, M. V., & Paknikar, K. M. (2011). Perspectives for nanobiotechnology enabled protection and nutrition of plants. *Biotechnology Advances, 29*(6), 792–803. https://doi.org/10.1016/j.biotechadv.2011.06.007

125. Perlatti, B., de Souza Bergo, & Luisa, P. (2013). Polymeric nanoparticle-based insecticides: A controlled release purpose for agrochemicals. In S. Trdan (Ed.), *Insecticides–development of safer and more effective technologies* (pp. 523–550). INTECH Open Access Publisher.
126. Kango, S., Kalia, S., Celli, A., Njuguna, J., Habibi, Y., & Kumar, R. (2013). Surface modification of inorganic nanoparticles for development of organic–inorganic nanocomposites—A review. *Progress in Polymer Science, 38*(8), 1232–1261. https://doi.org/10.1016/j.progpolymsci.2013.02.003
127. Pérez-Hernández, H., Pérez-Moreno, A., Sarabia-Castillo, C. R., García-Mayagoitia, S., Medina-Pérez, G., López-Valdez, F., Campos-Montiel, R. G., Jayanta-Kumar, P., & Fernández-Luqueño, F. (2021). Ecological drawbacks of nanomaterials produced on an industrial scale: Collateral effect on human and environmental health. *Water, Air, & Soil Pollution, 232*, 1–33. https://doi.org/10.1007/s11270-021-05370-2
128. Harish, V., Ansari, M. M., Tewari, D., Yadav, A. B., Sharma, N., Bawarig, S., et al. (2023). Cutting-edge advances in tailoring size, shape, and functionality of nanoparticles and nanostructures: A review. *Journal of the Taiwan Institute of Chemical Engineers, 149*, 105010. https://doi.org/10.1016/j.jtice.2023.105010
129. Kumari, A., Kumar, A., Seth, A., Guleria, I., Arya, V., & Shahi, S. K. (2024). Nanotechnology intervention for sustainable agriculture: Challenges and possibilities. *Nanotechnology: Agriculture, Environment and Health*, 1–51. https://doi.org/10.1007/978-981-97-6814-1_1
130. Sarmah, K., Borah, R., Boruah, S., & Sarmah, D. (2023). Nanopesticides: Revolutionizing Pest management with nanotechnology. *Biological Forum–An International Journal, 15*(7), 210–218.
131. Yang, F. L., Li, X. G., Zhu, F., & Lei, C. L. (2009). Structural characterization of nanoparticles loaded with garlic essential oil and their insecticidal activity against Tribolium castaneum (Herbst)(Coleoptera: Tenebrionidae). *Journal of Agricultural and Food Chemistry, 57*(21), 10156–10162. https://doi.org/10.1021/jf9023118

Plasmonic Nanosensors: Advancements and Challenges for Enhanced Agrotoxin Detection in Sustainable Agriculture

Amruta V. Shelar and Pravin N. Didwal

Contents

1 Introduction

Agrochemicals have significantly enhanced food production and address global nutritional needs but have also raised serious environmental and health concerns due to their extensive and prolonged use, jeopardizing ecological sustainability

A. V. Shelar
Department of Technology, Savitribai Phule Pune University, Pune, India

P. N. Didwal (✉)
Lawrence Berkeley National Laboratory, Berkeley, CA, USA

Department of Mechanical Engineering, MIT School of Engineering MIT Art, Design and Technology University, Pune, Maharashtra, India

R. S. Chaughule, A. V. Shelar (eds.), *Nanotechnology in Agriculture*,
https://doi.org/10.1007/978-3-031-97544-8_4

[1–4]. Despite their contributions to global sustainability, the extensive use of pesticides, fungicides, herbicides, insecticides, fertilizers, and soil conditioners has raised serious environmental and health concerns [5, 6]. Prolonged agrochemical use threatens ecological sustainability, leading to long-term negative consequences. Additionally, mycotoxins and harmful substances produced by molds and fungi pose significant risks to agricultural productivity, human health, and ecosystems [7, 8]. Common mycotoxins like alternariol, aflatoxin, and patulin, produced by *Alternaria*, *Aspergillus*, and *Penicillium* species, respectively, are frequently found in contaminated crops, fruits, grains, and vegetables [9, 10]. Together, agrochemical residues and mycotoxins (collectively termed agrotoxins) contaminate the environment, enter the food chain, and accumulate in humans and animals, causing serious health risks.

Agrotoxins are highly toxic, causing immune disorders, neurotoxicity, kidney damage, and diseases like cancer while disrupting plant growth, photosynthesis, and metabolic processes. They degrade soil, contaminate water, harm biodiversity, and promote pest resistance, requiring higher chemical use with reduced effectiveness [11, 12]. Advanced detection methods and sustainable practices like integrated pest management and organic farming are crucial to mitigating their impacts and ensuring environmental and public health protection. The ingestion of agrotoxin-contaminated food poses serious health risks. Prolonged exposure, even at minimal levels, has been associated with cancers, hormonal disruptions, reproductive and developmental issues, and neurological disorders [13, 14]. Vulnerable populations, including children, pregnant women, and farmworkers, are particularly susceptible due to their higher levels of exposure. Addressing these risks requires strict regulatory measures and the adoption of sustainable agricultural practices, such as integrated pest management, organic farming, and biopesticides. Continuous research and monitoring are vital to better understanding agrotoxin effects and developing safer, more effective pest control strategies.

Conventional methods like mass spectrometry and high-performance liquid chromatography (HPLC) effectively detect agrotoxins but are costly, time-consuming, and require specialized equipment [15–17]. Biosensors offer a more efficient and cost-effective alternative, providing quantitative data through biological recognition [18]. Advances in nanotechnology have further improved nanosensors, enabling precise detection with high sensitivity and adaptability. As shown in the Fig. 1, Nanosensors, nanoprobes, and other nanoscale technologies have revolutionized chemical and biosensing systems. These devices are engineered for swift response times and low power consumption, delivering efficient and high-performance solutions for modern sensing applications. Technologies such as electrochemical sensors, Surface Plasmon Resonance (SPR), and Surface-Enhanced Raman Spectroscopy (SERS) have shown excellent potential for agrotoxin monitoring, offering enhanced accuracy and versatility for agricultural applications [19, 20]. Noble metals exhibit distinctive plasmonic properties that enable efficient interaction with electromagnetic radiation, forming plasma environments that generate electromagnetic hot spots [21]. These hot spots are crucial for molecular sensing, allowing the detection of vibrational properties of compounds at nanomolar

Fig. 1 Figure provides an overview of the principles of nanoscale science, key materials used in nano-biosensors, and notable examples from the literature, highlighting their wide-ranging potential to address global challenges. This is a reprinted image of Ref. [78] with permission of MDPI

and femtomolar levels. Surface plasmons, occurring at the interface between noble metals and dielectric substances (e.g., in aqueous environments), are influenced by material-specific factors like wavelength and dielectric constants. SPR sensors leverage these phenomena to offer label-free identification and unique advantages through refractive index (RI) variations on plasmonic surfaces [22].

SPR sensors are valuable analytical tools for monitoring agrotoxins due to their ability to detect molecular interactions between analytes and plasmonic materials. However, despite their superior sensitivity compared to conventional biomonitoring methods, they are less effective at detecting agrotoxins in low concentrations [23, 24]. To overcome this limitation, researchers are exploring advanced SPR sensor designs incorporating innovative nanomaterials (NMs) and nanostructures enabled by modern nanofabrication technologies [25]. Plasmonic nanosensors, a class of optical biosensors, utilize plasmonic nanoparticles (NPs) to monitor molecular interactions. These nanosensors excel in detecting biological, chemical, environmental, and physical signals due to NPs unique properties, such as their high surface-to-volume ratio. Advances in nanomaterials and sensor configurations have resulted in novel designs and structures that significantly enhance field sensitivity. Noble metal NPs also demonstrate exceptional optical characteristics, including a broad UV-visible absorption band, absent in bulk metals [26]. When photons interact with their surfaces, these NPs produce enhanced SPR effects, making them highly effective at light absorption and scattering. Plasmonic nanosensors are composed of four main components: the analyte, the sensor, the transducer (which converts energy forms), and the detector [27]. When an analyte interacts with the plasmonic NPs surface, a specific reaction occurs, altering the transducer's physicochemical properties. These changes are then translated into electrical signals, enabling precise detection.

In recent years, advanced visual recognition techniques for biological and chemical sensing have seen significant growth, enabling the detection of a wide array of compounds [25]. Nanotechnology has played a crucial role in the evolution of SPR nanosensors over the last two decades, with plasmonic NPs demonstrating remarkable potential for detecting agricultural toxins [28, 29]. However, despite these advancements, there remain notable gaps in the literature and an absence of comprehensive reviews on the use of NMs in the design of SPR nanosensors specifically for agrotoxin detection. This chapter seeks to address these gaps by offering an overview of SPR nanosensors, including the types of nanosensors utilized for agrotoxin detection and the sensitive materials integrated into their construction. Furthermore, it examines the challenges and future prospects of applying nanosensors for agrotoxin monitoring, emphasizing the role of plasmonic NMs in enhancing detection efficiency and accuracy.

2 Agrotoxin Impacts and Advances in SPR Nanosensor Development

The need to address the advancements in SPR nanosensors for agrotoxin detection has become critical due to rising global agrochemical usage. Over 4.1 million tons of pesticides were consumed in 2020, leading to alarming agrotoxin residues and associated health risks like cancers, neurological disorders, and endocrine disruption [30]. The WHO reports 200,000 annual pesticide poisoning deaths, with 20% of global food exceeding residue limits. Current pesticide use, now at 4.2 million tons annually, has caused environmental damage, including water pollution, a 30–40% decline in bee populations, and increased pest resistance, resulting in higher crop losses [31]. Economically, pesticide exposure costs $15 billion annually in US healthcare, with global losses reaching $50 billion annually. Reliable detection technologies are urgently needed to mitigate these impacts [32].

This chapter examines recent advancements in nanotechnology for agrotoxin detection, with a focus on SPR nanosensors enhanced by NMs. Persistent and toxic agrotoxins, such as pesticide residues and mycotoxins, pose significant risks, while current monitoring methods fall short of addressing these challenges effectively. SPR nanosensors offer high specificity, sensitivity, and real-time monitoring in a compact, user-friendly design, outperforming traditional methods by enabling efficient, label-free detection, even in complex environmental conditions. It highlights the critical gaps in evaluating NMs used in SPR sensors for agrotoxin detection, particularly in soil environments, and explores potential solutions for on-site, rapid detection. It addresses key challenges such as sensor sensitivity, stability, and differentiation between agrotoxins, while offering recommendations for future research. By synthesizing recent advancements in NMs and SPR technology, this chapter emphasizes the need for innovative sensor designs to overcome current limitations. It provides a roadmap to advance nanosensor technology, aiming to protect human health, preserve ecosystems, and mitigate the growing impacts of agrotoxin contamination.

3 Revolutionizing SPR Sensors with Plasmonic NPs: Innovations in Detection and Applications

SPR is an effective method for detecting analytes, but challenges arise in identifying small quantities at low concentrations or molecular weights. This can be addressed by incorporating advanced NMs like plasmonic NPs, magnetic NPs, and carbon nanostructures such as graphene and carbon nanotubes (CNT) [26]. NMs enhance analyte adsorption and binding efficiency on sensor surfaces and the binding process through a sensor-gram [33]. These innovations significantly improve SPR sensor sensitivity and performance.

Metal NPs are crucial in advanced nanosensor development due to their unique optical, electronic, and catalytic properties. Gold NPs (AuNPs), in particular, stand out for their exceptional optical characteristics, biocompatibility, and ease of functionalization [34]. Their strong SPR properties make them ideal for sensing applications, including detecting biomolecules like DNA, proteins, and glucose with high sensitivity. El-Sayed et al. demonstrated AuNPs' effectiveness in colorimetric assays for precise biomolecule detection [35, 36]. Additionally, AuNPs have been successfully applied in environmental monitoring to detect toxic metals such as mercury and lead in water with outstanding sensitivity and selectivity. In addition, silver NPs (AgNPs) are widely used in biosensors for their antibacterial properties and ability to enhance Raman scattering in SERS sensors, enabling highly specific pathogen detection in food and water [36].

Platinum NPs (PtNPs), valued for their catalytic activity, are used in electrochemical sensors for glucose, hydrogen peroxide, and hydrogen gas, with studies highlighting their efficiency and sensitivity [37]. Copper NPs (CuNPs) offer cost-effectiveness and support surface plasmon resonances, as shown in hybrid SPR biosensors with graphene for enhanced sensitivity and in chemical sensors for detecting volatile organic compounds [38]. Iron oxide NPs (Fe_3O_4 NPs) are valued for their magnetic properties, making them suitable for detecting biomolecules and pollutants through easy separation and detection [39]. Some study demonstrated their effectiveness in pathogen detection using magnetic resonance techniques. Rhodium NPs (RhNPs), known for their catalytic properties, have been used in sensors for gases and chemicals [40]. Wang et al. showcased their high catalytic efficiency and improved performance in detecting environmental pollutants [41].

Metal NPs like Au, Ag, Pt, Cu, Fe_3O_4, and Rh play a crucial role in nanosensor applications due to their unique optical, electronic, and catalytic properties. They enable the creation of highly sensitive and versatile sensors for environmental monitoring, biomedical diagnostics, and chemical analysis. Ongoing research and optimization of these materials continue to drive advancements in nanosensor technology, unlocking new opportunities for precise and efficient detection across various fields. We have mainly focused on the Au and Ag NPs in the following sections.

AuNPs significantly enhance SPR sensor sensitivity, with their effectiveness influenced by size and shape. Commonly used as metallic films, AuNPs offer stability, low reactivity, and a narrow absorption band (500–550 nm), making them versatile for various applications. Studies, such as those by Kyeong Rim Ryu et al. [42], demonstrated that Au nanorods with varying aspect ratios improved SPR sensitivity through chemisorption-induced refractive index changes. At the same time, Rodrigo Calvo et al. [43] emphasized the role of precise metrological techniques in characterizing AuNPs. Early work by Brown and Natan achieved a 25-fold sensitivity improvement in immunoglobulin G detection using 11 nm AuNPs. Other advancements include glucose sensors, SPR magnetic field sensors, and ultrasensitive antigen detection. Additional innovations, such as polymer-templated AuNPs and Au nanorods, showcase the unique potential of AuNPs to amplify SPR signals [44]. Collectively, these findings underline the indispensable role of AuNPs in creating high-sensitivity SPR nanosensors for diverse applications, including diagnostics, environmental monitoring, and chemical analysis.

AgNPs, part of the plasmonic NPs family, offer exceptional optical properties, catalytic activity, and high extinction coefficients, making them valuable for advanced sensing applications. Green-synthesized AgNPs can detect heavy metals like copper, mercury, cobalt, and lead through distinct color changes and absorbance band shifts, demonstrating their high sensitivity [45]. AgNPs are increasingly used in optical and electrochemical biosensors for point-of-care applications. Studies, including those by Franziska Beck et al. [46], emphasize their ability to detect low-concentration analytes with enhanced performance. Advances in Ag nanostructures, such as those discussed by Ashank K. Gahlaut et al. [47], highlight their tunable resonances and excellent field enhancement for SPR and SERS applications.

Interestingly, stabilizers like polyvinyl alcohol (PVA) and polyvinylpyrrolidone (PVP) significantly sharpen SPR bands, improving signal strength and optical properties [48]. Additionally, silver nanoparticle (AgNP)-enhanced SPR biosensors have been employed for glucose detection, offering high sensitivity through the use of phase interrogation techniques. Xiang [49] found that noncircular Ag nanostructures, such as pentagonal and hexagonal nanoplates, further enhanced SPR sensitivity due to their sharp edges. AgNPs play a pivotal role in advancing SPR sensor technology. Their tunable properties, combined with improved stabilization and substrate designs, enable their use in diverse fields, including environmental monitoring, biomedical diagnostics, and chemical sensing.

4 Enhancing SPR Sensor Sensitivity with Carbon-Based Materials

Carbon-based NPs are categorized into fullerenes, graphene, carbon nanotubes, carbon nanofibers, and carbon black (Fig. 2). Among all carbon-based NPs, graphene and carbon nanotubes (CNTs) have transformed SPR sensor technology, delivering significant improvements in sensitivity and performance due to their exceptional electronic and structural properties.

Graphene, a single layer of carbon atoms in a hexagonal lattice, is celebrated for its excellent conductivity, high surface area, and mechanical strength, making it highly effective in enhancing SPR sensor performance [50]. Figure 3 shows that a significant majority (71.4%) of studies observed improved sensitivity with Gr-integrated SPR biosensors. Additionally, 14.3% reported substantial improvements, 10.7% noted enhanced sensitivity, and 3.6% recorded the highest sensitivity levels. These results underscore graphene's pivotal role in enhancing SPR biosensor performance, solidifying its value in biosensing applications. The graphene-coated Au sensor films significantly improved SPR sensitivity, with the number of graphene

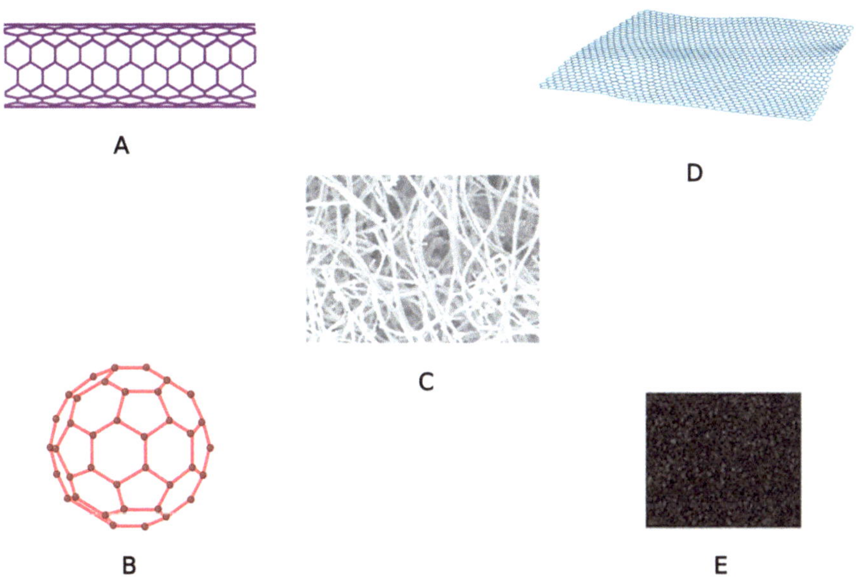

Fig. 2 Visual representation of carbon-based nanomaterials: (**a**) Carbon Nanotubes, (**b**) Fullerenes, (**c**) Carbon Nanofibers, (**d**) Graphene, and (**e**) Carbon Black. This is a reprinted image of Ref. [34] with permission of MDPI

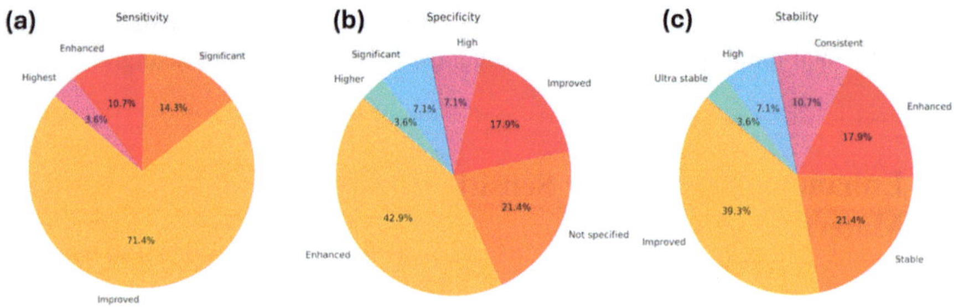

Fig. 3 Performance metrics in graphene-based SPR biosensors. This figure illustrates the distribution of qualitative evaluations across three key performance metrics in graphene-based SPR biosensors: (**a**) sensitivity, (**b**) specificity, and (**c**) stability. Each panel highlights the proportion of studies reporting various levels of enhancement, categorized as improved, enhanced, significant, highest, high, higher, ultra-stable, and not specified. The data emphasize the broad impact of graphene in advancing these critical biosensor performance metrics. This is a reprinted image of Ref. [51] with permission of MDPI

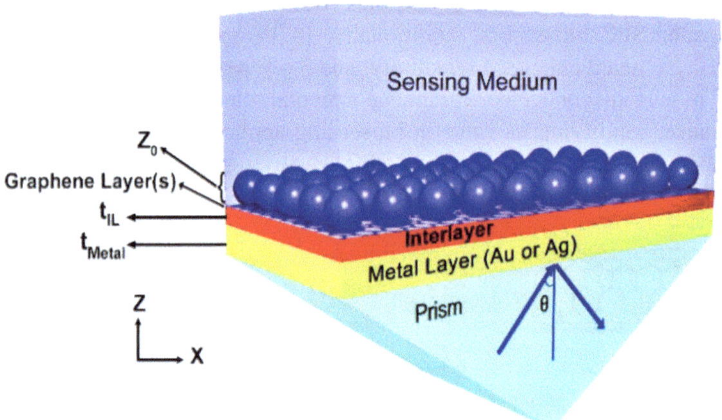

Fig. 4 Schematic representation of the SPR biosensor featuring a metal layer/interlayer/graphene layer structure integrated within the Kretschmann configuration. This is a reprinted image of Ref. [79] with permission of MDPI

layers directly influencing the sensor's ability to detect refractive index (RI) changes [50]. Interestingly, as illustrated in Fig. 4, a graphene-based multilayer structure comprising a prism, a metal layer (Ag or Au), an interlayer (IL), and a graphene layer can significantly enhance the sensitivity of the SPR biosensor by selecting an appropriate refractive index for the IL and optimizing its thickness. The advanced graphene-based SPR sensors by developing a system for salinity detection offer enhanced sensitivity and a broader detection range than conventional sensors [51]. The graphene's utility by incorporating it into nanocomposites enabling ultrasensitive detection of biological and chemical targets at femtomolar to atomic levels.

CNTs, with their cylindrical nanostructures and exceptional electrical and mechanical properties, are equally transformative in SPR sensing. CNT metasurfaces on silicon wafers are utilized to create plasmonic sensors capable of detecting pesticides at trace levels, achieving lower detection limits [52]. Developing a fiber optic SPR nanosensor using a CNT/Cu nanoparticle composite for ultra-trace nitrate detection demonstrates how changes in dielectric properties enhanced sensor performance. Additionally, the strong π-stacking interactions between graphene and analytes with aromatic ring structures enable graphene-based SPR sensors to efficiently detect complex biological molecules like DNA. The integration of graphene and CNTs into SPR sensors has significantly advanced their detection capabilities. Graphene's high conductivity and surface area enhance sensitivity and performance, while CNTs provide versatility and functionalization options for a broad range of analytes. These innovations have driven the development of highly sensitive, selective, and practical sensing technologies, expanding nanosensor applications across environmental monitoring, agriculture, and biomedical diagnostics.

5 Role of Two-Dimensional Nanomaterials in Enhancing Sensitivity and Performance

Two-dimensional nanomaterials (2D NMs) have revolutionized SPR sensor design and functionality due to their exceptional electronic, optical, and surface properties. Their unique single-layer or few-layer structures significantly enhance sensor sensitivity and performance. Graphene, a single layer of carbon atoms in a hexagonal lattice, is particularly valued for its excellent electrical conductivity, mechanical strength, and large surface area. The Au sensor films with graphene layers increased sensitivity by 25%, amplifying the plasmonic field and improving the detection of refractive index (RI) changes [53]. Graphene's large surface area also allows for the immobilization of more biomolecules, strengthening the SPR signal. Utilization of 2D NMs to develop an SPR device for biomedical diagnostics achieve a 300% increase in sensitivity by optimizing nanostructures [54]. This advancement highlights the potential of 2D NMs to detect low-concentration biomarkers with high precision, essential for accurate diagnostics. Similarly, 2D NMs into SPR sensors for antibiotic detection create a platform with exceptional sensitivity and broader detection capabilities for chemical and biological substances (Fig. 5). Yadvendra Singh et al. incorporated TiO_2, another notable 2D NM, into SPR sensors to detect bacteria-infected red blood cells, achieving an 18% sensitivity enhancement [55]. TiO_2's ability to modify the sensor's optical and electronic properties, combined with its high photocatalytic activity and large surface area, proved effective for biological detection. Nurzad Zakirov et al. developed a hybrid SPR biosensor by combining Cu substrates with graphene [56]. This approach leveraged Cu's superior conductivity and affordability alongside Gr's unique properties, resulting in enhanced sensitivity. By optimizing parameters like Cu thickness and graphene layers, they created an efficient, scalable, and cost-effective sensor solution.

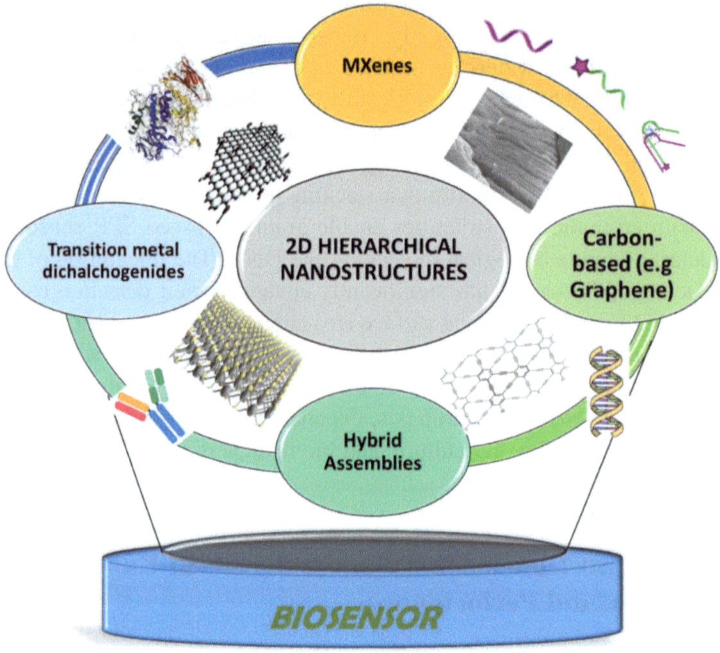

Fig. 5 Hierarchical nanostructures employed as electrode materials and bioimmobilization support in the development of electrochemical biosensors. These nanostructures enhance the electrochemical properties and facilitate the effective immobilization of biological recognition elements, enabling sensitive and selective biosensing applications. This is a reprinted image of Ref. [57] with permission of MDPI

Two-dimensional NMs such as graphene and TiO_2 are crucial for advancing SPR sensor technology. Their exceptional electronic properties, functionalization capabilities, and high surface area enable improved sensitivity and performance. These advancements have made SPR sensors invaluable for biomedical diagnostics, environmental monitoring, and chemical analysis. Ongoing research promises further innovations in highly sensitive and efficient sensing devices.

6 Nanosensors: Advancing Detection with Emerging Trends

Nanosensors are at the cutting edge of technology, providing exceptional sensitivity and specificity for detecting various analytes. Key innovations shaping their future include integrating multiple sensing modalities, leveraging machine learning for advanced data analysis, and developing portable, cost-efficient devices. These advancements promise to expand nanosensor applications in environmental monitoring, medical diagnostics, and industrial processes.

Combining multiple sensing mechanisms offers comprehensive detection capabilities. Traditional nanosensors often rely on a single technique like SPR or electrochemical sensing. By integrating methods such as SPR with fluorescence or Raman spectroscopy, researchers improve sensitivity and specificity. Liu et al. showed that combining SPR with fluorescence enhanced biomolecule detection by exploiting SPR's sensitivity and fluorescence quantification [58]. Similarly, Samuel et al. denote a hybrid sensor merging plasmonic and electrochemical detection, enabling effective monitoring of environmental pollutants [59]. These multimodal sensors are particularly valuable for multiplexed detection, where simultaneous analysis of multiple analytes is required. As shown in Fig. 6, nanosensors can be used to detect pesticide and mycotoxins. Machine learning is revolutionizing nanosensor data interpretation by addressing challenges with high-dimensional datasets. Machine learning algorithms like convolutional neural networks (CNNs) enhance pattern recognition, classification, and prediction. Cheng et al. used CNNs

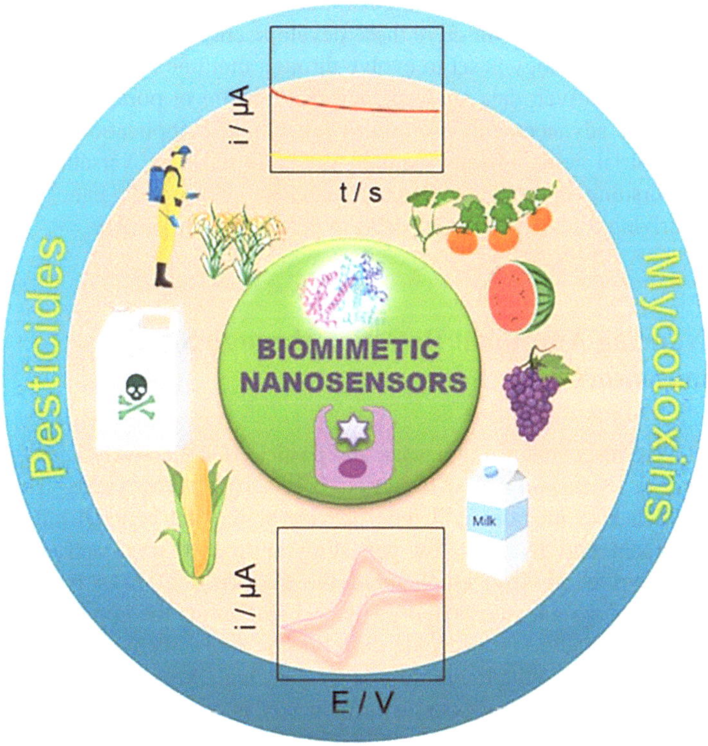

Fig. 6 As illustrated in Figure, nanosensors can be effectively utilized for the detection of pesticides and mycotoxins, offering high sensitivity, precision, and rapid analysis for enhanced food safety and environmental monitoring. This is a reprinted image of Ref. [60] with permission of MDPI

to analyze SPR sensor data, improving detection accuracy and reducing false positives [61]. Similarly, utilized support vector machines (SVMs) to analyze electrochemical signals, enabling accurate classification of structurally similar analytes. Real-time adaptability was demonstrated to integrate reinforcement learning into a nanosensor, optimizing performance based on environmental feedback. Machine learning integration enhances precision and efficiency, creating smarter nanosensing solutions.

To increase accessibility, recent efforts focus on portable, affordable nanosensors. Traditional devices require complex setups, limiting field use. Innovations like portable SPR nanosensors for real-time contaminant detection and cost-effective electrochemical sensors using screen-printed electrodes exemplify this trend [62]. Advances in materials science, such as carbon-based and metal-oxide platforms, further reduce costs without compromising performance. The future of nanosensors lies in hybrid systems combining multimodal sensing, Machine learning-based analytics, and portability. A hybrid sensor integrating SPR, electrochemical detection, and Machine learning could enable simultaneous detection of multiple analytes with unmatched sensitivity and real-time usability. Advances in nanomaterials and fabrication techniques, such as Zhang et al.'s exploration of multifunctional materials, will drive these developments [63].

Nanosensor technology is set to evolve through multimodal sensing integration, machine learning-driven data analysis, and the creation of portable, cost-effective solutions. These advancements promise to revolutionize applications across scientific and industrial fields, making nanosensors more sensitive, versatile, and accessible. By pursuing these innovations, researchers will unlock nanosensors' full potential, driving transformative progress in detection technologies.

7 Advancing Agrotoxin Detection through SPR Nanosensors

Chemicals like insecticides, fungicides, and pesticides, collectively known as agrochemicals, are extensively used in agriculture to boost crop yields. These compounds, which also include fertilizers, provide significant benefits but can harm the environment and livestock by persisting in treated areas and causing soil pollution. Talari et al. [64] emphasized the toxicological risks posed by high concentrations of agrochemicals to organisms, highlighting the importance of effective monitoring. Traditional pesticide detection methods, such as chromatography and spectroscopy, are reliable but limited by their high costs, complex setups, and lengthy processing times, as noted by Muenchen et al. [65]. To overcome these challenges, the development of rapid and highly sensitive detection technologies has become essential. SPR nanosensors have emerged as a promising solution over the past two decades, offering exceptional sensitivity, specificity, and adaptability for agrotoxin monitoring.

7.1 Innovations in Pesticide Detection with SPR Nanosensors

Pesticides, including fungicides, nematicides, herbicides, insecticides, and plant growth hormones, are essential for safeguarding crops from pests and weeds. However, these chemicals can harm nontarget species and the environment, emphasizing the need for advanced detection methods to mitigate their risks. SPR nanosensors have proven highly effective in pesticide detection due to their exceptional sensitivity, label-free detection capabilities, and miniaturization potential. Oymen et al. [62] developed a fast and efficient SPR nanosensor for detecting coumaphos, an organophosphate insecticide, using molecular imprinting. This nanosensor demonstrated excellent selectivity, reusability, and sensitivity in aqueous environments. Similarly, Wang et al. [66] utilized multi-walled carbon nanotube-based metasurfaces for pesticide sensors, achieving reliable detection with strong linear relationships between transmission amplitude and pesticide concentration.

Kant [67] advanced fiber-optic SPR nanosensors by embedding tantalum pentoxide nanostructures into a graphene oxide matrix. This setup effectively detected fenitrothion, an organophosphate pesticide, by exploiting refractive index changes. Dissanayake et al. [68] further enhanced plasmonic nanoparticle-based sensors' sensitivity by monitoring time-dependent shifts in localized surface plasmon resonance (LSPR) during pesticide interactions in natural conditions. Yilmaz et al. [69] introduced an SPR-based affinity sensor with molecularly imprinted NPs (plastic antibodies) for atrazine detection, showing superior selectivity and sensitivity. These advancements highlight SPR nanosensors as cutting-edge tools for efficient and precise pesticide detection, paving the way for improved agricultural practices and environmental protection.

7.2 Advancements in Fertilizer Residue Detection Using SPR Nanosensors

Fertilizers are essential in agriculture, supplying key nutrients like nitrogen, phosphorus, and potassium to improve soil fertility and boost crop yields. However, their overuse can harm ecosystems and pose risks to human health. SPR nanosensors have emerged as effective tools for detecting fertilizer residues in soil, offering enhanced environmental monitoring and agricultural management solutions. A thiol-mediated etching technique for gold nanorods to detect nitrite and nitrate ions resulted in distinct color changes at different ion concentrations, providing a simple yet reliable detection method. Similarly, Sharma and Gupta [70] developed a fiber-optic SPR potassium ion sensor for soil testing. While disposable due to its irreversible mechanism, the sensor enables real-time and accurate potassium detection. Verma and Gupta [71] designed a fiber-optic SPR probe for detecting phosphate ions in soil, emphasizing its affordability and suitability for on-site analysis. Rattanarat et al. [72] introduced a paper-based SPR sensor using Ag nanoplates to detect phosphate ions, offering a low-cost and visually accessible detection solution. Md. Azahar Ali et al.

[73] developed a microfluidic SPR nitrate sensor incorporating graphene oxide nanosheets and PEDOT nanofibers, effectively quantifying nitrate levels in soil samples with enhanced charge transfer resistance. Ye et al. proposed a streamlined method for NO_2^- detection based on aggregation-induced shifts in gold nanoparticle plasmon resonance, delivering an affordable and simplified nitrate detection approach. These advancements underscore SPR nanosensors' potential in fertilizer residue detection, enabling precise, cost-effective, and sustainable agricultural practices.

7.3 *Mycotoxin Detection with SPR Nanosensors: Ensuring Food Safety*

Mycotoxins, toxic by-products from fungi, such as *Penicillium, Aspergillus,* and *Fusarium,* pose significant threats to agriculture and human health. These contaminants affect food and feed during preharvest, harvest, and storage stages, necessitating robust detection systems for safety assurance. SPR nanosensors have emerged as a cutting-edge solution for real-time, sensitive detection of mycotoxins. Akgönüllü, Yavuz, and Denizli [74] designed an SPR nanosensor incorporating molecularly imprinted polymer films with AuNPs for detecting AFB1 in maize and peanut samples. This system showcased high sensitivity, specificity, and cost efficiency. Similarly, Bhardwaj et al. [75] employed multilayer AuNP-functionalized chips in microfluidic SPR biosensors, enabling precise AFB1 detection through self-assembled monolayer (SAM) formation and antibody grafting. Hossain and Maragos [76] developed imaging SPR (iSPR) assays using AuNPs for multiplexed detection of Fusarium mycotoxins in wheat, effectively identifying toxins such as zearalenone and deoxynivalenol with remarkable specificity. Hu et al. [77] further advanced SPR imaging by integrating antigen–antibody interactions with AuNPs, achieving enhanced sensitivity for detecting multiple mycotoxins. These advancements highlight the potential of SPR nanosensors in addressing mycotoxin challenges, ensuring food safety, and supporting sustainable agricultural practices. SPR nanosensors represent a transformative approach to agrotoxin detection, offering unparalleled sensitivity, specificity, and adaptability. Through advancements in nanomaterials, molecular imprinting, and multiplexed detection, these sensors are addressing critical challenges in agricultural monitoring. From pesticides and fertilizers to mycotoxins, SPR nanosensors are paving the way for safer, more sustainable agricultural practices.

8 Overcoming Challenges in SPR Nanosensors for Effective Agro-Toxin Detection

SPR nanosensors offer significant potential for detecting agro-toxins like pesticides, mycotoxins, and fertilizer residues, thanks to their exceptional sensitivity, specificity, and rapid detection capabilities. However, several challenges must be addressed to enable their practical application and widespread adoption.

(a) A primary challenge is achieving the necessary sensitivity and selectivity for detecting trace agro-toxins in complex agricultural environments. Background interference can obscure target signals, making it difficult to maintain a high signal-to-noise ratio. Specialized sensor coatings and receptor molecules are needed to selectively bind target toxins while minimizing cross-reactivity. These coatings must overcome matrix effects and accurately detect specific toxins among similar compounds present in agricultural samples. Other challenges include the potential toxicity of nanomaterials, sensor durability, and high production costs. Addressing these issues requires the development of safer, biocompatible materials, enhanced sensor robustness, and streamlined manufacturing processes.
(b) Cost is another major barrier to adopting SPR nanosensors, particularly in resource-limited areas. The reliance on expensive materials like gold and silver NPs, combined with complex fabrication processes and the need for advanced optical equipment, significantly increases costs. Scaling production for widespread use is further complicated by logistical challenges, including sensor housing, wireless communication systems, and maintenance requirements, particularly in remote locations. The environmental impact of SPR nanosensor production is also a pressing concern. NP's synthesis generates pollutants, consumes substantial energy and water, and produces waste. Hazardous materials used in synthesis pose risks to both human health and ecosystems. Moreover, the lifecycle of SPR sensors, including maintenance, calibration, and disposal, adds to their environmental footprint. Sustainable synthesis methods, environmentally friendly materials, and proper disposal protocols are essential to mitigate these impacts.
(c) Ensuring consistency in large-scale nanoparticle synthesis is critical for reliable sensor performance. Variations in NP's size and shape can affect optical properties, while long-term stability is required to prevent degradation or aggregation. Stringent quality control is essential to minimize variability and maintain reproducibility. Additionally, the synthesis process presents health and environmental risks, emphasizing the need for more sustainable practices. Durability is another challenge, as SPR nanosensors must withstand harsh agricultural conditions, including UV radiation, moisture, and varying soil compositions. Protective coatings and materials are essential to shield sensors from these factors while preserving their sensitivity and performance. These coatings can protect against UV exposure, moisture damage, and chemical interactions with soil, ensuring sensor reliability in field conditions.

Despite these challenges, SPR nanosensors have immense potential to transform agricultural practices and environmental monitoring. They enable real-time tracking of agro-toxin levels in soil, water, and crops, allowing for precise chemical application that reduces environmental impact and enhances productivity. In environmental monitoring, SPR nanosensors offer accurate detection of pollution sources and contamination, enabling early intervention to protect ecosystems. While obstacles such as sensitivity, cost, scalability, and environmental impact remain, advances in material science, sensor technology, and sustainable practices will be crucial in overcoming them. SPR nanosensors are positioned to revolutionize agricultural safety and environmental conservation, driving meaningful progress in food security and ecosystem health.

9 Conclusion

SPR nanosensors, combined with advanced nanomaterials, are revolutionizing agriculture by enhancing precision in agrochemical monitoring and management. These sensors mitigate environmental issues like water contamination and soil degradation while promoting sustainable farming practices. Integration with IoT and big data analytics enables real-time decision-making, optimizing resource use, and improving crop yields, making them indispensable for modern agriculture. Challenges remain, including the need for standardized protocols to ensure reliability and consistency. Collaboration among researchers, manufacturers, and regulatory bodies is essential. Technological integration with IoT, machine learning, and advanced imaging offers immense potential but requires addressing compatibility issues. Future research should focus on developing novel nanomaterials, improving manufacturing precision, and enhancing sensor sensitivity. Expanding applications in environmental monitoring and healthcare highlight the broad potential of nanosensors. Additionally, education and training are vital to equip professionals with the skills needed to implement and manage these innovations. With continued innovation, collaboration, and education, SPR nanosensors can drive progress in agriculture and beyond, addressing critical challenges while fostering sustainability.

References

1. Ahmad, M. F., Ahmad, F. A., Alsayegh, A. A., Zeyaullah, M., AlShahrani, A. M., Muzammil, K., Saati, A. A., et al. (2024). Pesticides impact on human health and the environment with their mechanisms of action and possible countermeasures. *Heliyon, 10 (7), e29128.*
2. Akgönüllü, S., Yavuz, H., & Denizli, A. (2020). SPR nanosensor based on molecularly imprinted polymer film with gold nanoparticles for sensitive detection of aflatoxin B1. *Talanta, 219*. Elsevier B.V.
3. Ali, M. A., Jiang, H., Mahal, N. K., Weber, R. J., Kumar, R., Castellano, M. J., & Dong, L. (2017). Microfluidic impedimetric sensor for soil nitrate detection using graphene oxide and conductive nanofibers enabled sensing interface. *Sensors and Actuators, B: Chemical, 239*, 1289–1299. Elsevier B.V.
4. Awuchi, C. G., Ondari, E. N., Ogbonna, C. U., Upadhyay, A. K., Baran, K., Okpala, C. O. R., Korzeniowska, M., et al. (2021). Mycotoxins affecting animals, foods, humans, and plants: Types, occurrence, toxicities, action mechanisms, prevention and detoxification strategies-a revisit. *Food, 10(6), 1279.*
5. Banerjee, A., Maity, S., & Mastrangelo, C. H. (2021). Nanostructures for biosensing, with a brief overview on cancer detection, IoT, and the role of machine learning in smart biosensors. *Sensors, 21(4), s21041253.*

6. Barathi, S., Sabapathi, N., Kandasamy, S., & Lee, J. (2024). Present status of insecticide impacts and eco-friendly approaches for remediation-a review. *Environmental Research, 240, 117432.*
7. Beaumelle, L., Tison, L., Eisenhauer, N., Hines, J., Malladi, S., Pelosi, C., Thouvenot, L., et al. (2023). Pesticide effects on soil fauna communities—A meta-analysis. *Journal of Applied Ecology, 60(7), 1-15.* John Wiley and Sons Inc.
8. Beck, F., Loessl, M., & Baeumner, A. J. (2023). Signaling strategies of silver nanoparticles in optical and electrochemical biosensors: Considering their potential for the point-of-care. *Microchimica Acta, 190(3), 91.* Springer.
9. Bhardwaj, H., Sumana, G., & Marquette, C. A. (2020). A label-free ultrasensitive microfluidic surface Plasmon resonance biosensor for aflatoxin B1 detection using nanoparticles integrated gold chip. *Food Chemistry, 307.* Elsevier Ltd.
10. Bharti, A., Jain, U., & Chauhan, N. (2024). Progressive analytical techniques utilized for the detection of contaminants attributed to food safety and security. *Talanta Open, 10.* Elsevier B.V.
11. Boedeker, W., Watts, M., Clausing, P., & Marquez, E. (2020). The global distribution of acute unintentional pesticide poisoning: Estimations based on a systematic review. *BMC Public Health, 20,* 1. BioMed Central Ltd.
12. Bolzonella, C., Lucchetta, M., Teo, G., Boatto, V., & Zanella, A. (2019). Is there a way to rate insecticides that is less detrimental to human and environmental health? *Global Ecology and Conservation, 20.* Elsevier B.V.
13. Brinco, J., Guedes, P., Gomes da Silva, M., Mateus, E. P., & Ribeiro, A. B. (2023). Analysis of pesticide residues in soil: A review and comparison of methodologies. *Microchemical Journal, 195, 109465.* Elsevier Inc.
14. Burlec, A. F., Corciova, A., Boev, M., Batir-Marin, D., Mircea, C., Cioanca, O., Danila, G., et al. (2023). Current overview of metal nanoparticles' synthesis, characterization, and biomedical applications, with a focus on silver and gold nanoparticles. *Pharmaceuticals, 16(10), 1410.* Multidisciplinary Digital Publishing Institute (MDPI).
15. Butt, M. A. (2025). Surface Plasmon resonance-based biodetection systems: Principles, Progress and applications—A comprehensive review. *Biosensors, 15(1), 150100035.* Multidisciplinary Digital Publishing Institute (MDPI).
16. Calvo, R., Thon, A., Saad, A., Salvador-Matar, A., Manso-Silván, M., Ahumada, Ó., & Pini, V. (2022). Size characterization of plasmonic nanoparticles with dark-field single particle spectrophotometry. *Scientific Reports, 12,* 1. Nature Research.
17. Campos, E. V. R., Proença, P. L. F., Oliveira, J. L., Bakshi, M., Abhilash, P. C., & Fraceto, L. F. (2019). Use of botanical insecticides for sustainable agriculture: Future perspectives. *Ecological Indicators, 105,* 483–495. Elsevier B.V.
18. Cheng, N., Chen, D., Lou, B., Fu, J., & Wang, H. (2021). A biosensing method for the direct serological detection of liver diseases by integrating a SERS-based sensor and a CNN classifier. *Biosensors and Bioelectronics, 186.* Elsevier Ltd.
19. Cuvelier, N., Avanasi, R., Grunenwald, M., Ramanarayanan, T., Wolf, D. C., & Bartell, S. M. (2024). A novel approach to integrate human biomonitoring data with model predicted dietary exposures: A crop protection chemical case study using lambda-Cyhalothrin. *Journal of Agricultural and Food Chemistry, 72*(20), 11663–11671. American Chemical Society.
20. Darwish, M. A., Abd-Elaziem, W., Elsheikh, A., & Zayed, A. A. (2024). Advancements in nanomaterials for nanosensors: A comprehensive review. *Nanoscale Advances, 6, 4015-4046.* Royal Society of Chemistry.
21. Das, S., Devireddy, R., & Gartia, M. R. (2023). Surface Plasmon resonance (SPR) sensor for cancer biomarker detection. *Biosensors, 13(3), 396* . MultidisciplinaryDigital Publishing Institute (MDPI).
22. Dissanayake, N. M., Arachchilage, J. S., Samuels, T. A., & Obare, S. O. (2019). Highly sensitive plasmonic metal nanoparticle-based sensors for the detection of organophosphorus pesticides. *Talanta, 200,* 218–227. Elsevier B.V.
23. Dreaden, E. C., Alkilany, A. M., Huang, X., Murphy, C. J., & El-Sayed, M. A. (2012). The golden age: Gold nanoparticles for biomedicine. *Chemical Society Reviews, 41*(7), 2740–2779.

24. El-Sayed, R. A., Jebur, A. B., Kang, W., & El-Demerdash, F. M. (2022). An overview on the major mycotoxins in food products: Characteristics, toxicity, and analysis. *Journal of Future Foods, 2(2), 91-102*. Beijing Academy of Food Sciences.
25. Fernández-Cruz, M. L., Mansilla, M. L., & Tadeo, J. L. (2010). Mycotoxins in fruits and their processed products: Analysis, occurrence and health implications. *Journal of Advanced Research, 1(2), 119-122.*
26. Gahlaut, S. K., Pathak, A., & Gupta, B. D. (2022). Recent advances in silver nanostructured substrates for plasmonic sensors. *Biosensors, 12(9), 12090713*. MDPI.
27. Haleem, A., Javaid, M., Singh, R. P., Suman, R., & Rab, S. (2021). Biosensors applications in medical field: A brief review. *Sensors International, 2(3), 100100*. KeAi Communications Co.
28. Hernández-Mesa, M., & Moreno-González, D. (2022). Current role of mass spectrometry in the determination of pesticide residues in food. *Separations, 9(6), 9060148*. MDPI.
29. Hinman, S. S., McKeating, K. S., & Cheng, Q. (2018). Surface plasmon resonance: Material and interface design for universal accessibility. *Analytical Chemistry, 90(1), 19-39*. American Chemical Society.
30. Ho, V., Pelland-St-Pierre, L., Gravel, S., Bouchard, M. F., Verner, M. A., & Labrèche, F. (2022). Endocrine disruptors: Challenges and future directions in epidemiologic research. *Environmental Research, 204, 111969*. Academic Press Inc.
31. Hossain, M. Z., & Maragos, C. M. (2018). Gold nanoparticle-enhanced multiplexed imaging surface plasmon resonance (iSPR) detection of Fusarium mycotoxins in wheat. *Biosensors and Bioelectronics, 101*, 245–252. Elsevier Ltd.
32. Hu, D., Ji, X., Zhu, J., & Xu, J. (2024). Crystallization-dictated assembly of block copolymers and nanoparticles under three-dimensional confinement. *Chemical Communications, 60, 10854-10865*. Royal Society of Chemistry (RSC).
33. Hu, W., Chen, H., Zhang, H., He, G., Li, X., Zhang, X., Liu, Y., et al. (2014). Sensitive detection of multiple mycotoxins by SPRi with gold nanoparticles as signal amplification tags. *Journal of Colloid and Interface Science, 431*, 71–76. Academic Press Inc.
34. Ibrahim, N. H., Taha, G. M., Hagaggi, N. S. A., & Moghazy, M. A. (2024). Green synthesis of silver nanoparticles and its environmental sensor ability to some heavy metals. *BMC Chemistry, 18*, 1. BioMed Central Ltd.
35. Jin, C., Wu, Z., Molinski, J. H., Zhou, J., Ren, Y., & Zhang, J. X. J. (2022). Plasmonic nanosensors for point-of-care biomarker detection. *Materials Today Bio, 16, 100263*. Elsevier B.V.
36. Kant, R. (2020). Surface plasmon resonance based fiber–optic nanosensor for the pesticide fenitrothion utilizing Ta2O5 nanostructures sequestered onto a reduced graphene oxide matrix. *Microchimica Acta, 187*, 1. Springer.
37. Kassotis, C. D., Vandenberg, L. N., Demeneix, B. A., Porta, M., Slama, R., & Trasande, L. (2020). Endocrine-disrupting chemicals: Economic, regulatory, and policy implications. *The Lancet Diabetes and Endocrinology*. Lancet Publishing Group.
38. Khan, I., Saeed, K., & Khan, I. (2019). Nanoparticles: Properties, applications and toxicities. *Arabian Journal of Chemistry*. Elsevier B.V.
39. Khan, R., Radoi, A., Rashid, S., Hayat, A., Vasilescu, A., & Andreescu, S. (2021). Two-dimensional nanostructures for electrochemical biosensor. *Sensors*. MDPI AG.
40. Koya, A. N., Zhu, X., Ohannesian, N., Yanik, A. A., Alabastri, A., Proietti Zaccaria, R., Krahne, R., et al. (2021). Nanoporous metals: From plasmonic properties to applications in enhanced spectroscopy and photocatalysis. *ACS Nano*. American Chemical Society.
41. Kumar, M., Sarma, D. K., Shubham, S., Kumawat, M., Verma, V., Prakash, A., & Tiwari, R. (2020). Environmental endocrine-disrupting chemical exposure: Role in non-communicable diseases. *Frontiers in Public Health*. Frontiers Media S.A.
42. Kumar, S., Singh, R., Wang, Z., Li, M., Liu, X., Zhang, W., Zhang, B., et al. (2023). (Invited) Advances in 2D nanomaterials-assisted plasmonics optical fiber sensors for biomolecules detection. *Results in Optics, 10*. Elsevier B.V.
43. Lakavath, K., Kafley, C., Sajeevan, A., Jana, S., Marty, J. L., & Kotagiri, Y. G. (2024). Progress on electrochemical biomimetic nanosensors for the detection and monitoring of mycotoxins and pesticides. *Toxins*. Multidisciplinary Digital Publishing Institute (MDPI).

44. Liu, C., Geng, Q., & Geng, Z. (2024). Strategies to improve performances of fluorescent biosensors based on smartphones: Sensitivity, high throughput, and smart detection. *Sensors and Actuators A: Physical*. Elsevier B.V.
45. Mbayachi, V. B., Ndayiragije, E., Sammani, T., Taj, S., Mbuta, E. R., & Khan, A. U. (2021). Graphene synthesis, characterization and its applications: A review. *Results in Chemistry*. Elsevier B.V.
46. Mitra, S., Saran, R. K., Srivastava, S., & Rensing, C. (2024). Pesticides in the environment: Degradation routes, pesticide transformation products and ecotoxicological considerations. *Science of the Total Environment*. Elsevier B.V.
47. Mousavi-Kiasari, S. M. G., Rashidi, K., Fathi, D., Taleb, H., Mirjalili, S. M., & Faramarzi, V. (2022). Computational design of highly-sensitive graphene-based multilayer SPR biosensor. *Photonics, 9*, 10. MDPI.
48. Muenchen, D. K., Martinazzo, J., Brezolin, A. N., de Cezaro, A. M., Rigo, A. A., Mezarroba, M. N., Manzoli, A., et al. (2018). Cantilever functionalization using peroxidase extract of low cost for glyphosate detection. *Applied Biochemistry and Biotechnology, 186*(4), 1061–1073. Humana Press Inc.
49. Muthukrishnan, L. (2022). An overview on the nanotechnological expansion, toxicity assessment and remediating approaches in agriculture and food industry. *Environmental Technology and Innovation*. Elsevier B.V.
50. Mutunga, T., Sinanovic, S., & Harrison, C. S. (2024). Integrating wireless remote sensing and sensors for monitoring pesticide pollution in surface and groundwater. *Sensors*. Multidisciplinary Digital Publishing Institute (MDPI).
51. N'Diaye, J., Poorahong, S., Hmam, O., Izquierdo, R., & Siaj, M. (2017). Facile synthesis rhodium nanoparticles decorated single layer graphene as an enhancement hydrogen peroxide sensor. *Journal of Electroanalytical Chemistry, 789*, 85–91. Elsevier B.V.
52. Oymen, B., Jalilzadeh, M., Yılmaz, F., Aşır, S., Türkmen, D., & Denizli, A. (2023). Simple and fast pesticide nanosensors: Example of surface Plasmon resonance Coumaphos Nanosensor. *Micromachines, 14*, 4. MDPI.
53. Pathak, V. M., Verma, V. K., Rawat, B. S., Kaur, B., Babu, N., Sharma, A., Dewali, S., et al. (2022). Current status of pesticide effects on environment, human health and it's eco-friendly management as bioremediation: A comprehensive review. *Frontiers in Microbiology*. Frontiers Media S.A.
54. Pinyorospathum, C., Rattanarat, P., Chaiyo, S., Siangproh, W., & Chailapakul, O. (2019). Colorimetric sensor for determination of phosphate ions using anti-aggregation of 2-mercaptoethanesulfonate-modified silver nanoplates and europium ions. *Sensors and Actuators, B: Chemical, 290*, 226–232. Elsevier B.V.
55. Ramli, N. H., Mohamad Nor, N., Abu Bakar, A. H., Zakaria, N. D., Lockman, Z., & Abdul Razak, K. (2024). Platinum-based nanoparticles: A review of synthesis methods, surface functionalization, and their applications. *Microchemical Journal*. Elsevier Inc.
56. Roper, D. K. (2007). Determining surface plasmon resonance response factors for deposition onto three-dimensional surfaces. *Chemical Engineering Science, 62*(7), 1988–1996.
57. Ryu, K. R., Kim, G. W., & Ha, J. W. (2021). Localized surface plasmon resonance inflection points for improved detection of chemisorption of 1-alkanethiols under total internal reflection scattering microscopy. *Scientific Reports, 11*, 1. Nature Research.
58. Samuel, V. R., & Rao, K. J. (2022). A review on label free biosensors. *Biosensors and Bioelectronics: X*. Elsevier Ltd.
59. Shah, M. M., Ahmad, K., Boota, S., Jensen, T., La Frano, M. R., & Irudayaraj, J. (2023). Sensor technologies for the detection and monitoring of endocrine-disrupting chemicals. *Frontiers in Bioengineering and Biotechnology*. Frontiers Media S.A.
60. Sharma, A., Kumar, V., Shahzad, B., Tanveer, M., Sidhu, G. P. S., Handa, N., Kohli, S. K., et al. (2019). Worldwide pesticide usage and its impacts on ecosystem. *SN Applied Sciences, 1(11), 1446*. Springer Nature.
61. Sharma, S., & Gupta, B. D. (2021). Surface plasmon resonance based fiber optic potassium ion disposable sensing probe for soil testing. *Optical Fiber Technology, 64*. Academic Press Inc.

62. Shelar, A., Salve, S., Shende, H., Mehta, D., Chaskar, M., Nile, S. H., & Patil, R. (2024). Recent advances on highly sensitive plasmonic nanomaterial enabled sensors for the detection of agrotoxins: Current progress and future perspective. *Computers and Electronics in Agriculture, 227, 109545*. Elsevier B.V.
63. Singh, R., & Singh, G. S. (2017). Traditional agriculture: A climate-smart approach for sustainable food production. *Energy, Ecology and Environment, 2(5), 296-316*. Joint Center on Global Change and Earth System Science of the University of Maryland and Beijing Normal University.
64. Singh, Y., & Raghuwanshi, S. K. (2021). Titanium dioxide (TiO2) coated optical fiber-based SPR sensor in near-infrared region with bimetallic structure for enhanced sensitivity. *Optik, 226*. Elsevier GmbH.
65. Talari, F. F., Bozorg, A., Faridbod, F., & Vossoughi, M. (2021). A novel sensitive aptamer-based nanosensor using rGQDs and MWCNTs for rapid detection of diazinon pesticide. *Journal of Environmental Chemical Engineering, 9*, 1. Elsevier Ltd.
66. Tene, T., Bellucci, S., Arias Arias, F., Carrera Almendariz, L. S., Flores Huilcapi, A. G., & Vacacela Gomez, C. (2024). Role of graphene in surface Plasmon resonance-based biosensors. *Sensors, 24, s24144670*. Multidisciplinary Digital Publishing Institute (MDPI).
67. Verma, A., & Gupta, B. D. (2021). Fiber optic surface plasmon resonance based disposable probe for the detection of phosphate ion in soil. *Optik, 243*. Elsevier GmbH.
68. Vu, X. H., Dien, N. D., Ha Pham, T. T., Trang, T. T., Ca, N. X., Tho, P. T., Vinh, N. D., et al. (2020). The sensitive detection of methylene blue using silver nanodecahedra prepared through a photochemical route. *RSC Advances, 10*(64), 38974–38988. Royal Society of Chemistry.
69. Wang, L., Huang, X., Wang, C., Tian, X., Chang, X., Ren, Y., & Yu, S. (2021). Applications of surface functionalized Fe3O4 NPs-based detection methods in food safety. *Food Chemistry, 16, 128343*. Elsevier Ltd.
70. Wang, R. Z., Huang, D. L., Liu, Y. G., Zhang, C., Lai, C., Wang, X., Zeng, G. M., et al. (2019). Recent advances in biochar-based catalysts: Properties, applications and mechanisms for pollution remediation. *Chemical Engineering Journal, 16, 2022117*. Elsevier B.V.
71. Wang, Y., Cui, Z., Zhang, X., Zhang, X., Zhu, Y., Chen, S., & Hu, H. (2020). Excitation of surface plasmon resonance on multiwalled carbon nanotube metasurfaces for pesticide sensors. *ACS Applied Materials and Interfaces, 12*(46), 52082–52088. American Chemical Society.
72. Wei, C., Li, M., & Zhao, X. (2018). Surface-enhanced raman scattering (SERS) with silver nano substrates synthesized by microwave for rapid detection of foodborne pathogens. *Frontiers in Microbiology, 9*(DEC) Frontiers Media S.A.
73. Xiong, Y., McLellan, J. M., Chen, J., Yin, Y., Li, Z. Y., & Xia, Y. (2005). Kinetically controlled synthesis of triangular and hexagonal nanoplates of palladium and their SPR/SERS properties. *Journal of the American Chemical Society, 127*(48), 17118–17127.
74. Yesudasu, V., Pradhan, H. S., & Pandya, R. J. (2021). Recent progress in surface plasmon resonance based sensors: A comprehensive review. *Heliyon, 7(3), e06321*. Elsevier Ltd.
75. Yılmaz, E., Azgir, E., Bereli, N., Trkmen, D., & Denizli, A. (2017). Plastic antibody based surface plasmon resonance nanosensors for selective atrazine detection. *Materials Science and Engineering C, 73*, 603–610. Elsevier Ltd.
76. Zain, M. E. (2011). Impact of mycotoxins on humans and animals. *Journal of Saudi Chemical Society, 15*(2), 129–144.
77. Zakirov, N., Zhu, S., Bruyant, A., Lérondel, G., Bachelot, R., & Zeng, S. (2022). Sensitivity enhancement of hybrid two-dimensional nanomaterials-based surface Plasmon resonance biosensor. *Biosensors, 12*, 10. MDPI.
78. Zhang, Z., Zhou, R., Ke, L., Li, J., Jayan, H., El-Seedi, H. R., Zou, X., et al. (2024). Development of multifunctional metal-organic frameworks (MOFs)-based nanofiller materials in food packaging: A comprehensive review. *Trends in Food Science and Technology, 154, 14771*. Elsevier Ltd.
79. Zhou, W., Li, M., & Achal, V. (2025). A comprehensive review on environmental and human health impacts of chemical pesticide usage. *Emerging Contaminants, 11(1-30), 100410*. KeAi Communications Co.

Nanotechnology and Plant Defense: Molecular Insights into Immune Response Activation

Kalpana Dabhade and Navjeet Kaur

Contents

K. Dabhade (✉)
Department of Life Sciences and Biotechnology, Chhatrapati Shivaji Maharaj University, Panvel, Navi Mumbai, India
e-mail: kalpanadabhade@csmu.ac.in

N. Kaur
Department of Biological Sciences and Biotechnology, Institute of Chemical Technology, Matunga, Mumbai, India

R. S. Chaughule, A. V. Shelar (eds.), *Nanotechnology in Agriculture*,
https://doi.org/10.1007/978-3-031-97544-8_5

1 Introduction

Plants, unlike animals, are immobile and cannot escape from threats in their environment, such as pathogens and herbivores. Plants face numerous biotic stressors, including bacteria, fungi, viruses, nematodes, and herbivores, which can significantly impact their growth, productivity, and survival. To combat these threats, plants have developed a highly effective innate immune system composed of two layers: Pattern-triggered immunity (PTI) and effector-triggered immunity (ETI). Plants rely entirely on these innate defenses unlike animals, which have adaptive immune systems. In the context of modern agriculture, strengthening these defense mechanisms is vital to enhance crop resilience, reduce dependence on chemical pesticides, and improve yields sustainably. The interplay between PTI and ETI allows plants to mount rapid, specific, and potent responses to pathogen attacks, making these mechanisms essential for survival in a pathogen-rich environment [1].

In modern agriculture, where crop loss due to pathogens poses severe economic and food security risks, understanding and enhancing plant immunity has become increasingly important. Improving natural plant defenses can reduce dependency on chemical pesticides, lower environmental impacts, and contribute to sustainable agricultural practices. As agriculture faces the escalating challenges of climate change, pest infestations, and evolving plant pathogens, it has become essential to leverage cutting-edge technologies to secure food production. Traditional crop improvement methods have their limitations, especially in responding to dynamic biotic stress factors like pathogens and abiotic stressors such as drought or salinity [2]. Nanotechnology and CRISPR (clustered regularly interspaced short palindromic repeats)/Cas gene editing seem to be the next-age solutions for enhancing plant immunity and resilience. In this chapter, we have discussed various aspects including the plant immune system, the role of nanotechnology and the integration of the CRISPR/Cas system to overcome existing challenges in achieving high-yielding plants produce with enhanced nutritional qualities.

2 Plant Immune System: Molecular Insights

2.1 Complexity of the Plant Immune System

Plants rely on an intricate, multilayered immune system to defend themselves against a diverse range of pathogens, including bacteria, fungi, viruses, and herbivorous insects. This immune system is inherently complex, involving cellular and molecular mechanisms that work together to identify and neutralize threats. Plants cannot actively move to evade infections, so they have evolved a comprehensive, two-layered immune system. Each layer plays a distinct but complementary role in recognizing and responding to pathogens. This two-tier plant defense mechanism can be divided in the following categories, viz., pattern-triggered immunity (PTI) and effector-triggered immunity (ETI). Together, PTI and ETI establish a sophisticated line of defense, which can be reinforced and optimized through advanced genetic and nanotechnology-based interventions [3].

2.1.1 Pattern-Triggered Immunity (PTI)

PTI is the first line of defense in plants. This immune response is activated when plant cells recognize common pathogen structures, known as Pathogen-Associated Molecular Patterns (PAMPs). PAMPs are conserved molecules found in broad groups of pathogens and are recognized by Pattern Recognition Receptors (PRRs) located on the plant cell surface. Examples of well-characterized PAMPs include Flagellin, Chitin, and Lipopolysaccharides (LPS).When a PRR recognizes a PAMP, it initiates a signaling cascade that leads to several defensive responses. The rapid generation of ROS (e.g., hydrogen peroxide) creates a toxic environment for pathogens, damaging their cells and signaling nearby plant cells to initiate defenses. Activation of MAPK pathways amplifies the immune signal, leading to changes in gene expression and the synthesis of antimicrobial proteins. The deposition of lignin and callose around the infection site strengthens the cell wall, limiting pathogen entry and spread. Plants produce small molecules like phytoalexins that are toxic to pathogens and prevent their growth.

These responses create an inhospitable environment for a wide range of pathogens, offering broad-spectrum protection. PTI is a kin to a "general immune response" that is quick to activate and effective against diverse pathogens [4].

2.1.2 Effector-Triggered Immunity (ETI)

While PTI provides an initial barrier, many pathogens have evolved to produce effectors—molecules that suppress PTI responses, allowing the pathogen to invade the plant more effectively. ETI represents the plant's counter-defense mechanism, where it detects these effectors and initiates a strong immune response. ETI is mediated by specialized intracellular receptors, mainly Nucleotide-Binding Leucine-Rich Repeat (NLR) Proteins, which detect pathogen effectors inside plant cells [5].When an effector is detected by an NLR receptor, the plant activates a set of potent immune responses, effector-triggered immunity (ETI) often results in a rapid, localized cell death at the site of infection, a phenomenon known as the hypersensitive response (HR). This "scorched earth" strategy sacrifices infected cells to limit the pathogen's spread. Additionally, ETI can activate systemic acquired resistance (SAR), a long-lasting defense mechanism that primes distant plant tissues to resist future infections. SAR is mediated by signaling molecules like salicylic acid, which help communicate the immune state across the plant. Furthermore, ETI enhances the production of defense-related hormones such as jasmonic acid, ethylene, and salicylic acid, amplifying immune signaling pathways and preparing the plant for sustained resistance [6]. ETI is typically faster and more robust than PTI because it targets specific pathogen effectors, allowing plants to mount a tailored response against particular pathogens. This specificity is both an advantage and a limitation, as it provides effective resistance to specific pathogens but may leave the plant vulnerable to new or rapidly evolving pathogens [7].

2.2 *Evolutionary Dynamics of Plant–Pathogen Interactions*

The relationship between plants and pathogens can be described as an evolutionary "arms race." As plants evolve mechanisms like PTI and ETI to recognize and counteract pathogens, pathogens, in turn, evolve new strategies to evade or suppress these defenses. This constant cycle of adaptation and counteradaptation shapes the immune responses in plants and the infection strategies in pathogens [8]. For example, some pathogens produce effectors that can suppress PTI responses, prompting plants to evolve NLR receptors that can detect these specific effectors and activate ETI. In response, pathogens may evolve new effectors or modify existing ones to avoid recognition, continuing the cycle. This coevolutionary process is a powerful driver of diversity in plant immune systems and pathogen virulence [9]. Improving plant innate immunity is critical in modern agriculture, where crop losses due to diseases can result in economic instability and threaten food security. Enhanced immunity in crops offers a sustainable path toward resilient agriculture with reduced reliance on chemical inputs [10]. Enhancing crop resilience through the genetic improvement of PTI (pattern-triggered immunity) and ETI (effector-triggered immunity) offers significant potential for modern agriculture. By integrating disease-resistance genes, particularly those associated with NLR (nucleotide-binding leucine-rich repeat) proteins, breeding programs can develop crop varieties capable of withstanding intense pathogen pressures, thereby reducing disease outbreaks and minimizing yield losses [11]. Advanced molecular tools like CRISPR/Cas genome editing enable precise modifications or insertions of immune-related genes, boosting both PTI and ETI responses. These innovations also support a reduction in the reliance on chemical pesticides, which have long been a cornerstone of disease management but pose risks to human health and ecosystems due to overuse. Strengthening plant immunity allows crops to resist a broader range of pathogens through PTI. At the same time, ETI enhances defense against specific, virulent threats, enabling farmers to adopt more sustainable and environmentally friendly practices. This contributes to preserving biodiversity by minimizing pesticide applications, which often disrupt ecosystems, and aligns with global goals for sustainable agriculture by fostering low-input farming systems that conserve natural resources [12]. Furthermore, crops with optimized immune responses are vital in addressing the challenges posed by climate change, as shifting environmental conditions are likely to increase pathogen spread and virulence [13]. Drought- and heat-tolerant crops with robust PTI and ETI mechanisms can better endure these stresses, offering resilience against pathogen-induced losses in unpredictable climates. Together, these strategies bolster agricultural productivity and ensure ecological balance and long-term sustainability [14].

3 CRISPR/Cas Gene-Editing System in Agriculture

CRISPR (clustered regularly interspaced short palindromic repeats) and associated Cas (CRISPR-associated), an innovative gene-editing technology, has rapidly transformed plant sciences by allowing precise, efficient, and flexible genetic modifications. Its use in enhancing plant immunity has been a major area of focus, with researchers targeting genes related to defense mechanisms, including PTI and ETI. CRISPR/Cas technology has emerged as a revolutionary tool in plant science, allowing for precise gene edits that improve disease resistance, drought tolerance, and crop yield. CRISPR/Cas systems, initially discovered as part of the adaptive immune system in prokaryotes, have become invaluable tools for targeted genome editing across diverse organisms, including plants. In plants, CRISPR/Cas proteins enable precise modifications to DNA sequences, allowing for precise genetic interventions. This technology has been transformative in enhancing plant resilience to various stresses, including pathogens, by enabling the modification of genes directly involved in immune response pathways [15]. The CRISPR/Cas system leverages a cell's DNA repair mechanisms to insert, delete, or modify genes associated with desired traits [16]. A general mechanism of CRISPR/Cas gene editing system has been illustrated in Fig. 1.

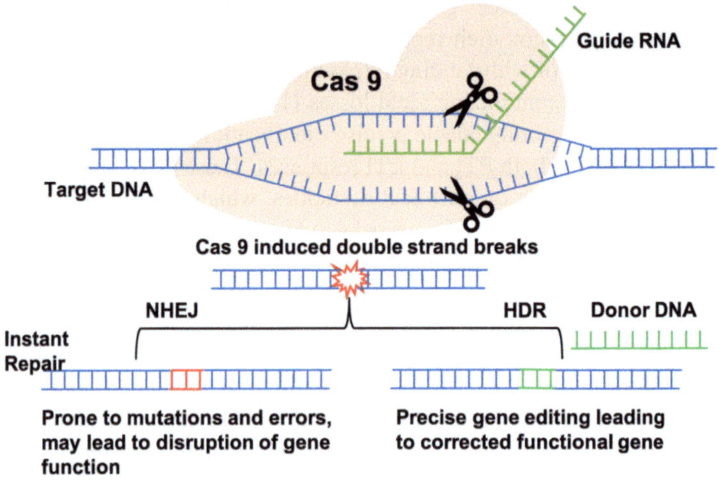

Fig. 1 The Cas9 protein, guided by a single-guide RNA (sgRNA), creates a double-strand break (DSB) at a specific DNA target, enabling repair for genetic modifications. Non-homologous end joining (NHEJ) and homology-directed repair (HDR) are common repair pathways wherein NHEJ is a quick mechanism used for gene knockouts, introducing random indels, whereas HDR is comparatively less frequently used and is time-consuming but, it enables precise editing using donor templates with complex modifications

3.1 Applications in Plant Immunity

3.1.1 Enhancing PTI and ETI Mechanisms

The CRISPR/Cas system seems like an attractive tool with several distinct advantages in plant genome editing, particularly for enhancing defense mechanisms. It can provide controlled precision and efficiency in gene editing, allowing for targeted modifications while minimizing unintended genetic alterations compared to traditional methods. This precision is particularly beneficial for rapidly adapting crops to evolving pathogens by editing genes associated with immunity to develop resistant varieties. Its multiplexing capability enables simultaneous targeting of multiple genes and is especially advantageous for modifying complex immune response pathways involving various genetic components. CRISPR/Cas can be applied to modify or enhance genes involved in PTI and ETI, leading to stronger or more durable immunity in plants. Some potential gene targets include PRR genes, such as FLS2 (Flagellin-Sensing 2) and EFR (EF-Tu Receptor), which are essential for initiating pattern-triggered immunity (PTI) [17]. By editing these genes to improve their sensitivity or durability, plants can enhance their initial immune responses, providing stronger protection against a broad range of pathogens. NLR genes, pivotal for effector-triggered immunity (ETI), recognize specific pathogen effectors and trigger robust resistance mechanisms. Modifying or enhancing NLR genes can lead to more targeted and effective resistance against particular pathogens.

Additionally, genes involved in signal transduction, including those encoding kinases and transcription factors, play a crucial role in immune signaling. Enhancing the stability or activity of these signaling components can improve plant immunity's overall strength and coordination, making crops more resilient to pathogen attacks [18]. However, it is important to acknowledge that there is limited study on CRISPR/Cas-mediated precision targeting in plant genes since the plant cell wall has challenging barriers to overcome as compared to animal or bacterial cells, where this technology has been widely experimented with.

3.1.2 CRISPR/Cas-Based Editing of Disease-Resistant Genes (e.g., R-Genes, NLRs)

Resistance (R) genes, particularly those encoding NLRs, are pivotal in the plant defense system. CRISPR/Cas enables targeted editing of these genes, potentially providing the following advantages proving disease resistance in crops can be achieved by editing R-genes, which are responsible for recognizing and responding to specific pathogens. By targeting these genes, plants can be engineered to resist diseases that affect economically important crops, reducing yield losses. Additionally, modifying NLR genes can expand the plant's ability to recognize a broader range of pathogen effectors, enhancing its overall immunity and protecting against diverse threats. CRISPR/Cas technology also offers the potential to stack multiple R-genes, helping to maintain resistance and durability. This strategy prevents resistance breakdown by addressing the risk of pathogens evolving new effectors, ensuring long-term protection for crops [19, 20]. In rice, genes such as Xa13 and Xa21 have been targeted through CRISPR editing to develop resistance to bacterial blight, a significant disease that impacts rice yields. Similarly, in wheat, specific R-genes have been modified to confer resistance to powdery mildew, a fungal disease that affects wheat crops. These examples highlight how CRISPR technology can be used to tailor pathogen resistance in crops, providing crop-specific solutions to combat diseases, thereby improving crop resilience and reducing yield losses [21, 22].

3.2 Advanced Multiplex Gene Editing Strategies

Given the complex nature of plant immune responses, single-gene editing often does not suffice for effective disease resistance. Multiplex CRISPR systems enable the simultaneous editing of multiple genes, facilitating a more comprehensive modification of the immune network [23, 24]. By editing multiple immunity-related genes, multiplexing allows for the customization of immunity profiles tailored to meet specific agricultural needs and address varying pathogen pressures. Examples of multiplexing in plant immunity demonstrate its potential to enhance disease resistance across various crops. In tomatoes, CRISPR has been used to simultaneously target multiple genes associated with susceptibility to bacterial wilt, resulting in plants with broad-spectrum resistance to the disease. Similarly, in potatoes, multiplex gene editing has been employed to increase resistance against late blight by modifying specific susceptibility (S) genes. These examples highlight how multiplexing can be leveraged to improve plant immunity by targeting several genes at once, offering efficient and tailored solutions for combating crop diseases [25, 26].

4 Role of Nanotechnology in Enhancing Agriculture

Nanotechnology, the science and engineering of manipulating matter at an atomic or molecular scale, has emerged as a transformative field with broad applications across sectors, including agriculture [27]. Nanotechnology has evolved from a theoretical concept to a powerful tool in agriculture, capable of revolutionizing plant protection and enhancing immune responses. By enabling precise interventions and reducing environmental impacts, nanotechnology supports sustainable agricultural practices and strengthens food security. As researchers continue to explore its potential, nanotechnology is set to play an essential role in building resilient, disease-resistant crop systems, paving the way for a new era in agriculture [28].

Nanotechnology has opened new avenues in agriculture by enabling precise and controlled delivery of agrochemicals, nutrients, and genetic materials. In the context of plant immunity, nanotechnology offers innovative tools for early disease detection, efficient delivery of immunity-boosting molecules, and environmentally friendly pesticides. Nanoparticles can improve the bioavailability and uptake of these compounds, thus enhancing plant health and resilience [29].

4.1 Historical Background and Development of Nanotechnology

The origins of nanotechnology trace back to the mid-twentieth century, with the famous 1959 lecture by physicist Richard Feynman titled "There's Plenty of Room at the Bottom." Feynman proposed the possibility of manipulating individual atoms to build extremely small devices, laying the theoretical groundwork for the field.

In the 1980s, with the invention of the scanning tunneling microscope (STM) and atomic force microscope (AFM), scientists gained the tools to visualize and manipulate materials at the atomic level. By the 1990s, nanotechnology research gained momentum, leading to the formation of dedicated research institutions and funding programs. Key developments during this period have significantly shaped the landscape of nanotechnology and its applications, particularly in agriculture and bioengineering. One of the most notable milestones was the formation of the

National Nanotechnology Initiative (NNI) in 2000 in the United States. The NNI played a pivotal role in fostering research and collaboration across federal agencies, academic institutions, and the private sector, accelerating advancements in nanotechnology. This initiative not only spurred innovation in materials science and biotechnology but also laid the groundwork for the development of nanotechnology applications in agriculture, including precision farming, nano-pesticides, and plant disease diagnostics. In addition to the NNI, introducing new nanomaterials such as fullerenes, carbon nanotubes, and quantum dots revolutionized the field of bioengineering. Fullerenes, with their unique spherical structure, and carbon nanotubes, known for their exceptional strength and conductivity, opened new possibilities for creating more efficient, durable, and lightweight materials. These nanomaterials have since been explored for their potential in enhancing agricultural productivity, such as improving nutrient delivery systems and developing advanced sensors for monitoring crop health. Quantum dots, with their tunable optical properties, have also shown promise in bio-imaging and detecting plant pathogens, offering a new approach for real-time crop health monitoring. Together, these innovations have paved the way for the integration of nanotechnology into agricultural practices, promising improvements in efficiency, sustainability, and disease management [30]. A comprehensive overview of the various applications of nanotechnology within the agricultural sector, highlighting its potential to enhance productivity, improve resource efficiency, and address challenges in crop management, pest control, and sustainable farming practices is depicted in Fig. 2 [30].

4.2 *Early Agricultural Applications*

The initial exploration of nanotechnology in agriculture began in the early 2000s, with studies focusing on improving soil health, nutrient delivery, and pest control. Scientists recognized that nanoscale materials could enhance the efficiency and effectiveness of agricultural inputs. This early period focused on applications such as nano-encapsulated fertilizers and pesticides that offer controlled release systems that minimize the leaching and runoff of agrochemicals, enhancing efficiency and reducing environmental impact. Additionally, early prototypes of nano-biosensors enable precise soil and crop health monitoring by detecting specific pathogens and nutrient levels [31]. These advancements support sustainable agriculture by optimizing resource use and ensuring timely interventions for crop management [32]. As research progressed, nanotechnology applications in agriculture evolved to address specific challenges, such as plant disease, nutrient deficiencies, and environmental sustainability. Advances in nanomaterials and nano-delivery systems fuelled innovations, allowing for precise interventions that target plants at the cellular level. The following developments highlight the milestones in agricultural nanotechnology [33].

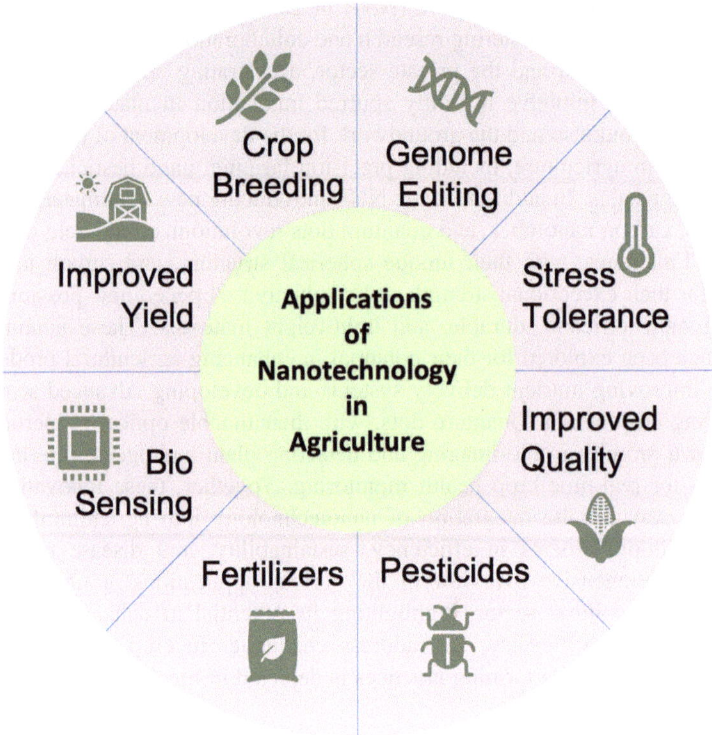

Fig. 2 An overview of various applications of nanotechnology in agriculture

4.2.1 Controlled-Release Fertilizers and Pesticides

The development of controlled-release fertilizers and pesticides was a significant step forward. By using nanoparticles as carriers, scientists created formulations that gradually release nutrients or pesticides, minimizing waste and environmental impact. Nanoparticles enhance nutrient bioavailability, allowing plants to absorb nutrients more efficiently and thereby reducing the need for excessive fertilizer application. Additionally, their targeted delivery systems minimize chemical runoff into water and soil, decreasing environmental contamination and promoting sustainable agricultural practices [34].

4.2.2 Nano-Biosensors for Disease and Pest Detection

As the demand for sustainable pest control grew, nano-biosensors became crucial tools for early detection of pathogens and pests. Nano-biosensors utilize materials like gold nanoparticles (AuNPs), graphene, and carbon nanotubes to detect specific pathogens based on biomarkers or environmental changes. Nanotechnology facilitates rapid and precise diagnostics, enabling farmers to monitor crop health in real time and implement timely interventions. Early detection of diseases or nutrient deficiencies allows for targeted treatments, reducing the reliance on broad-spectrum pesticides. This not only minimizes pesticide use but also helps prevent the development of pesticide resistance, supporting sustainable and efficient agricultural practices [35].

4.2.3 Nano-Carriers for Plant Immune Activation

With the growing understanding of plant immune systems, researchers have begun using nanotechnology to enhance plant immunity directly. Nano-carriers deliver immunity-boosting molecules like salicylic acid, jasmonic acid, and chitosan to plants, activating defense pathways more efficiently than traditional methods. Nano-carriers offer targeted immune activation by delivering molecules directly to plant immune pathways, maximizing their efficacy and precision. Furthermore, nanoparticles can be designed as slow-release systems, providing prolonged defense responses and enhancing long-term immunity in plants. These advancements strengthen the overall resilience of crops against pathogens while optimizing resource use [36].

4.3 Emergence of Nanotechnology in Enhancing Plant Protection and Activating Immune Responses

With advancements in nanotechnology, the field has moved beyond traditional uses in agriculture toward innovative applications that enhance plant protection and immune response activation. Nanotechnology's role in plant immunity primarily focuses on creating novel materials that stimulate defense mechanisms, improve pathogen resistance, and support sustainable pest control [37].

4.3.1 Nano-Pesticides and Their Advantages

Traditional pesticides have significant environmental drawbacks, such as water pollution, soil degradation, and harm to beneficial organisms. Nano-pesticides, in contrast, offer targeted and controlled release, minimizing these side effects. Chitosan nanoparticles, derived from natural sources, exhibit potent antifungal, antiviral, and antibacterial properties, making them effective as both treatments for plant pathogens and activators of the plant's innate immune system. Similarly, metal-based nano-pesticides, such as nanoparticles of silver, copper, and zinc oxide, possess strong antimicrobial capabilities. For instance, silver nanoparticles can disrupt bacterial and fungal cell membranes, providing a robust, natural defense against a wide range of pathogens. These nanoparticle-based solutions offer innovative, sustainable approaches to plant protection [38].

4.3.2 Nanoparticles as Immune Activators

Nanotechnology enables the development of particles that mimic natural signals to stimulate plant immune responses. These nano-based immune activators are particularly useful in activating pattern-triggered immunity (PTI) and effector-triggered immunity (ETI) pathways, strengthening plant defenses against specific pathogens. Silica nanoparticles (SiNPs) play a vital role in fortifying plant cell walls, thereby enhancing resistance to fungal infections. They also stimulate pattern-triggered immunity (PTI) by mimicking a pathogen attack, bolstering the plant's natural defense mechanisms. Additionally, nanoparticles loaded with salicylic acid—a critical signaling molecule in plant immunity—enable controlled delivery to optimize systemic-acquired resistance (SAR). This targeted approach ensures prolonged and effective immune activation, offering sustainable solutions for improved crop protection [39].

4.3.3 Nanoparticle-Based Delivery Systems for RNA Interference (RNAi)

RNA interference (RNAi) is a promising method for silencing specific genes in pathogens, thus preventing infection. Nanoparticles can effectively deliver double-stranded RNA (dsRNA) to plant cells, triggering RNAi pathways that inhibit pathogen development. This technology is particularly promising for viral diseases, which are difficult to control through traditional methods. Carbon Nanotubes and Gold Nanoparticles: Both materials are effective carriers for dsRNA, allowing for precise targeting of pathogen genes without altering the plant's genetic material [40].

While nanotechnology holds immense promise, several challenges must be addressed for widespread agricultural adoption. These include environmental and safety concerns, high costs of nanoparticle production, and the need for regulatory frameworks. However, the future of nanotechnology in agriculture is bright, with ongoing research focused on developing sustainable, cost-effective, and scalable solutions.

5 Integration of Nanotechnology with CRISPR/Cas Systems

5.1 Need for Integrated Systems

CRISPR/Cas gene editing has intrigued plant biotechnologists, offering its potential to create crops with enhanced resilience, higher nutritional value, and increased adaptability to changing environmental conditions. However, the path to realizing its full potential is fraught with obstacles. Traditional delivery methods like

Agrobacterium-mediated transformation, though widely utilized, are limited by their host range and often result in random transgene integration, leading to gene silencing or unpredictable expression. Similarly, biolistic particle bombardment, while bypassing host range constraints, frequently causes tissue damage, multiple unstable insertions, and low editing efficiency. The tough and rigid nature of the plant cell wall further complicates the delivery process, making it significantly more challenging than in mammalian or bacterial cells. Beyond these technical challenges, the large size and structural complexity of CRISPR/Cas components add to spatial barriers to effective delivery. Cas proteins, such as Cas9, are difficult to package and transport into plant cells due to their high charge density and low stability in ionic solutions. Additionally, the reliance on protospacer adjacent motif (PAM) sequences for site-specific targeting limits the editing scope in plant genomes, which are often rich in AT regions, especially in noncoding regions critical for gene regulation. Furthermore, successful editing frequently requires advanced tissue culture techniques, such as somatic embryogenesis, which are not only time-consuming but also have low efficiency and are not universally applicable across plant species. Another significant challenge lies in the incomplete understanding of plant genomics. Effective genome editing depends on identifying precise target sites, which requires a comprehensive understanding of transcriptomic, proteomic, and metabolomic interactions. Unfortunately, the functions of many plant genes remain unknown, increasing the risk of off-target effects and unintended phenotypic changes [41–45].

5.2 Nanomaterials Suitable for CRISPR/Cas Systems

Nanotechnology, with no doubt, is a field with immense potential in various biological applications [46, 47]. Nanomaterials act as versatile carriers that protect gene-editing components, enhance their stability, and improve their cellular uptake. These materials are particularly advantageous in plant biotechnology, where the rigid plant cell wall and large genome sizes present unique challenges [48, 49]. In the case of gene editing, especially in applying CRISPR/Cas technology to plant cells, researchers are exploring nanoscale materials to overcome traditional challenges associated with gene editing, such as inefficient delivery, off-target effects, and cellular barriers. Various types of nanomaterials have been explored for their suitability as delivery vehicles for CRISPR/Cas systems. Each type offers distinct advantages depending on the application, target, and biological environment. A broad categorization of nanomaterials is shown in Fig. 3.

5.2.1 Metallic Nanoparticles

Metallic nanoparticles, such as gold and iron oxide nanoparticles, are increasingly being explored in plant genetic engineering to deliver gene-editing tools like

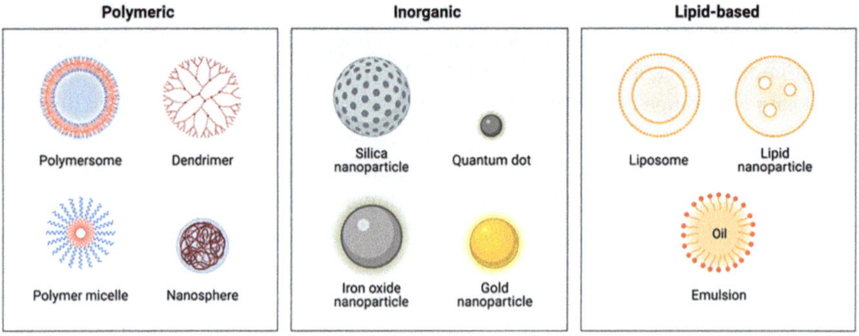

Fig. 3 Major categories of nanoparticles used in the nanomedicine sector based on the material composition [50]

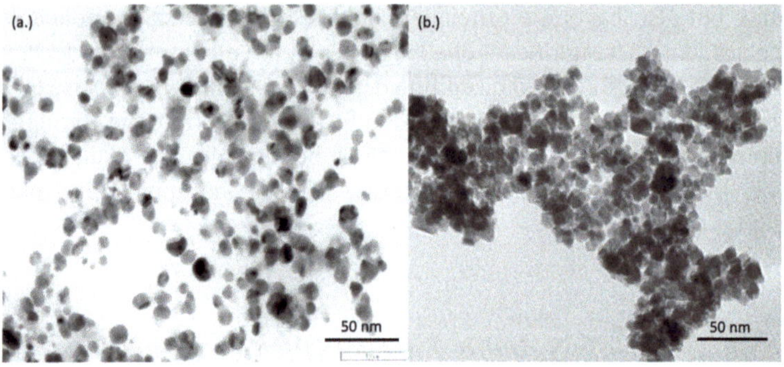

Fig. 4 (**a**) Gold nanoparticles, scale 50 nm and (**b**) Iron oxide nanoparticles, scale 50 nm

CRISPR/Cas systems effectively. These nanoparticles can enhance the stability of genetic material, improve cellular uptake, and offer controlled release, making them ideal candidates for precise and efficient genome editing in plants. They are easily customized into a variety of morphological shapes and sizes to suit the purpose of the application. A transmission electron microscopy image of gold and iron oxide nanoparticles captured from our lab has been shown in Fig. 4.

Gold Nanoparticles (AuNPs): Known for their biocompatibility and ease of functionalization, AuNPs are widely used for delivering nucleic acids and proteins. Their ability to conjugate with biomolecules via thiol or amine linkages makes them highly adaptable for CRISPR components. Their surface can be easily functionalized with biomolecules, including CRISPR/Cas proteins and guide RNA (gRNA), enabling targeted delivery. Moreover, their optical properties allow them to be tracked in biological systems using imaging techniques. The ability of AuNPs to cross biological barriers, including the plant cell wall and mammalian plasma membranes, further enhances their versatility.

Magnetic Iron Oxide Nanoparticles (MIONPs): These nanoparticles provide magnetic properties that allow for precise targeting using external magnetic fields. Their stability and biocompatibility make them suitable for intracellular delivery. Magnetic iron oxide nanoparticles are specifically explored for gene editing due to their magnetic responsiveness and stability. These nanoparticles can be directed to specific locations using external magnetic fields, ensuring precise delivery. Additionally, MIONs are nontoxic and can be metabolized by natural pathways, making them safe for biological applications. Their unique ability to combine imaging and therapeutic functions enables real-time tracking of delivery and editing efficiency [51–53].

5.2.2 Polymeric Nanoparticles

These are synthesized from natural or synthetic polymers such as polyethylene glycol (PEG) or polylactic-co-glycolic acid (PLGA). Polymeric nanoparticles are biodegradable, reducing concerns about long-term toxicity, and can encapsulate both hydrophilic and hydrophobic molecules. Polymeric nanoparticles, particularly those made from biodegradable polymers like PLGA and PEG, offer exceptional versatility for gene editing purposes. These nanoparticles can encapsulate CRISPR components, protecting them from enzymatic degradation and premature release. Polymeric nanoparticles are tuneable in size, charge, and hydrophobicity, allowing for optimization based on the target system. They exhibit sustained release properties, ensuring that the gene-editing components remain effective over time. Importantly, their biocompatibility reduces the risk of immune responses, making them a popular choice for both plant and mammalian applications [54, 55].

5.2.3 Other Nanomaterials

Several other nanomaterials are also explored for their individual unique properties. Lipid nanoparticles provide excellent protection against enzymatic degradation and facilitate endosomal escape. Carbon-based materials like quantum dots, graphene oxide, and carbon nanotubes are being investigated for their ability to adsorb and transport genetic material. Their high surface area and ability to penetrate cellular membranes are advantageous for gene editing. Structurally robust, silica nanoparticles provide a rigid, porous structure that can encapsulate CRISPR components and release them in a controlled manner. Zinc-based nanoparticles are engineered to specifically recognize and bind to targeted DNA sequences, offering precise genome editing capabilities for applications in plant genetic engineering [52–56]. Some of the key advantages and limitations have been summarized in Fig. 5 and have been discussed in detail in the sections following subsequently.

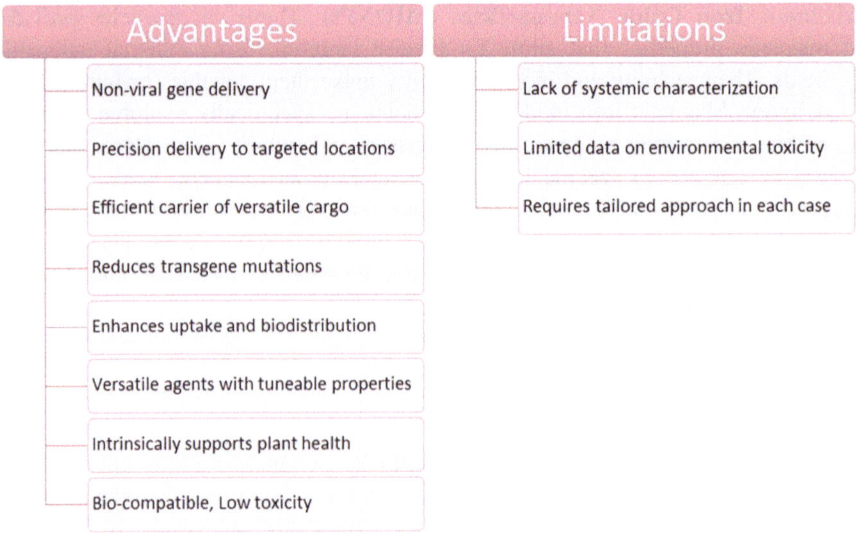

Fig. 5 Key advantages and limitations of nanomaterials used for CRISPR/Cas gene editing in plant systems

5.3 Case Studies of Nanotech-Enabled CRISPR Cas Gene Editing in Plants

5.3.1 Nonviral Delivery Agent

Viral delivery carries risks such as unwanted genetic mutations, insertional mutagenesis, and immunogenicity. Viral vectors have limitations in the size of expression cassettes they can carry, which restricts their use in delivering large or complex cargoes that hold greater potential in plant biotechnology. Additionally, viral vectors are unable to directly transport proteins, rendering them unsuitable for gene editing techniques such as CRISPR, which rely on Cas proteins. Furthermore, the production of viral vectors is highly regulated due to their pathogenic origins and the potential for integrating segments of their genetic material into the plant host genome [56–58]. Nonviral vectors for gene transfection in plants offer a significant advantage for overcoming species-specific transformation limitations associated with traditional methods like *Agrobacterium*-mediated transformation, which does not work effectively in many monocots. By using nonviral carriers such as cationic liposomes, polycations, and cell-penetrating peptides (CPPs), researchers can bypass the cell wall barrier and deliver genetic material to diverse plant species without the need for specialized equipment or reliance on pathogen-based systems. Nanotechnology plays a crucial role in this context by enhancing the precision and efficiency of these carriers. Nanoparticles, including carbon nanotubes, can be combined with CPPs to form stable complexes that effectively condense and deliver nucleic acids, facilitating penetration and uptake by plant cells. This approach not only broadens the scope of genetic transformation across plant species but also allows for targeted delivery within specific cellular structures, potentially transforming crop breeding and plant biotechnology [59, 60].

Magnetic nanoparticles (MNPs) have been extensively used as drug-delivery agents in mammalian cells [61, 62]. These unique properties can be translated into gene editing such as employing a magnetofection technique utilizing an external magnetic force. In one such reported method by Dulam et al. Fe3O4 nanoparticles coated with polyethyleneimine (positively charged) form complexes with negatively charged DNA vectors or CRISPR/Cas9 components, which are then applied to pollens under a magnetic field for direct delivery. This approach enables vector/DNA-free genome editing, allowing non-transgenic crop production by delivering transcribed sgRNA and Cas9 mRNA or Cas9 protein with sgRNA directly to plant reproductive tissues like stigma or protoplasts. Notably, magnetofection bypasses the lengthy tissue culture process and avoids transgene integration, making it efficient for editing crops such as maize, tobacco, lettuce, and fruit crops like apple and grapevine. Although yet to be reported, pollen-mediated magnetofection could enable faster, direct genome editing by transferring CRISPR/Cas9 ribonucleoproteins (RNPs) into pollens for quicker crop improvement [63, 64].

A review article by Mousa et al. discusses the potential of exosome- and liposome-like nanoparticles as carriers for CRISPR/Cas genome editing in plants, offering a promising alternative to traditional viral vectors, which face challenges such as immunogenicity, insertional mutagenesis, and limited cargo capacity. Nonviral vectors like exosomes and liposomes provide advantages in terms of biocompatibility, low cytotoxicity, and enhanced targeting capabilities. These nanoparticles improve the stability and bioavailability of CRISPR/Cas components, enabling more precise genome editing with minimal immune response. Their small size and versatility in surface modifications make them particularly effective for efficient plant cell penetration [65].

Another research highlights the use of hybrid and multilayer nanoparticles, which combine viral-like properties with the safety of nonviral systems for more effective delivery. Hybrid nanoparticles have been known to improve the efficacy of the overall delivery and systemic circulation in various biological applications. Suitable hybridization techniques can improve targeting specificity, endosomal escape, and nuclear localization of nanoparticles to increase the editing efficiency in plants [52, 66].

5.3.2 Precision Gene Editing

Advancements in precision gene-editing technologies like CRISPR/Cas systems are staggered by challenges like off-target effects, inconsistent editing outcomes, and inefficiencies in delivery mechanisms that hinder their broader application in agriculture. These issues compromise the reliability of genetic modifications. Nanoscale materials have proven to mitigate these issues to a certain extent by providing solutions for precise delivery, controlled release, and enhanced targeting of gene-editing components in plants.

In one study by Haroon et al., Strigolactones (SLs) are a key determinant of plant architecture, known for their role in inhibiting shoot branching (tillering). CRISPR/Cas9-based manipulation of SL biosynthesis in crops holds promise for improving yield and reducing susceptibility to parasitic plants like *Striga*. However, this approach faced challenges such as disrupting SLs to prevent *Striga* germination can inadvertently hinder beneficial mycorrhizal associations, and gene editing outcomes can vary significantly across different crop cultivars, impacting consistency in traits like plant height and tillering [67]. Nanocarriers can enable the controlled release of synthetic SL analogs to target *Striga* without fully disrupting SL levels, preserving symbiotic relationships. Targeted delivery of CRISPR tools can minimize off-target effects and ensure consistent gene editing across cultivars. Stimuli-responsive nanocarriers can regulate SL activity dynamically, inhibiting parasitic plants like Striga while preserving mycorrhizal associations. Additionally, nanosensors may be used to monitor SL levels for real-time phenotype adjustments, ensuring optimized crop performance.

Feng Chen et al. have discussed two case studies of CRISPR/Cas-based plant modification: (i) Rice—for enhancing traits such as yield, stress resistance, and nutritional value and (ii) Maize—for enhancing its tolerance to various abiotic and biotic stresses. The main challenges faced in these cases were off-target effects, which may culminate in adverse consequences, including undesirable traits and reduced yield. Efforts to minimize off-target effects include the development of high-fidelity Cas variants and optimizing gRNA design, as well as exploring alternative systems like Cas12b, which show promise for precise targeting with minimal off-target activity. Delivery methods also present a major hurdle, particularly for recalcitrant crops like maize, where traditional methods like *Agrobacterium*-mediated transformation and biolistic delivery are inefficient [68]. Specific nanomaterials, such as silica NPs, are known to induce stress-resilience in plants and may be particularly suited for monocots like rice due to their ease of surface functionalization, biocompatibility, and controlled release capabilities. The stable and predictable behavior of silica NPs in rice, where tissue regeneration and transformation are generally less efficient compared to dicots, makes them a better option for this crop [69, 70]. Maize is a larger and more recalcitrant monocot, which presents specific challenges for gene editing, particularly in terms of delivery. Carbon nanotubes, with their strong mechanical properties and efficient penetration ability, are especially advantageous for overcoming the thick cell walls in maize. They have shown high efficacy in translocating cargo even in difficult-to-transform species [70–73].

Another review by Vinay Kumar et al. discusses the issue of inconsistent Cas9 activity, which can lead to variable mutation patterns, and certain CRISPR constructs may result in off-target effects or require transgene-free plants for consumer acceptance [74]. Nanoparticles can be designed to carry CRISPR/Cas9 components, such as Cas9 proteins and guide RNAs (gRNAs), directly into plant cells without integrating foreign DNA. This approach would produce nontransgenic, genome-edited plants by avoiding the permanent integration of the CRISPR construct into the plant genome, potentially reducing off-target mutations. Delivering RNPs to whole plants through nano-mediated, non-biolistic methods presents a promising strategy to circumvent issues associated with transgene integration and plant regeneration. Furthermore, nano-based delivery of CRISPR components may facilitate the rapid screening and validation of sgRNA-endonuclease candidates, accelerating the planta testing of various sgRNAs [70, 72]. In the same review, Vinay et al. also discuss the potential for improved nutrient uptake and metabolism arising due to genetic changes in plant pathways altering nutrient uptake due to changes in root structure and function. Alterations in carotenoid biosynthesis can affect overall plant health and pigmentation, impacting agricultural and market value. Nanoparticles can be engineered to release biosynthesis inhibitors or activators in a controlled manner, specifically targeting carotenoid pathways. For instance, encapsulating fluridone (a carotenoid biosynthesis inhibitor) in nanocarriers could allow precise control over carotenoid levels, reducing unwanted side effects in gene-edited plants. Nanoparticles containing essential micronutrients or growth-promoting agents can enhance nutrient uptake efficiency, compensating for any disruptions in root morphology. These nano-fertilizers can also be used to maintain root health and optimize growth in modified plants [75].

5.3.3 Effective Carrier

An interesting study by Zhen Liang et al. demonstrates the use of nanoparticles, specifically through particle bombardment, to deliver CRISPR/Cas9 ribonucleoproteins (RNPs) into wheat immature embryos, enabling precise gene editing of TaGW2 and TaGASR7 without integrating foreign DNA, resulting in transgene-free mutant plants. The method involves pre-incubating Cas9 protein and sgRNA with gold nanoparticles, which are then used for biolistic bombardment. This nanoparticle-mediated delivery ensures high specificity and minimizes off-target effects by promoting rapid RNP degradation, making it an efficient alternative for genome editing in cereal crops like wheat, maize, and barley. The approach enhances public acceptance of genome-edited crops due to their transgene-free nature and offers a promising tool for crop improvement [76].

Another work presented by Zhang Y et al. on the genome-editing method mentions the use of gold nanoparticles to deliver CRISPR/Cas9 components directly into wheat cells via biolistic bombardment, allowing precise, transient editing without transgene integration. This nanotechnology-based approach can be extended for gene editing in difficult-to-transform plants, advancing molecular breeding in other varieties of wheat and similar crops [77].

Another reported research uses gold nanoparticles to improve the delivery of CRISPR/Cas9 into potato protoplasts through GNOME laser transfection. By targeting gold nanoparticles with a laser, temporary pores are created in the cell membrane, allowing CRISPR/Cas9 ribonucleoproteins to enter more efficiently. This method addresses challenges specific to plant cells, like sensitivity to osmotic stress and weak nanoparticle adhesion, by using membrane-binding lectins and linker molecules to enhance binding. As a proof-of-concept, the project aims to reduce the toxic compound α-solanine in potatoes, showcasing the potential of GNOME laser transfection for plant gene editing [78].

Bettina Nagy et al. demonstrate the use of the novel cationic polymer poly (2-hydroxypropylene imine) (PHPI) for CRISPR/Cas9-mediated genome editing in maize protoplasts, targeting the EGFP gene. PHPI facilitated the co-introduction of Cas9 (pFGC-pcoCas9) and gRNA (pZmU3-gRNA) plasmids, with DsRed (red fluorescent protein marker gene) as a transformation marker. Fluorescence microscopy identified DsRed-positive cells as transformed, and GFP signal loss indicated successful gene editing, with an average transformation frequency of 4.1% and mutation frequency of 1.65%. Sanger sequencing of DsRed-positive microcalli confirmed targeted multi- and single-base deletions, as well as nucleotide insertions near the PAM site (5′-NGG-3′), typical of Cas9 activity. This approach, demonstrating high-efficiency DNA delivery and targeted mutation, suggests PHPI's potential for broader application in plant genome editing.

5.3.4 Overcoming Transgene Mutations

Demirer et al. present an elegant study on the nanomaterial-based delivery platform using carbon nanotubes (CNTs) for efficient, nontoxic, and species-independent DNA delivery in plants, enabling transient gene expression without transgene integration. By developing dialysis and electrostatic grafting methods to load DNA onto CNTs, the platform was successfully tested on various plants such as *Nicotiana benthamiana*, arugula, wheat, and cotton, achieving strong expression of target proteins. Unlike traditional methods like *Agrobacterium* or gene gun transformations, which can damage cells or result in random transgene integration, CNT-mediated delivery is minimally invasive and circumvents strict GMO regulations, especially important for complex species (e.g., cassava and cacao). When paired with genome editing, CNTs support high-throughput screening and stable gene edits without genetic integration, positioning them as a scalable and regulatory-friendly tool for plant biotechnology and rapid phenotype mapping [79, 80].

In a work reported by Lamiaa M. et al. cationic lipid nanoparticles have been used to enable the precise delivery of the Cas9/crRNA ribonucleoprotein (RNP) complex into citrus protoplast cells. These nanoparticles enhance cellular uptake and protect the RNP complex from degradation, ensuring efficient, DNA-free genome editing. This approach not only avoids integration of foreign DNA, resulting in transgene-free, edited citrus plants, but also supports regulatory compliance and the advancement of marker-free, precision breeding techniques for improved citrus varieties [81]. In another study by same group, cationic lipid nanoparticles, particularly Lipofectamine LTX, were used to enhance CRISPR/Cas9 DNA delivery into citrus protoplasts by protecting the DNA from enzymatic degradation and enabling it to bypass cellular barriers. This nanoparticle-mediated approach improved DNA stability and delivery efficiency, achieving high gene expression with minimal cytotoxicity—critical for woody plants like citrus, where traditional PEG-based transfection methods are often toxic and less effective. This method holds promise for more efficient genome editing in citrus, facilitating the production of homozygous edited plants in vegetatively propagated varieties [82].

5.4 Challenges Associated with Nanomaterials in Agriculture

The cellular uptake and intracellular fate of nanoparticles (NPs) are crucial for the advancement of nanoparticle-mediated delivery systems, particularly in plant genetic engineering. Nanoparticle-mediated cellular internalization occurs through various mechanisms, such as endocytosis, where nanoparticles are engulfed by cell membranes and internalized into the cytoplasm. The size, surface charge, and shape of nanoparticles significantly influence this process, determining their ability to cross biological barriers. However, for every application, a tailored approach is needed. For instance, studies on the translocation of gold nanoparticles (AuNPs) within barley plant roots presented by Anna Milewska and colleagues revealed that AuNPs did not translocate effectively within the plant. This lack of translocation was attributed to the size and surface charge variations of the nanoparticles, which impacted their behavior and led to off-target accumulation in the cell wall. Such accumulation altered the chemical composition of the host tissue, demonstrating the importance of nanoparticle characteristics in their biological interactions and suggesting potential barriers to effective delivery [83]. Nanoparticles such as carbon nanotubes (CNTs) have been observed to inhibit root elongation in crops like tomatoes and lettuce. Despite their potential in plant genetic engineering, the toxic effects of NPs on plant growth must be thoroughly studied to ensure that their application does not harm the plants or hinder their development [84].

Further, the need for systematic characterization of nanoparticle properties, including their interactions with various plant tissues, is essential to understanding the mechanistic details of their uptake and intracellular fate. For example, Dr. Strano's research demonstrated that the transport of single-walled carbon nanotubes across pollen cell membranes is primarily influenced by their surface zeta potential. Additionally, the diversity of plant materials, such as protoplasts, leaves, and root hairs, requires tailored approaches for nanoparticle delivery. Such comprehensive studies on the interaction of nanoparticles with different plant cell types are lacking in our current knowledge base, making it difficult to implement strategies for practical applications in plant biotechnology [85, 87].

There is no doubt that nanoparticle-mediated delivery systems hold great promise for advancing plant genetic engineering, but several challenges remain. From understanding the fundamental mechanisms of nanoparticle uptake to improving the stability and cytotoxicity of these systems, continued research is needed for the wide usage of nanoparticles in agricultural and biotechnological applications. Acknowledging these challenges empowers our research on nanoparticle-based strategies that could revolutionize plant transfection and genetic manipulation, toward the development of more resilient and high-yielding crops.

6 Potential of CRISPR/Cas System as Advanced Biosensors

The integration of nanotechnology with CRISPR/Cas systems has opened new horizons for real-time monitoring. By addressing existing challenges and harnessing emerging innovations, nano-based biosensors can significantly enhance the reliability and scope of CRISPR/Cas applications, marking a transformative step in molecular diagnostics and therapeutics. CRISPR/Cas systems have revolutionized the field of molecular biology by enabling precise genome editing with unprecedented accuracy and efficiency. These systems, originally discovered as an adaptive immune mechanism in bacteria and archaea, have been repurposed for various biotechnological applications. The ability to monitor CRISPR/Cas activity in real-time is critical for ensuring the safety, accuracy, and efficiency of genome-editing experiments. Real-time monitoring provides insights into the dynamics of CRISPR/Cas interactions, enhances experimental reproducibility, and aids in identifying off-target effects [88].

Biosensors, devices that combine biological recognition elements with physicochemical transducers, have emerged as indispensable tools for real-time monitoring of biomolecular interactions, including CRISPR/Cas activity. Nano-based biosensors, in particular, have garnered significant attention due to their sensitivity, specificity, and scalability. This section explores the design and application of nano-based biosensors for real-time CRISPR/Cas monitoring, highlighting their mechanisms, benefits, and challenges.

6.1 Designing Nano-Based CRISPR/Cas Monitoring Systems

CRISPR/Cas systems rely on RNA-guided nucleases, such as Cas9 and Cas12, to recognize and cleave target DNA sequences. The CRISPR/Cas system functions through a series of precise steps to target and modify DNA. Initially, the CRISPR/Cas complex binds to the target DNA through the guide RNA, which directs the Cas protein to the specific sequence. Upon binding, the Cas protein, typically Cas9, performs cleavage of the target DNA, making a double-strand break at the desired location. This break activates the DNA repair mechanisms within the cell, which can be harnessed for gene editing. Additionally, some CRISPR proteins like Cas12 or Cas13 are capable of generating secondary signals, such as collateral cleavage. This occurs when Cas12, after cleaving the target DNA, indiscriminately cleaves nearby nontarget DNA, amplifying the signal for detection in diagnostic applications, or in the case of Cas13, cleaving RNA molecules, providing further versatility for gene editing and diagnostics [89].

Nano-based biosensors are ideal for detecting these events due to their high surface-to-volume ratio, tunable properties, and ability to integrate with various transduction mechanisms, including optical, electrochemical, and mechanical.

Nano-based CRISPR/Cas biosensors consist of several key components that work together to enable sensitive and precise detection of specific DNA or RNA sequences. The biorecognition elements include the guide RNA (gRNA), which provides sequence specificity, and the Cas proteins, which act as the catalytic components responsible for cleaving the target nucleic acids. DNA or RNA reporters are engineered oligonucleotides that produce measurable signals, such as fluorescence or color changes, upon cleavage. Nanomaterials play a pivotal role in amplifying these signals. Gold nanoparticles (AuNPs) are used for colorimetric and plasmonic sensing due to their optical properties, while carbon nanotubes (CNTs) and graphene enhance electron transfer in electrochemical sensors. Quantum dots (QDs) offer high-resolution fluorescence, and metal-organic frameworks (MOFs) improve sensor efficiency with their high surface area and tunable chemical properties. Transducers convert the biochemical interactions into measurable physical signals. Optical transducers detect changes in fluorescence or surface plasmon resonance, electrochemical transducers measure changes in current, voltage, or impedance, and mechanical transducers detect shifts in mass or resonance frequency. Together, these components enable the development of highly sensitive and accurate biosensors with applications ranging from pathogen detection to genetic analysis [90].

Fluorescent Biosensors: These biosensors utilize the cleavage of fluorophore-quencher pairs by Cas proteins. A fluorophore is attached to one end of a DNA or RNA reporter, and a quencher is attached to the other. When the Cas protein cleaves the target nucleic acid, the fluorophore is separated from the quencher, resulting in a measurable fluorescence signal. To enhance signal sensitivity, quantum dots or fluorescent dyes may be incorporated into the system, allowing for highly sensitive detection of the target [91].

Electrochemical Biosensors: Electrochemical biosensors measure changes in electrical properties upon target recognition or cleavage. Conductive nanomaterials, such as carbon nanotubes (CNTs) or graphene, are integrated into electrodes. When the target binds to the sensor, it induces changes in conductivity or generates redox reactions, producing measurable electrical signals. These sensors offer high sensitivity and are ideal for portable applications [92].

Colorimetric Biosensors: Colorimetric sensors use visible color changes to indicate CRISPR/Cas activity. Gold nanoparticles are commonly employed in these sensors due to their plasmonic properties. When the Cas protein cleaves a DNA-linked nanoparticle, the nanoparticles either aggregate or disperse, causing a visible color shift. This color change can be easily detected by the naked eye, providing a simple and rapid method for detecting CRISPR activity.

Surface Plasmon Resonance (SPR) Biosensors: SPR biosensors detect changes in the refractive index on a metal surface when biomolecules bind or undergo conformational changes. CRISPR/Cas activity alters the mass or conformation of the biomolecules on the sensor surface, which results in measurable shifts in the SPR signal. These sensors offer high sensitivity and real-time detection capabilities, making them suitable for detailed molecular analyses [93].

6.2 Applications of CRISPR/Cas in Agro-Biosensing

CRISPR/Cas biosensors have a wide range of applications in plant sciences, particularly in areas like molecular diagnostics, genome editing monitoring, pathogen detection, and plant trait analysis.

Detection of Plant Pathogens: CRISPR/Cas biosensors can detect bacterial, viral, or fungal pathogens in plants at an early stage. By targeting pathogen-specific nucleic acid sequences, they enable rapid and sensitive diagnostics. Quick detection helps farmers implement timely interventions, reducing crop losses. Aman et al. presented a novel diagnostic method for detecting *Xanthomonas oryzae*, the pathogen responsible for bacterial blight in rice, and have developed a detection system utilizing the CRISPR/Cas12a enzyme, which is guided by specific CRISPR RNAs (crRNAs) to recognize and bind to target DNA sequences unique to *Xanthomonas oryzae*. Upon binding, Cas12a exhibits collateral cleavage activity, indiscriminately cutting nearby single-stranded DNA (ssDNA). This property is harnessed to cleave reporter molecules, producing a detectable signal indicative of the pathogen's presence [94].

Monitoring Genome Editing in Plants: Biosensors are used to verify the accuracy of CRISPR-mediated genome editing in plants. They can confirm successful edits or detect off-target mutations which ensures precision in genome editing and reduces unintended effects. Han et al. (2019) applied CRISPR/Cas9 technology to create a targeted mutation in the E1 gene to regulate flowering in soybeans. The resulting truncation of the E1 protein disrupted its ability to inhibit the GmFT2a/5a gene, leading to enhanced expression of these genes and earlier flowering under long-day conditions. This genetic modification produced a photo-insensitive soybean variant, making it a promising candidate for cultivation in higher latitude regions. CRISPR/Cas9 biosensors monitor the presence or absence of specific DNA sequences post-editing in crops like maize or wheat [95].

Identification of Plant Genetic Markers: CRISPR/Cas systems detect single nucleotide polymorphisms (SNPs) or specific genetic markers associated with desirable traits like drought resistance or high yield which enhances the efficiency of breeding programs through precise marker-assisted selection. *Zhang* et al. discussed the application of CRISPR technology in detecting single nucleotide polymorphisms (SNPs) to enhance plant breeding and functional genomics. CRISPR-based biosensors, particularly CRISPR/Cas9 and CRISPR/Cas12, were engineered to identify specific SNPs within plant genomes to identify SNPs in genes regulating stress tolerance in rice [96].

Quality Control in Agricultural Products: Biosensors can be used to test for genetic modifications or quality traits in agricultural produce. For example, they identify specific alleles linked to higher nutritional content, which ensures compliance with regulatory standards for genetically modified organisms (GMOs). *Tang* et al. developed novel Cas12 and Cas13 Systems for harnessing collateral cleavage activity, enabling signal amplification for ultrasensitive detection. It detects RNA sequences in genetically modified soybeans to confirm their genetic makeup [97].

Detection of Environmental Stress Responses: Biosensors have been designed to measure changes in gene expression in plants under environmental stresses like drought, salinity, or nutrient deficiency which aids in developing stress-resistant crop varieties. Ghosh focused on CRISPR/Cas systems as functional genomics tools to dissect plant responses to biotic and abiotic stresses through gene knockouts or activations and targeting transcription factors like DREB or WRKY to enhance plant stress resilience. CRISPR/Cas13 biosensors detected RNA transcripts of stress-responsive genes in wheat under drought conditions [98].

Detection of Residual Herbicides or Pesticides: By targeting specific genes involved in herbicide or pesticide resistance, CRISPR biosensors can confirm the presence of resistance traits in weeds or crops which helps in managing resistance and optimizing herbicide use. Shan et al. programmed Cas9 to recognize mutations in the EPSPS gene that confer glyphosate resistance. The system distinguishes resistant and susceptible weed populations in *Amaranthus palmeri* (Palmer amaranth) [99].

Identification of Plant Viruses: CRISPR biosensors rapidly identify plant viruses by detecting their DNA or RNA which helps in early detection to prevent widespread outbreaks and reduce crop damage. Fozouni et al. utilized CRISPR/Cas13 for its RNA-targeting ability and developed a CRISPR/Cas13a biosensor to detect RNA of the *Tomato Yellow Leaf Curl Virus (TYLCV)* Potato Virus Y [100].

Monitoring of Transgene Integration: Ensuring the successful integration of transgenes during genetic engineering in plants which reduces time in confirming successful genetic modifications. Kim et al. programmed Cas12 and Cas13 systems to identify specific transgene sequences which can validate the insertion of desired transgenes in genetically engineered cotton plants [101].

Studying Plant Epigenetics: CRISPR-based biosensors can identify specific epigenetic modifications, such as DNA methylation or histone modifications, in plants which provides insights into gene regulation mechanisms in plants. *Liang* et al. discussed that CRISPR-dCas9 (dead Cas9) can be used to study and manipulate plant epigenomes. Site-specific DNA methylation or demethylation using fusion proteins like dCas9-TET or dCas9-DNMT were used to study epigenetic changes in *Arabidopsis thaliana* during developmental stages [102].

Field-Deployable Diagnostics: CRISPR/Cas biosensors can be adapted for portable, on-site detection of plant diseases and genetic traits using simple devices like lateral flow strips which enables real-time decision-making in agricultural management. *Jiao* et al. *developed a* CRISPR/Cas12 biosensor coupled with a paper based device which was used for field detection of *Phytophthora infestans*, the pathogen causing late blight in potatoes [103].

Future advancements may focus on integrating biosensors with microfluidics for point-of-care applications and developing fully autonomous systems powered by artificial intelligence. Additionally, the emergence of novel Cas variants with unique properties may pave the way for next-generation biosensors with enhanced capabilities for enhancing defense mechanisms in plants.

7 Future Perspectives and Ethical Considerations

As discussed in the above sections, nanoparticles like gold and lipid-based carriers enable the targeted delivery of tools such as CRISPR/Cas9, bypassing traditional methods that may involve viral vectors or chemicals. Advanced innovations like quantum dots allow real-time tracking of gene-editing complexes, ensuring precision and minimizing off-target effects, while enzymes and devices like nanoneedles help overcome biological barriers, such as plant cell walls. These breakthroughs are paving the way for multigene editing, climate-resilient crops, and reduced dependence on chemical fertilizers and pesticides, making agriculture more sustainable [104, 105].

However, these advancements bring ethical and environmental challenges that require urgent attention. Accessibility remains a pressing issue, as the unequal availability of nanotechnology-enhanced gene editing could deepen global disparities in food security and agricultural productivity. The possibility of off-target edits, with unintended genetic alterations potentially impacting ecosystems or human health, raises significant ethical concerns. Environmental risks, such as the persistence of nanoparticles in soil and water ecosystems, further complicate matters, as their accumulation could harm nontarget organisms and disrupt soil health. Genetically modified crops also pose ecological risks, including potential crossbreeding with wild relatives or outcompeting native species, necessitating long-term monitoring. Public scepticism about these technologies underscores the need for transparent communication, public education, and stakeholder engagement to build trust and ensure responsible deployment [106–108].

To address these challenges, robust and harmonized regulatory frameworks are crucial. International organizations like ISO, FAO, and the Codex Alimentarius Commission must collaborate to establish guidelines for nanoparticle use, risk assessment, and environmental monitoring. Regional bodies such as the OECD, ECHA, and EFSA in Europe, as well as national agencies like the EPA, USDA, and FDA in the United States, must align their efforts to minimize regulatory gaps. Mechanisms for public consultation and inclusion of stakeholders such as farmers, scientists, and policymakers can strengthen governance and foster public trust. Regulatory policies must remain flexible, with periodic updates informed by ongoing research and technological advancements, to ensure nanotechnology's safe and equitable integration into agriculture [109–111].

8 Conclusion

This chapter has demonstrated how the integration of these transformative tools addresses critical challenges in plant immunity and paves the way for sustainable crop production. Nanomaterials such as gold nanoparticles, carbon nanotubes, and lipid-based carriers have emerged as efficient tools for overcoming the structural barriers presented by plant cells. These nanoparticles not only facilitate the precise delivery of CRISPR components but also protect these components from enzymatic degradation, reducing off-target effects. Their role in activating both PTI and ETI pathways has opened new avenues for engineering disease-resistant crops. Furthermore, the co-delivery capabilities of nanomaterials, combined with advancements in biosensors, have enabled real-time monitoring of CRISPR activity, enhancing the reliability of gene-editing processes. The successful integration of these technologies has profound implications, from reducing reliance on chemical pesticides to improving crop resilience against climate-induced stresses. This approach aligns with global goals for sustainable agriculture and food security, addressing critical environmental and socioeconomic concerns.

We have addressed the challenges, especially regarding the scalability of nanoparticle-mediated delivery systems, the optimization of CRISPR components for diverse plant species, and the environmental safety of nanomaterials that require focused research and regulatory oversight. The ethical implications of genome-edited crops, particularly regarding biodiversity and equitable access, also warrant careful consideration. Looking ahead, the fusion of nanotechnology with CRISPR/Cas systems promises to revolutionize agriculture. As we move forward, continued collaboration among scientists, policymakers, and stakeholders will be essential in realizing the vision of a sustainable agricultural future.

References

1. Jones, J. D. G., & Dangl, J. L. (2006). The plant immune system. *Nature, 444*(7117), 323–329. https://doi.org/10.1038/nature05286
2. Boller, T., & Felix, G. (2009). A renaissance of elicitors: Perception of microbe-associated molecular patterns and danger signals by pattern-recognition receptors. *Annual Review of Plant Biology, 60*, 379–406. https://doi.org/10.1146/annurev.arplant.57.032905.105346
3. Hammond-Kosack, K. E., & Parker, J. E. (2003). Deciphering plant-pathogen communication: Fresh perspectives for molecular resistance breeding. *Current Opinion in Biotechnology, 14*(2), 177–193. https://doi.org/10.1016/S0958-1669(03)00030-4
4. Dixon, R. A. (2001). Natural products and plant disease resistance. *Nature, 411*(6839), 843–847. https://doi.org/10.1038/35081178
5. Balmer, A., & Mauch-Mani, B. (2013). The role of salicylic acid and jasmonic acid in pathogen defense. *Plant Signaling & Behavior, 8*(9), e25772. https://doi.org/10.4161/psb.25772
6. Chisholm, S. T., Coaker, S. A., Day, B. P., & Staskawicz, J. B. (2006). Host-microbe interactions: Shaping the evolution of the plant immune response. *Cell, 124*(4), 803–814. https://doi.org/10.1016/j.cell.2006.02.008

7. Zipfel, C. (2008). Pattern-recognition receptors in plant innate immunity. *Current Opinion in Immunology, 20*(1), 10–16. https://doi.org/10.1016/j.coi.2007.11.003
8. Dodds, P. N., & Rathjen, J. P. (2010). Plant immunity: Towards an integrated view of plant–pathogen interactions. *Nature Reviews Genetics, 11*(8), 539–548. https://doi.org/10.1038/nrg2812
9. Glazebrook, J. (2005). Contrasting mechanisms of defense against biotrophic and necrotrophic pathogens. *Annual Review of Phytopathology, 43*(1), 205–227. https://doi.org/10.1146/annurev.phyto.43.040204.135923
10. Heath, M. C. (2000). Hypersensitive response-related death. *Plant Molecular Biology, 44*(3), 321–334. https://doi.org/10.1023/A:1026592509060
11. Schwessinger, B., & Ronald, P. C. (2012). Plant innate immunity: Perception of conserved microbial signatures. *Annual Review of Plant Biology, 63*, 451–482. https://doi.org/10.1146/annurev-arplant-042811-12
12. Wang, Y., Wei, L., & Ji, B. (2021). CRISPR-based microbial genome editing: Methods and applications. *Biotechnology Advances, 47*, 107696. https://doi.org/10.1016/j.biotechadv.2020.107696
13. Khot, L. R., Sankaran, S., Schuster, G. L., & Ehsani, R. (2012). Applications of nanomaterials in agricultural production and crop protection: A review. *Crop Protection, 35*, 64–70. https://doi.org/10.1016/j.cropro.2012.01.007
14. Chhipa, H. (2017). Nanofertilizers and nanopesticides for agriculture. *Environmental Chemistry Letters, 15*(1), 15–22. https://doi.org/10.1007/s10311-016-0581-3
15. Majeed, Z., Ali, K., Khan, S. S., Bukhari, A. A. P., & Ahmad, F. (2021). Advanced nanotechnology: A sustainable solution for crop protection. *Agronomy, 11*(6), 1118. https://doi.org/10.3390/agronomy11061118
16. Eberl, F., Moser, G. N. H. S., López, D. M., Attallah, A. A. M., & Preger, D. M. L. C. (2020). Designing pathogen-specific nanocarriers for precision plant disease management. *Nature Nanotechnology, 15*, 30–39. https://doi.org/10.1038/s41565-019-0595-3
17. Chen, Y., & Tang, Y. (2020). Application of nanotechnology in sustainable crop protection. *Journal of Agricultural and Food Chemistry, 68*(40), 11248–11260. https://doi.org/10.1021/acs.jafc.0c05476
18. Rai, M., & Ingle, A. P. (2012). Role of nanotechnology in agriculture with special reference to management of insect pests. *Applied Microbiology and Biotechnology, 94*(2), 287–293. https://doi.org/10.1007/s00253-012-3969-4
19. Kumar, S. (2021). *Plant nanotechnology: Principles, practices, and prospects*. Springer Nature. https://doi.org/10.1007/978-3-030-69353-0
20. Ahmed, B., Ali, F., Shah, M. W., Bakhsh, A. H., Khan, F. G., Ali, M. A., & Jamil, M. (2017). Nanotechnology-based approaches for enhancing plant growth and disease resistance. *Plant Science, 256*, 34–46. https://doi.org/10.1016/j.plantsci.2016.12
21. Subramanian, K. S., Saha, S. K., Raghavendra, P. R., & Srivastava, S. K. (2015). Nanofertilizers for balanced crop nutrition. *Current Science, 108*(1), 25–28.
22. Sundaram, R. M., Choi, S. R., & Kim, B. J. (2014). Functional genomics for plant innate immunity: Challenges and opportunities. *Trends in Biotechnology, 32*(4), 224–233. https://doi.org/10.1016/j.tibtech.2014.02.001R
23. Gogos, A., Meier, N. J. M. J., & McDonald, R. L. (2012). Nanomaterials in plant protection and fertilization: Current state, foreseen applications, and research priorities. *Journal of Agricultural and Food Chemistry, 60*(39), 9781–9792. https://doi.org/10.1021/jf302154y
24. Bai, X., Yadav, D. S., & Sharma, R. K. (2019). Mechanisms of nanoparticle-mediated gene delivery in plants. *Plant Biotechnology Journal, 17*(3), 656–674. https://doi.org/10.1111/pbi.13020
25. Savary, S., Willocquet, P. F., & Elings, J. (2019). The global burden of plant diseases and its implications for food security. *Nature Sustainability, 2*(9), 767–777. https://doi.org/10.1038/s41893-019-0291-9

26. Gao, X., Chen, F. J., & Chen, Y. (2020). Enhanced pathogen resistance through targeted delivery of nanoparticles in plants. *Advanced Functional Materials, 30*(7), 1910445. https://doi.org/10.1002/adfm.201910445
27. Popp, J., Birol, F. L. L. R., & Sims, C. C. L. (2013). The effect of bioenergy expansion: Food, energy, and environment. *Renewable and Sustainable Energy Reviews, 32*, 559–578. https://doi.org/10.1016/j.rser.2014.01.056
28. Parisi, C., Malek, M., & Richardson, J. J. (2015). Potential impact of nanomaterials on the agro-food sector. *Journal of Agricultural and Food Chemistry, 63*(16), 3980–3999. https://doi.org/10.1021/jf506000h
29. Zhang, H., O'Connell, L. G., & Li, C. L. (2016). Engineering nanoscale structures for precision agriculture. *ACS Nano, 10*(9), 8978–8982. https://doi.org/10.1021/acsnano.6b04181
30. Peters, R. J. B., Mulder, A. M., & van Rijen, M. A. W. H. (2016). Nanomaterials for agricultural applications: Benefits and challenges. *Environmental Science: Nano, 3*(4), 682–702. https://doi.org/10.1039/C6EN00076H
31. Singh, S., Gupta, M. N. K., & Ghosh, R. K. (2017). Biological synthesis of nanomaterials and their role in sustainable agriculture. *Applied Microbiology and Biotechnology, 101*(1), 60–68. https://doi.org/10.1007/s00253-016-8009-6
32. Azzawi, A. I., Hussen, Z. M. A., & Ali, A. F. (2021). Applications of nanotechnology in plant growth and protection: A review. *Environmental Technology & Innovation, 24*, 101872. https://doi.org/10.1016/j.eti.2021.101872
33. Bai, Y., Brown, R. J. A., & Zhang, S. L. (2013). Functional characterization of a plant immunity regulator. *Plant Cell, 25*(7), 2438–2453. https://doi.org/10.1105/tpc.113.114959
34. Khan, M. R., & Rizvi, T. F. (2021). Nanotechnology and its application in plant science: A review. *Journal of Plant Physiology and Biochemistry, 162*, 23–36. https://doi.org/10.1016/j.plaphy.2021.02.009
35. Duraimurugan, P., Rajendran, K. S., & Srinivasan, M. S. (2022). Bioengineered nanoparticles for crop protection: Recent trends. *Nanotechnology Reviews, 11*(1), 1537–1554. https://doi.org/10.1515/ntrev-2022-0072
36. Cardi, T., Murovec, J., Bakhsh, A., et al. (2023). CRISPR/Cas-mediated plant genome editing: Outstanding challenges a decade after implementation. *Trends in Plant Science, 28*, 1144–1165.
37. Son, S., & Park, S. R. (2022). Challenges facing CRISPR/Cas9-based genome editing in plants. *Frontiers in Plant Science, 13*, 902413.
38. Pak, S., Li, C., Pak, S., & Li, C. (2022). Progress and challenges in applying CRISPR/Cas techniques to the genome editing of trees. *Forestry Research, 1*, 6. https://doi.org/10.48130/FR-2022-0006
39. Singh, S., Chaudhary, R., Deshmukh, R., & Tiwari, S. (2022). Opportunities and challenges with CRISPR/Cas mediated homologous recombination based precise editing in plants and animals. *Plant Molecular Biology, 111*, 1–20.
40. Cunningham, F. J., Goh, N. S., Demirer, G. S., Matos, J. L., & Landry, M. P. (2018). Nanoparticle-mediated delivery towards advancing plant genetic engineering. *Trends in Biotechnology, 36*, 882–897.
41. Ahmar, S., Mahmood, T., Fiaz, S., Mora-Poblete, F., Shafique, M. S., Chattha, M. S., & Jung, K. H. (2021). Advantage of nanotechnology-based genome editing system and its application in crop improvement. *Frontiers in Plant Science*. https://doi.org/10.3389/FPLS.2021.663849/FULL
42. Fincheira, P., Tortella, G., Seabra, A. B., Quiroz, A., Diez, M. C., & Rubilar, O. (2021). Nanotechnology advances for sustainable agriculture: Current knowledge and prospects in plant growth modulation and nutrition. *Planta*. https://doi.org/10.1007/s00425-021-03714-0
43. Pathania, P. K., Lotey, N. K., Kashyap, D., & Sabherwal, P. (2023). Nanomaterial-based sensing platforms for food-borne pathogen detection. *Applications of Nanotechnology in Microbiology, 193–228.*

44. Khan, I., Saeed, K., & Khan, I. (2019). Nanoparticles: Properties, applications and toxicities. *Arabian Journal of Chemistry, 12*, 908–931.
45. Elahi, N., Kamali, M., & Baghersad, M. H. (2018). Recent biomedical applications of gold nanoparticles: A review. *Talanta, 184*, 537–556.
46. Lotey, N. K., Pednekar, S., & Chaughule, R. (2022). Hybrid nanoparticles in biomedical applications. In *Nanomaterials for cancer detection using imaging techniques and their clinical applications* (pp. 365–400). Springer International Publishing.
47. Lotey, N. K., Chaughule, R., & Pednekar, S. (2022). Low concentration PAMAM dendrimer coated superparamagnetic iron oxide nanoparticles—potentialsin MR imaging and theranostics. *Res Sq*. https://doi.org/10.21203/rs.3.rs-2269226/v1
48. Haladjova, E., Rangelov, S., Tsvetanov, C. B., Posheva, V., Peycheva, E., Maximova, V., Momekova, D., Mountrichas, G., Pispas, S., & Bakandritsos, A. (2014). Enhanced gene expression promoted by hybrid magnetic/cationic block copolymer micelles. *Langmuir, 30*, 8193–8200.
49. Lu, A. H., Salabas, E. L., & Schüth, F. (2007). Magnetic nanoparticles: Synthesis, protection, functionalization, and application. *Angewandte Chemie - International Edition, 46*, 1222–1244.
50. Sung, Y. K., & Kim, S. W. (2020). Recent advances in polymeric drug delivery systems. *Biomaterials Research*. https://doi.org/10.1186/s40824-020-00190-7
51. Deng, S., Gigliobianco, M. R., Censi, R., & Di Martino, P. (2020). Polymeric nanocapsules as nanotechnological alternative for drug delivery system: Current status, challenges and opportunities. *Nanomaterials*. https://doi.org/10.3390/nano10050847
52. Mohanty, A., Uthaman, S., & Park, I. K. (2020). Utilization of polymer-lipid hybrid nanoparticles for targeted anti-cancer therapy. *Molecules*. https://doi.org/10.3390/molecules25194377
53. Riley, M. K., & Vermerris, W. (2017). Recent advances in nanomaterials for gene delivery—A review. *Nanomaterials, 7*, 94.
54. Gad, M. A., Li, M. J., Ahmed, F. K., & Almoammar, H. (2020). Nanomaterials for gene delivery and editing in plants: Challenges and future perspective. *Multifunctional Hybrid Nanomaterials for Sustainable Agri-food and Ecosystems*, 135–153.
55. Mostafavi, E., & Zare, H. (2022). Carbon-based nanomaterials in gene therapy. *OpenNano, 7*, 100062.
56. Lv, Z., Jiang, R., Chen, J., & Chen, W. (2020). Nanoparticle-mediated gene transformation strategies for plant genetic engineering. *The Plant Journal, 104*, 880–891.
57. Vats, S., Kumawat, S., & Brar, J. (2022). Opportunity and challenges for nanotechnology application for genome editing in plants. *Plant Nano Biology, 1*, 100001.
58. Midorikawa, K., Kodama, Y., & Numata, K. (2019). Vacuum/compression infiltration-mediated permeation pathway of a peptide-pDNA complex as a non-viral carrier for gene delivery in planta. *Scientific Reports, 9*(1 9), 1–10.
59. Malzahn, A., Lowder, L., & Qi, Y. (2017). Plant genome editing withTALEN and CRISPR. *Cell & Bioscience, 7*, 21.
60. Belhaj, K., Chaparro-Garcia, A., Kamoun, S., Patron, N. J., & Nekrasov, V. (2015). Editing plant genomes with CRISPR/Cas9. *Current Opinion in Biotechnology, 32*, 76–84.
61. Li, L., Hu, S., & Chen, X. (2018). Non-viral delivery systems for CRISPR/Cas9-based genome editing: Challenges and opportunities. *Biomaterials, 171*, 207.
62. Lotey, N. K. B., Chaughule, R., & Pednekar, S. (2022). Dendrimer coated SPIONs: Synthesis, characterization and potentials in biomedical applications. *Acta Scientific Medical Sciences*, 118–123.
63. Lotey, N. K., Sabherwal, P., Chaughule, R., Patkar, D., & Pednekar, S. (2024). Ultra—Low concentration dendrimer coated SPIONs: Synthesis, characterization and potentials in MR imaging. *BioNanoScience*. https://doi.org/10.1007/s12668-024-01409-w
64. Bortesi, L., & Fischer, R. (2015). The CRISPR/Cas9 system for plant genome editing and beyond. *Biotechnology Advances, 33*, 41–52.

65. Sandhya, D., Jogam, P., Allini, V. R., Abbagani, S., & Alok, A. (2020). The present and potential future methods for delivering CRISPR/Cas9 components in plants. *Journal of Genetic Engineering & Biotechnology, 18*, 25.
66. Alghuthaymi, M. A., Ahmad, A., Khan, Z., Khan, S. H., Ahmed, F. K., Faiz, S., Nepovimova, E., Kuča, K., & Abd-Elsalam, K. A. (2021). Exosome/liposome-like nanoparticles: New carriers for CRISPR genome editing in plants. *International Journal of Molecular Sciences, 22*, 7456.
67. Takeshita, V., Campos, E. V. R., Rodrigues, J. S., & Fraceto, L. F. (2023). Opinion: Hybrid nanoparticle systems—Two-way delivery approach for agriculture. *Plant Nano Biology, 6*, 100053.
68. Butt, H., Jamil, M., Wang, J. Y., Al-Babili, S., & Mahfouz, M. (2018). Engineering plant architecture via CRISPR/Cas9-mediated alteration of strigolactone biosynthesis. *BMC Plant Biology, 18*, 1–9.
69. Chen, F., Chen, L., & Yan, Z. (2024). Recent advances of CRISPR-based genome editing for enhancing staple crops. *Frontiers in Plant Science, 15*, 1478398.
70. Du, J., Liu, B., & Zhao, T. (2022). Silica nanoparticles protect rice against biotic and abiotic stresses. *Journal of Nanobiotechnology, 20*, 197.
71. Yan, G., Huang, Q., Zhao, S., Xu, Y., He, Y., Nikolic, M., Nikolic, N., Liang, Y., & Zhu, Z. (2024). Silicon nanoparticles in sustainable agriculture: Synthesis, absorption, and plant stress alleviation. *Frontiers in Plant Science, 15*, 1393458.
72. Safdar, M., Kim, W., Park, S., Gwon, Y., Kim, Y. O., & Kim, J. (2022). Engineering plants with carbon nanotubes: A sustainable agriculture approach. *Journal of Nanobiotechnology, 20*(1 20), 1–30.
73. Chen, Z., Zhao, J., & Cao, J. (2022). Opportunities for graphene, single-walled and multi-walled carbon nanotube applications in agriculture: A review. *Crop Design, 1*, 100006.
74. Khan, M. N., Ali, W., Grillo, R., Wu, H., & Nie, L. (2024). Carbon dots and single-walled carbon nanotubes enhances maize shading stress tolerance. *bioRxiv*, 2024.07.11.603111.
75. Bari, V. K., Nassar, J. A., Kheredin, S. M., Gal-On, A., Ron, M., Britt, A., Steele, D., Yoder, J., & Aly, R. (2019). CRISPR/Cas9-mediated mutagenesis of CAROTENOID CLEAVAGE DIOXYGENASE 8 in tomato provides resistance against the parasitic weed Phelipanche aegyptiaca. *Scientific Reports, 9*(1 9), 1–12.
76. Squire, H. J., Tomatz, S., Voke, E., González-Grandío, E., & Landry, M. (2023). The emerging role of nanotechnology in plant genetic engineering. *Nature Reviews Bioengineering, 1*(5 1), 314–328.
77. Liang, Z., Chen, K., & Li, T. (2017). Efficient DNA-free genome editing of bread wheat using CRISPR/Cas9 ribonucleoprotein complexes. *Nature Communications, 8*(1 8), 1–5.
78. Zhang, Y., Liang, Z., Zong, Y., Wang, Y., Liu, J., Chen, K., Qiu, J. L., & Gao, C. (2016). Efficient and transgene-free genome editing in wheat through transient expression of CRISPR/Cas9 DNA or RNA. *Nature Communications, 7*, 12617.
79. Nagy, B., Öktem, A., Ferenc, G., Ungor, D., Kalac, A., Kelemen-Valkony, I., Fodor, E., Nagy, I., Dudits, D., & Ayaydin, F. (2023). CRISPR/Cas9 mutagenesis through introducing a nanoparticle complex made of a cationic polymer and nucleic acids into maize protoplasts. *International Journal of Molecular Sciences, 24*, 16137.
80. Demirer, G. S., Silva, T. N., Jackson, C. T., Thomas, J. B., Ehrhardt, D. W., Rhee, S. Y., Mortimer, J. C., & Landry, M. P. (2021). Nanotechnology to advance CRISPR–Cas genetic engineering of plants. *Nature Nanotechnology, 16*, 243.
81. Demirer, G. S., Zhang, H., Goh, N. S., González-Grandío, E., & Landry, M. P. (2019). Carbon nanotube–mediated DNA delivery without transgene integration in intact plants. *Nature Protocols, 14*, 2954.
82. Mahmoud, L. M., & Dutt, M. (2024). Cationic lipid nanoparticle-mediated delivery of a Cas9/crRNA ribonucleoprotein complex for transgene-free editing of the citrus plant genome. *Plant Cell Reports*. https://doi.org/10.1007/S00299-024-03254-3

83. Mahmoud, L. M., Kaur, P., Stanton, D., Grosser, J. W., & Dutt, M. (2022). A cationic lipid mediated CRISPR/Cas9 technique for the production of stable genome edited citrus plants. *Plant Methods, 18*, 1–14.
84. Milewska-hendel, A., Sala, K., Gepfert, W., & Kurczyńska, E. (2021). Gold nanoparticles-induced modifications in cell wall composition in barley roots. *Cells*. https://doi.org/10.3390/CELLS10081965
85. Cañas, J. E., Long, M., Nations, S., Vadan, R., Dai, L., Luo, M., Ambikapathi, R., Lee, E. H., & Olszyk, D. (2008). Effects of functionalized and nonfunctionalized single-walled carbon nanotubes on root elongation of select crop species. *Environmental Toxicology and Chemistry, 27*, 1922–1931.
86. Lew, T. T. S., Park, M., Wang, Y., Gordiichuk, P., Yeap, W. C., Mohd Rai, S. K., Kulaveerasingam, H., & Strano, M. S. (2020). Nanocarriers for transgene expression in pollen as a plant biotechnology tool. *ACS Materials Letters, 2*, 1057–1066.
87. Hu, P., An, J., Faulkner, M. M., Wu, H., Li, Z., Tian, X., & Giraldo, J. P. (2020). Nanoparticle charge and size control foliar delivery efficiency to plant cells and organelles. *ACS Nano, 14*, 7970–7986.
88. Chen, J. S., Ma, E., & Harrington, L. B. (2018). CRISPR/Cas12a target binding unleashes indiscriminate single-stranded DNase activity. *Science, 360*(6387), 436–439.
89. Doudna, J. A., & Charpentier, E. (2014). The new frontier of genome engineering with CRISPR/Cas9. *Science, 346*(6213), 1258096.
90. Jinek, M., Chylinski, K., & Fonfara, I. (2012). A programmable dual-RNA-guided DNA endonuclease in adaptive bacterial immunity. *Science, 337*(6096), 816–821.
91. Li, J., Wang, J., & Yang, L. (2019). Nano-based biosensors for monitoring CRISPR/Cas systems. *Biosensors and Bioelectronics, 123*, 59–68.
92. Wang, Q., Zhang, D., & Li, H. (2020). Nanomaterials for real-time CRISPR/Cas detection. *Advanced Functional Materials, 30*(12), 1908915.
93. Zhang, Z., Liu, X., & Cheng, L. (2021). Advances in biosensors for CRISPR/Cas systems. *Trends in Biotechnology, 39*(5), 523–535.
94. Aman, R., Mahas, A., & Mahfouz, M. M. (2020). CRISPR/Cas12a-based detection of *Xanthomonas oryzae* causing bacterial blight in rice. *Molecular Plant Pathology, 21*, 1725–1733. https://doi.org/10.1111/mpp.12902
95. Li, J., Meng, X., Sun, Y., & Liang, Y. (2021). CRISPR technology in plant genome editing: Biosensing and detection tools for precision editing. *Journal of Integrative Plant Biology, 63*, 1245–1260. https://doi.org/10.1111/jipb.13100
96. Zhang, X., Wang, L., & Xu, X. (2020). CRISPR-based SNP detection for plant breeding and functional genomics. *Trends in Biotechnology, 38*, 1325–1333. https://doi.org/10.1016/j.tibtech.2020.02.009
97. Tang, X., Liu, B., Ma, H., & Wang, Z. (2021). CRISPR/Cas biosensors for genetic detection in agricultural products. *Biosensors and Bioelectronics, 175*, 112857. https://doi.org/10.1016/j.bios.2021.112857
98. Ghosh, R., & Chakraborty, S. (2020). CRISPR/Cas systems as tools to study stress-related gene expression in plants. *Plant Science, 299*, 110618. https://doi.org/10.1016/j.plantsci.2020.110618
99. Shan, Q., Zhang, F., & Wang, K. (2021). CRISPR-mediated detection of glyphosate resistance in weed species. *Weed Science, 69*, 281–289. https://doi.org/10.1017/wsc.2021.22
100. Fozouni, P., Knott, G. J., & Doudna, J. A. (2021). CRISPR-based detection of RNA viruses in crops. *Nature Biotechnology, 39*, 292–296. https://doi.org/10.1038/s41587-020-0682-1
101. Kim, D., Lee, J., & Kim, J. (2019). Biosensors for transgene detection in genetically modified crops. *Plant Biotechnology Journal, 17*, 1924–1933. https://doi.org/10.1111/pbi.13161
102. Liang, C., Zhou, W., & Li, X. (2021). CRISPR/Cas systems for studying plant epigenetic modifications. *Plant Journal, 106*, 1183–1192. https://doi.org/10.1111/tpj.15167

103. Jiao, X., Zhang, X., & Wang, Y. (2021). Paper-based CRISPR/Cas12 biosensor for field detection of *Phytophthora infestans* in potatoes. *Biosensors and Bioelectronics, 186*, 113292. https://doi.org/10.1016/j.bios.2021.113292
104. Zhang, L., Gu, F., & Chan, C. T. (2020). Nanotechnology in biosensor design for gene-editing applications. *Nature Nanotechnology, 15*(3), 219–231.
105. Kumar, S., & Gupta, R. (2021). Functionalization of nanomaterials for biosensing. *Biosensors and Bioelectronics, 170*, 112707.
106. Kessler, R. (2019). Advances in optical biosensors for monitoring gene editing. *Trends in Biotechnology, 37*(12), 1189–1203.
107. Scott, N. R., & Chen, H. (2020). Electrochemical approaches to CRISPR/Cas detection. *ACS Nano, 14*(8), 9819–9830.
108. Smith, A. M., & Jones, R. (2021). Signal amplification strategies in nanobiosensors. *Analytical Chemistry, 93*(5), 2448–2460.
109. Rajan, R., & Sinha, A. (2022). Quantum dots in biosensing applications. *Journal of Nanobiotechnology, 20*, 67.
110. Environmental Protection Agency. (2021). *Safety and environmental impact of nanotechnology*. EPA Publications.
111. World Health Organization. (2021). *Guidelines on biosensor technology in healthcare*. WHO.

Nano-Priming Technology to Improve Seed Germination and Seedling Development

Sanyukta Salve, Amruta V. Shelar, and Vishnu Mishra

Contents

S. Salve
Department of Biotechnology (previously Institute of Bioinformatics and Biotechnology), Savitribai Phule Pune University, Pune, MH, India

A. V. Shelar
Department of Technology, Savitribai Phule Pune University, Pune, MH, India

V. Mishra (✉)
Department of Plant and Soil Sciences, Delaware Biotechnology Institute, University of Delaware, Newark, DE, USA

National Institute of Plant Genome Research (NIPGR), Aruna Asaf Ali Marg, New Delhi, India
e-mail: vmishra@udel.edu

R. S. Chaughule, A. V. Shelar (eds.), *Nanotechnology in Agriculture*,
https://doi.org/10.1007/978-3-031-97544-8_6

1 Introduction

Agriculture and ecological restoration are essential for sustaining life and preserving ecological balance. These fields depend on the correct germination of seeds and the desired growth of seedlings [1]. The phases of plant growth are crucial for determining the vitality, efficiency, and adaptability of crops and natural plant species. The significance of optimizing these first stages of growth cannot be exaggerated, considering their direct influence on efforts to provide food security and restore ecosystems [2]. The input of global agriculture in 2020 stood roughly at a staggering value of $ 3.5 trillion, insight into its economy [3]. It is stressing for the industry confronting growing challenges about environmental pressures and climate change. The Food and Agriculture Organisation (FAO) estimated that, by 2050, agriculture should increase production to meet the needs of an anticipated world population of 9.3 billion [4]. Simultaneously, the success of ecological restoration initiatives, like the Bonn Challenge which seeks to rehabilitate 350 million hectares of damaged land by 2030, depends on the achievement of good plant establishment [5]. Therefore, it is crucial to comprehend and enhance the process of seed germination and the growth of seedlings to tackle these issues effectively and get lasting results.

Seed germination is the transition from dormancy to young plant growth. It includes the physiologically and biochemically more complex water imbibition, enzyme activation, and transfer of nourishment [6]. Germination is a key step in seedling development. Root and shoot systems of the animal plant grow and differentiate during this stage, hence allowing the acquisition of water, nutrients, and sunlight. Successful seed germination and vigorous seedling growth are important, as they are management points through which crop production increases and food security is achieved [6]. Robust seedlings have greater resistance to environmental pressures such as drought, salt, and temperature changes, which are intensified by climate change. Studies have shown that improving the strength and vitality of young plants may increase crop production by as much as 30%. This highlights the important connection between the strong early growth of plants and the overall productivity of agriculture [7]. Both seed germination and seedling establishment are equally crucial in ecological restoration. Efficiently germinating and nurturing native plant species is crucial for the restoration of damaged landscapes, the promotion of biodiversity, and the recovery of ecosystem processes [8]. Enhanced seedling establishment significantly enhances survival rates in restoration projects, so aiding in soil stabilization, water filtering, and carbon sequestration [9]. These processes are very important in maintaining ecosystem balance. Their infancy, that is, the uniformity of germination, seedling survival, and growth, is impaired by different hurdles coming from diverse environmental and biological pressures [10]. Environmental stresses like high values of temperature, shortage of water, salinity, and soil pollution have always impeded germination, weakened seedlings, and increased rates of mortality. Climate change exacerbates these difficulties by amplifying the frequency and magnitude of detrimental weather phenomena. For example, varying temperatures can interrupt the absorption process, while elevated saline levels can impede the essential enzyme activity needed for germination [11].

Seeds may become infected with fungi, bacteria, and viruses, which often hamper germination and lead to weak plantlets [12]. Animals such as insects and rodents, considered pests, may cause significant damage to crops at the early stages of growth, thus compromising establishment. Robust techniques are required to safeguard seeds and seedlings from biological obstacles, so maintaining their health and viability throughout crucial early growth stages [10]. Seed dormancy is a significant barrier that prevents seeds from germinating, even when the environmental circumstances are favorable. Dormancy processes, whether physiological or physical, need certain treatments or circumstances to be overcome, hence complicating the process of germination. Overcoming seed dormancy is essential for attaining consistent germination rates and effective restoration of indigenous plant species into deteriorated ecosystems [13]. Insufficient seed quality also hampers the success of germination and the vigor of seedlings. Seeds may experience diminished viability and vigor due to factors such as unfavorable storage conditions, mechanical harm during harvesting, and genetic variability [14]. Optimizing germination rates and boosting early plant growth requires ensuring excellent seed quality through good handling, storage techniques, and the selection of superior genetic material. Conventional methods of seed priming have been created to improve the process of seed germination and the first development of seedlings [15]. Hydropriming is the process of immersing seeds in water to stimulate the initial phases of germination before planting. The objective of this procedure is to decrease the duration of germination and enhance consistency in the emergence of seedlings. Osmo-priming uses osmotic solutions, commonly composed of salts or polyethylene glycol, to regulate the absorption of water by seeds. This technique facilitates the coordination of germination and can enhance the strength of seedlings, particularly under less-than-ideal environmental circumstances [16]. Biopriming is a process where seeds are treated with advantageous microorganisms, such as rhizobacteria or fungus that can improve germination and the growth of seedlings through several methods. These bacteria have the capacity to synthesize growth-promoting compounds, increase the availability of nutrients, and provide protection against infections [17]. Although traditional seed priming methods have advantages, they frequently encounter issues with consistency, scalability, and efficiency. Although hydropriming is a straightforward and affordable method, it might lead to inconsistent absorption of water and unpredictable germination results [16]. Osmo-priming, while offering more control, necessitates significant labor and meticulous regulation of osmotic conditions. The efficiency of biopriming relies on the compatibility and potency of the microorganisms employed, which might exhibit significant variation. The constraints mentioned impede the widespread use of conventional priming methods and emphasize the necessity for more sophisticated and dependable procedures that may consistently and efficiently enhance seed germination and seedling growth.

Nanotechnology, the manipulation of materials at the nanoscale (1–100 nanometers), has become a potential field in agriculture. The use of nanoparticles in seed treatment, also known as nano-priming, takes advantage of the distinct characteristics of nanoparticles to improve the performance of the seed. Nanoparticles possess unique physical and chemical properties, stemming from their diminutive size and expansive surface area, which can enhance seed germination and seedling growth through several mechanisms. Nano-priming technology provides several benefits compared to conventional priming methods [15]. Nanoparticles can more efficiently penetrate the outer layer of seeds, allowing them to directly supply nutrients and chemicals that promote growth to the developing tissues inside. This precise delivery method can optimize the total metabolic activity of the seed, facilitating accelerated and consistent germination [18]. In addition, nanoparticles can be designed to have protective benefits against environmental stresses and infections. As an illustration in Fig. 1, nanoparticles may be engineered to progressively release growth-promoting

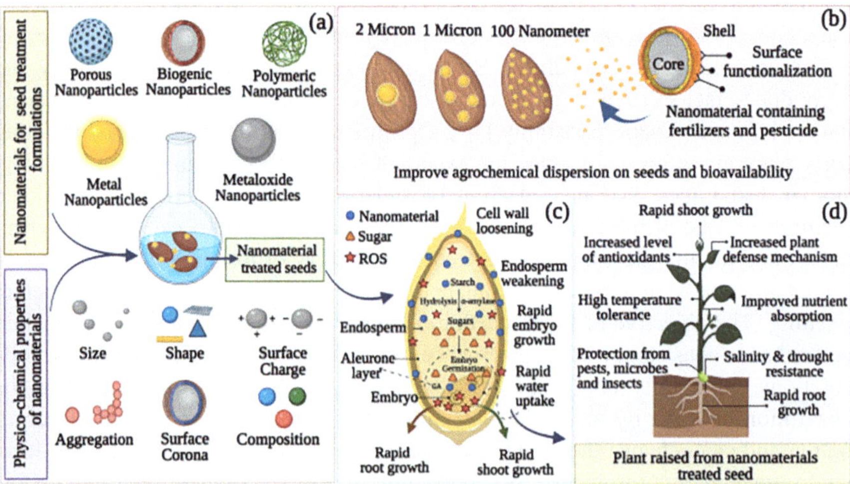

Fig. 1 Role and interaction of nanomaterials in seed treatment for enhanced germination and seedling growth. (**a**) Types of Nanomaterials: Porous, biogenic, polymeric, metal, and metal oxide nanoparticles with key properties like size, shape, and surface charge. (**b**) Seed Surface Functionalization: Nanomaterials with fertilizers and pesticides improve dispersion and bioavailability at different sizes (2 micron, 1 micron, 100 nano-meters). (**c**) Mechanism of Interaction: Nanomaterials enhance endosperm weakening, cell wall loosening, water uptake, and rapid root and shoot growth through sugar and ROS release. (**d**) Benefits: Rapid growth, increased antioxidants, improved defences, nutrient absorption, temperature tolerance, pest resistance, and enhanced salinity and drought resistance. (This is a reprinted image of Ref. [18] With permission)

chemicals, guaranteeing persistent advantages during the initial phases of seedling development [19]. The implementation of this regulated discharge mechanism has the potential to greatly enhance the strength and adaptability of young plants. In addition, nano-priming is easily scalable and can be adjusted to different seed types and climatic circumstances, providing a flexible alternative to the constraints of conventional priming methods [15]. The possibility of standardizing nano-priming procedures guarantees consistent results, making it a viable method for improving seed germination and seedling growth on a wide scale.

This chapter aims to examine the mechanics of nano-priming technology, investigate the functions of various types of nanoparticles, and provide empirical data to support their effectiveness. In addition, we will analyze the environmental and safety factors, practical uses, and future prospects of this groundbreaking method, emphasizing its capacity to transform the process of seed germination and seedling growth in agricultural and ecological settings. This chapter seeks to offer a full knowledge of how nano-priming technology may address critical elements, overcome the limits of existing approaches, and contribute to sustainable agricultural and ecological restoration activities.

2 Mechanisms Underlying Nano-Priming Technology

The use of nano-priming technology is revolutionizing seed treatment, providing a hitherto unknown opportunity to improve seed germination and seedling growth. This revolutionary methodology exceeds standard methods in terms of efficiency and efficacy because it takes advantage of nanoparticles' unique features. Research on the basic ideas of nano-priming—the interactions between nanoparticles and seeds—as well as their major effects on enzyme activity and seed metabolism would help one to completely understand its great consequences [20].

2.1 Basic Principles of Nano-Priming

Nano-priming capitalizes on nanoparticles' exceptional characteristics to optimize seedlings' germination and development. Nano-priming is a novel method of priming that employs nanoparticles with a size range of 1–100 nm, in contrast to conventional priming methods such as hydropriming, osmo-priming, and biopriming. These particles are capable of interacting with seed tissues in a manner that is not possible with traditional agents due to their high surface area-to-volume ratio and extreme reactivity [21]. Nanoparticles can rapidly penetrate seed coverings due to their diminutive size. The penetration process is essential because the seed coat serves as a substantial protective barrier, preventing the seed from absorbing water and protecting it from harmful microorganisms and external factors that may induce stress [22]. It is typically challenging for conventional techniques to provide consistent penetration and reliable results.

Nanoparticles can get around these problems by delivering chemicals precisely, which help embryonic cells grow. Nanoparticles are also very reactive, which makes it easier for them to connect with parts of cells. This helps important physiological processes that are needed for seedling growth and germination. Because of their reactivity, nanoparticles can also be used to control the release of substances. This guarantees that young plants receive an adequate supply of growth stimulants and minerals.

2.2 Nanoparticle Interaction with Seeds

An essential process for the efficacy of nano-priming technology is the interaction between nanoparticles and seeds. Entering the seed coat, nanoparticles allow essential nutrients, hormones, and protective molecules to be directly transported to the embryonic tissues Fig. 1b. This exact distribution technique ensures that the young seedling obtains the necessary help for strong development right from the start. As needed for many metabolic functions, nanoparticles can carry macronutrients such as potassium, phosphorous, and nitrogen as well as micronutrients including iron and zinc [15]. Moreover, one can add phytohormones regulating seed germination and seedling development: auxins, cytokinins, gibberellins, and so on. Thanks to nanoparticles, these substances may be provided exactly and under control, guaranteeing a constant and balanced supply and promoting uniform and fast development. Additionally, nanoparticles can be specifically engineered to possess antibacterial properties, thereby offering an additional layer of defence against the diseases that are transmitted by seeds. This dual function not only preserves the seedlings' health but also enhances their resilience during the early stages of development to environmental stresses and disease attacks [23].

2.3 Improvement of Seed Metabolism and Enzymatic Activity

Beyond merely supplying nutrients, nano-priming technology significantly affects enzyme activity and seed metabolism, two processes essential to effective germination and seedling development. Seed metabolic pathways include the synthesis of essential chemicals, the mobilization of nutrients, and the production of energy. Nanoparticles can help these pathways by delivering readily available nutrients and growth regulators [24]. As shown in Fig. 1c, nanoparticles can increase the activity of enzymes involved in nutrition mobilization, such as lipases, amylases, and proteases. These enzymes degrade previously stored macromolecules like as proteins, lipids, and carbohydrates into simpler forms that the developing embryo may easily consume [25]. By raising enzyme activity, nanoparticles enable more efficient conversion of nutrients kept in reserve into the building blocks and energy needed for development. Moreover, nanoparticles can affect the generation and action of reactive oxygen species (ROS) in seeds. ROS can be a signaling molecule during seed germination even if they are linked with cellular damage. Enzymatic activity can be induced and seed dormancy can be disrupted by increasing ROS levels. ROS levels are reduced by nanoparticles, which in turn affect the expression of stress-responsive genes in seeds, thereby facilitating germination. These genes produce proteins that protect the embryo and seedling from environmental challenges like dehydration, salt, and temperature fluctuations [26]. Nanoparticles augment seedling resilience by elevating the expression of specific genes, hence enhancing their probability of successful establishment in diverse demanding settings [27]. Furthermore, the interaction between nanoparticles and seed tissues may improve hormonal balance. Phytohormones strongly influence seed germination and seedling growth. Nanoparticles can be used to control how these hormones are delivered, making sure that the right amounts are always present to support healthy growth. Gibberellins, for instance, help seeds germinate and come out of their dormant state. Auxins and cytokinins, on the other hand, control cell division and growth as seedlings grow [23].

3 Types of Nanoparticles and Their Specific Roles

Nanotechnology has made huge strides in making seeds healthier and encouraging plants to grow quickly in agriculture. Very small nanoparticles with a lot of surface area and high reactivity have shown a lot of promise in making plants stronger, helping seeds sprout, and making plants more resistant to external stresses. This is because their chemistry and physical properties are different. As shown in Table 1, this section gives a full rundown of different nanoparticles, such as zinc, silica, silver, iron oxide, copper oxide, and gold. It explains what their specific roles are and how they improve the quality of seeds. Every topic includes numerous case studies that illustrate the practical applications and advantages in agricultural contexts.

3.1 Zinc Oxide Nanoparticles

Zinc is essential for various physiological and biochemical processes in seeds. According to current research, zinc oxide nanoparticle priming of seeds can significantly enhance zinc content, hence enhancing yield and growth rate. Brown seaweed extract (*Turbinaria ornata*) was used in a study to create biogenic nanozinc for priming rice seeds [29]. Both crop output and seed quality significantly increased as a result of the treatment. Moreover, zinc oxide nanoparticles have been shown to increase zinc concentrations in the roots, shoots, and grains while simultaneously promoting wheat development and lowering cadmium accumulation [28]. This ultimately enhanced wheat biomass and nutrient retention while mitigating cadmium toxicity. It has been reported that the use of bio-fabricated nanozinc is an effective

Table 1 Impact of nanoparticle characteristics on seed germination and seedling development: a detailed evaluation of nanoparticle size, concentration, and seed type outcomes

Nanoparticle	Size of the nanoparticle (nm)	Concentration (mg/L)	Type of seed	Outcomes	Reference
Zinc oxide (ZnO)	20	50	Wheat (*Triticum aestivum L.*)	Increased growth, decreased cadmium accumulation	[28]
	20	40	Rice (*Oryza sativa L.*)	Improved seed quality, higher crop yield	[29]
	15	60	Bean (*Phaseolus vulgaris*)	Enhanced germination, slow zinc release, lower toxicity	[30]
	10	70	Lupinus termis (*Lupinus albus*)	Prevented germination loss under salinity stress	[31]
	25	80	Pearl millet (*P. glaucum*)	Growth promotion, mildew control	[32]
	387.7 ± 4 nm	100–1600	Maize seeds (*Zea mays L.*)	Exhibited antifungal properties and promoted seedling growth	[33]
Silica (SiO_2)	~100	2	Pea seeds (*P. sativum* L.)	Improved seed and seedlings vigor	[34]
	10–50	75	*Stevia rebaudiana Bertoni*	Enhanced seed germination and physiological properties	[35]
	10–17	1000	Tomato (*Solanum lycopersicum*)	Enhance tomato seed germination by improving water uptake, antioxidant enzyme activity, and stress resistance. Also boost seedling vigor and growth by mitigating abiotic stress effects.	[36]
	30	120	Wheat (*Triticum aestivum L.*)	Increased growth, reduced cadmium accumulation	[37]
	40	500	Soybean (*Glycine max* L.)	Enhanced the seed germination percentage, speed of emergence and seedling vigour under drought stress condition	[38]
	35	160	Pearl millet (*Pennisetum glaucum*)	Improved growth and disease resistance	[32]
Silver (Ag)	10	30	Rice (*Oryza sativa L.*)	Enhanced seed germination, increased starch metabolism	[39]
	20	40	Onion (*Allium cepa*)	Improved growth and yield, no toxic effects	[40]
	15	50	Watermelon (*Citrullus lanatus*)	Improved seed germination, enhanced fruit quality	[41]
	5	25	Soybean (*Glycine max*)	Increased seed germination, effective as seed protection agent	[42]
	30	35	Wheat (*Triticum aestivum L.*)	Enhanced growth promoter, reduced toxic effects	[43]

Iron oxide (Fe_3O_4)	19–30	20–160 ppm	Flax seeds (*Linum usitatissimum*)	Increased the activity of plant growth regulator	[44]
	6–20	30 μg/mL	Rice (*Oryza sativa L.*)	High antimicrobial activity	[45]
	80	25–1000 μg/mL	Wheat (*Triticum aestivum L.*)	Increased harvest yield	[46]
	33.8 ± 3.59	10–160 mg/L	Rice (*Oryza sativa L.*)	Improved water uptake	[47]
	25–100	50 μg/mL	Rice (*Oryza sativa L.*)	Improved enzymatic activity	[48]
Copper oxide (CuO)	40–60	4	Indian mustard (*Brassica juncea*)	Improved shoot and root length, enhanced net photosynthetic rate, increased internal CO_2 concentration, and elevated proline content	[49]
	25	60	Wheat (*Triticum aestivum L.*)	Enhanced metabolic processes, better nutrient availability	[50]
	19	0.12% and 0.16% (w/v)	Wheat (*Triticum aestivum L.*)	Improved germination and seedling vigor under normal and stress conditions (PEG-induced hyperosmotic stress and salt stress)	[51]
	30–25 mg/L	25	Mung bean (*Vigna radiata*)	Enhanced seed germination, seedling growth, chlorophyll content, and antioxidant enzyme activities. Improved tolerance to oxidative stress by reducing lipid peroxidation	[52]
Gold (Au)	15–25	100	*Gloriosa superba*	The treatment with gold nanoparticles significantly enhanced the seed germination rate and overall growth of *Gloriosa superba*. The nanoparticles improved the antioxidant defense mechanism and promoted better root and shoot development	[53]
	2–19	10 μg/mL	Barley (*Hordeum vulgare L.*)	Uptake of macronutrients and micronutrients, as well as on biomass production of exposed plants	[54]

seed-priming agent for promoting growth and controlling mildew in pearl millet [32]. The researcher developed zinc–chitosan nanoparticles and evaluated their effectiveness in maize through seed priming and foliar application [33]. These nanoparticles stimulated seedling development and showed antifungal action. Investigated have been the effects of various zinc nanoparticle concentrations and sizes on bean (*Phaseolus vulgaris*) germination [30]. They found most of the zinc maintained in the seed coat with a smaller percentage accessing the cotyledons by detecting the biotransformation of zinc oxide nanoparticles using X-ray absorption spectroscopy. This indicated that zinc nanoparticles could be valuable agrochemicals given their slow zinc release and lower toxicity relative to zinc sulfate. Another study indicated that nanozinc might lower germination loss in *Lupinus termis* (*Lupine*) seeds under salinity stress, suggesting that zinc nanoparticles could boost plant development and yield in salted soils [31]. Still mostly unknown are the precise processes by which zinc oxide nanoparticles reduce salt stress in seedlings. This research taken together shows the great possibilities for nanozinc in agriculture. To completely maximize their advantages, nevertheless, one must first grasp their interactions with seeds.

3.2 Silicon Nanoparticles

There is a lot of interest in silicon nanoparticles (Si NPs) because of their potential to improve plant development and stress tolerance. Despite not being a necessary vitamin, silica has several advantages, particularly during trying times. Plant cell walls that have been strengthened by silica nanoparticles (Si NPs) are stronger. They enhance cell wall integrity and activate stress-responsive mechanisms, thereby improving a plant's resilience to biotic and abiotic stresses, including infections, high salinity, and water deficiency. Furthermore, modifying the architecture of the root system and augmenting the surface area of the roots allows Si nanoparticles to improve nutrient absorption.

A recent study provides a good example of the practical application of Si NPs to enhance seed quality. A study indicates that silica nanoparticles can enhance the physiological characteristics and seed germination of *Stevia rebaudiana Bertoni* [35]. The promising results indicate that silica nanoparticle nano-priming could serve as an effective strategy in modern agriculture to improve crop resilience and performance. The presence of silicon nanoparticles (Si NPs) greatly improves resistance to drought. It has been demonstrated that nano silica priming significantly enhances seed germination, growth, and stress tolerance. It achieved this by reducing oxidative damage and bolstering the antioxidant defence system [38]. Furthermore, under drought stress, nano silica enhanced osmotic adjustment, which is essential for preserving cellular turgor and water uptake. The data imply that utilizing microsilica to reduce drought stress in soybeans may improve crop tolerance and production. Exogenous silicon has been shown to strengthen bud seedlings' antioxidant defence system, increasing seed germination and minimizing oxidative stress. The effectiveness of cinnamon essential oil against seed diseases was enhanced by encasing it in nanoparticles, as reported in [55]. The combination of sodium alginate seed coating and cinnamonaldehyde-mesoporous silica nanoparticles was a novel approach that effectively treated seed-borne diseases. A study that examined the effects of nanosilica on wheat discovered that it enhanced cadmium accumulation, yield, and growth [37].

3.3 *Silver Nanoparticles*

Due to their potent antibacterial properties, silver nanoparticles (Ag NPs) can be utilized to protect seeds and seedlings from detrimental infections. Silver nanoparticles (Ag NPs) safeguard seed health by their extensive antibacterial properties against bacteria, fungi, and viruses. Silver nanoparticles (Ag NPs) possess antibacterial properties that protect seeds and seedlings from pathogens, enhancing germination rates and promoting vigorous plant development [56].Ag NPs not only inhibit infections but also improve nutrient absorption and boost plants' resistance to stress by regulating reactive oxygen species (ROS) levels and maintaining a balance between vital signaling and potential oxidative damage [57]. By creating nanopores in the seed coat, acting as nanocatalysts to increase the activity of enzymes that break down starch, and inducing moderate stress or reactive oxygen species, silver nanoparticles aid in seed germination [39]. Silver nanoparticles can significantly enhance the germination and starch metabolism of old rice seeds when used as nanopriming agents. This enhancement is demonstrated in Fig. 2, where increased amounts of soluble carbohydrates that promote seedling growth result from amylase activation caused by silver nanopriming. This technique enhances water absorption via osmosis by reducing the osmotic potential and water potential of the cells. Moreover, it is stated that reactive oxygen species, especially hydroxyl radicals, are essential for the relaxation of the cell wall and the weakening of the seed sheath, both of which facilitate the emergence of the radicle [39]. Nano-priming elevated reactive oxygen species (ROS) levels in growing seeds compared to unprimed controls, indicating that both ROS and aquaporins enhance seed germination. Furthermore, the infiltration of silver nanoparticles into seed coats may result in the formation of nanopores that enhance water absorption and accelerate starch digestion. The proposed mechanisms for nanopriming-induced seed germination include the generation of hydroxyl radicals for cell wall weakening, the rejuvenation of ROS/antioxidant systems, and the creation of nanopores [58].

While nanoparticles lower ROS levels under stress conditions by increasing enzyme activities such as superoxidasedismutase, catalases, and guaiacol-peroxidase, hence minimizing seed cell damage, nanopriming keeps ROS levels within an oxidative window that promotes germination. In both greenhouse and field studies, it was found that the internalization of nanosilver derived from onion extract greatly improved the germination, growth, and yield of onion seedlings [40]. Researchers have recently looked into the possibility of using biogenic nanosilver as a wheat growth promoter without the negative side effects associated

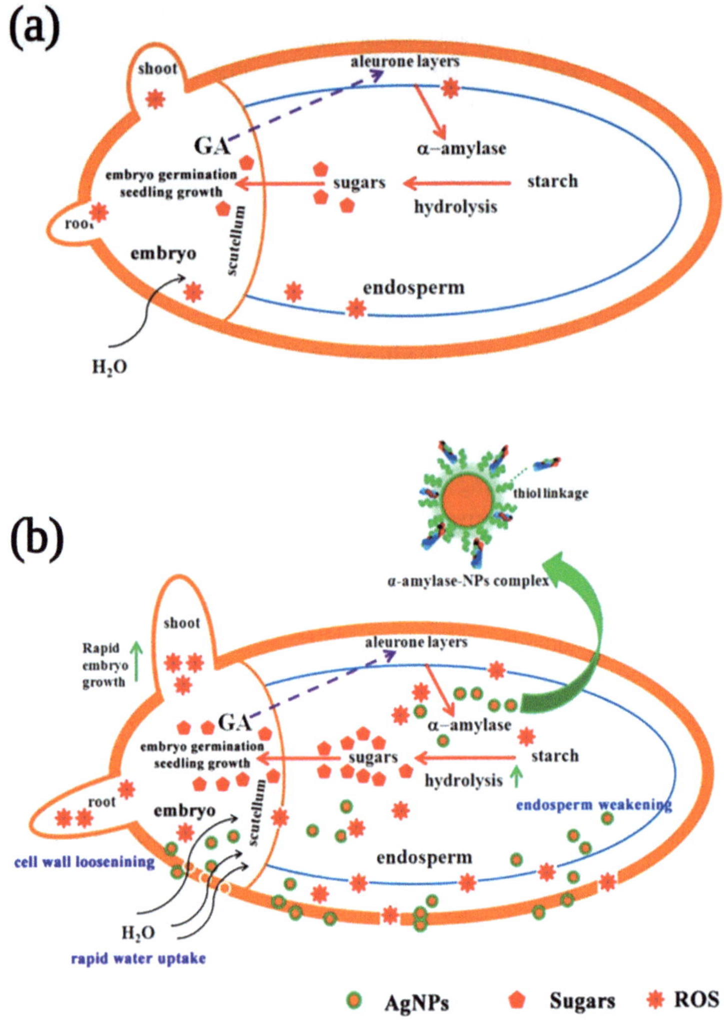

Fig. 2 Photosynthesized silver nanoparticles enhance aged rice seeds' germination and starch metabolism. (**a**) Seeds without silver nanoparticle treatment: illustrates natural seed germination. Gibberellic acid (GA) induces α-amylase production in aleurone layers, hydrolysing starch into sugars that fuel embryo growth, forming the shoot and root. (**b**) Seeds with silver nanoparticle treatment: shows enhanced germination with silver nanoparticles (AgNPs). Nanoparticles boost water uptake, cell wall loosening, and α-amylase activity. Interaction with ROS accelerates sugar production and endosperm weakening, promoting rapid embryo and seedling growth. (This is a reprinted image of Ref. [18] with permission)

with chemically synthesized nanosilver [43]. Additionally, employing agro-industrial by-products, silver nanoparticles were synthesized as nano-priming agents for diploid and triploid watermelon seeds [41]. Their findings demonstrated improved fruit quality, growth, and germination. A cost-effective, sustainable synthesis method was utilized to generate stable silver nanoparticles employing the soil fungus *Macrophomina phaseolina* [42]. The effectiveness of silver nanoparticles as a protective agent for seeds was assessed by analyzing their dosage in soybean seed germination.

3.4 Iron Oxide Nanoparticles

Iron is vital for plant metabolism; hence, FeO nanoparticles are necessary to enhance seed quality. Iron oxide nanoparticles have been proven to promote seed germination, stimulate plant growth, and boost environmental resistance in plants because they are also a source of iron and assist the plant in the absorption and utilization of this element which is an important nutrient. Iron is a precursor to many metabolic processes occurring in seedlings such as respiration, photosynthesis, and DNA synthesis. The large surface area of FeO nanoparticles promotes the absorption of iron by the plants; thus, enhancement of their growth and development through increased iron uptake and utilization [59].

Nanoparticles of iron oxide, when present in low concentrations, were found to stimulate the growth and development of seeds. Using soaking and priming of seed in salinity conditions it has been observed that nano-iron oxide can promote sorghum (*Sorghum bicolor*) germination and seedling growth [60]. This was proved through the use of the seed. Iron oxide nanoparticles were manufactured through the utilization of onion extract [61]. The application of non-toxic iron oxide nanoparticles to watermelon seeds in a sustainable manner was demonstrated to be capable of increasing the anti-inflammatory characteristics of the seeds and enhancing their defenses. In a study it was evaluated the phytopathological effects of ferrous sulfide nanoparticles that were distributed in water against the rice-born fungus *Fusarium verticillioides* [45]. Rice seeds that were treated with nano-iron exhibited impressive levels of seed germination and inhibited the growth of the fungus *F. verticillioides*. To treat anemia caused by iron deficiency, biofortifying wheat with iron through seed priming has proven to be an effective approach. This method resulted in a significant increase in grain iron content and a greater accumulation of iron in the grains [46]. A study demonstrated that fragrant rice cultivars could benefit from the application of nanopriming, which involved the utilization of zerovalent iron [47]. It has been suggested a seed dressing method for rice, and their research has shown that the nano-pyrite seed dressing stimulates the creation of rice that is equal in nitrogen, phosphorous, and potassium without affecting yield [48]. As a result, the iron oxide nanoparticles that were discussed earlier have the potential to serve as a foundation for the creation of additional asset delivery systems. Nanoparticles made of iron oxide have the ability to dramatically limit the amount of iron that is present in the cotyledon, which ultimately prevents nanoparticles from being taken up and moved about [62]. In addition, these iron oxides are aggregated on the seeds' exterior, which makes them an excellent choice for applications involving seed priming and pest management [63].

3.5 Copper Oxide Nanoparticles

Due to their unique physicochemical properties and ability to enhance seed germination and seedling development, copper oxide nanoparticles (CuO NPs) act as useful tools for seed priming. Copper, as an essential mineral, plays a very vital role in photosynthesis, respiration, and metabolic processes of proteins and carbohydrates. The nanoscale copper oxide possesses greater reactivity and function against bulk copper complexes as it enjoys certain advantages compared with the latter. The use of copper oxide nanoparticles in seed priming increases seed viability and percentage germination in its resistance to varied environmental conditions.

These nanoparticles improve nutrient absorption, elevate antioxidant enzyme activity, and promote root and shoot development. CuO nanoparticles enhance plant health and yield by mitigating the adverse impacts of biotic and abiotic stresses, including drought and pathogen infestations.

Copper is responsible for the activation of several enzymes, which in turn contributes to the creation of RNA and strengthens the performance of photosystems. The germination and growth of *P. vulgaris* seedlings are affected by a variety of copper oxide nanoparticle sizes and concentrations. Following the findings, X-ray absorption spectroscopy was utilized to determine that the majority of copper was in its unaltered state [64]. Although copper nanoparticles did not affect the germination of seeds, they did improve the weight gain of seedlings at low concentrations, but they reduced the growth of seedlings at high concentrations. Copper oxide nanoparticles that were biosynthesized exhibited a significant upregulation and enhancement of antioxidant enzyme activity. Moreover, research has been done to investigate the possible role that nanocopper could play in wheat yield [65]. The poisonous action of the biosynthesized copper oxide nanoparticles against two wheat grain-damaging insects, *Sitophilus granarius* and *Rhyzopertha dominica*, was discovered to be dependent on both the size of the nanoparticles and the dose at which they were administered. The results indicated that smaller doses of copper oxide nanoparticles should be applied as pesticides in agricultural fields so as not to impede wheat growth. Researchers investigated the effects of copper oxide nanoparticles on the germination, growth, and physiological responses of *Brassica pekinensis L.* about changing concentrations [50]. Taking into consideration these findings, nanocopper has the potential to act as an effective pesticide that can facilitate the storage of a wide variety of seeds. Furthermore, at lower concentrations, it has the ability to considerably improve the rate at which seeds germinate.

3.6 Gold Nanoparticles

Gold nanoparticles (Au NPs) have garnered attention for their distinctive physical and chemical properties, as well as their capacity to enhance seed germination and seedling growth. Gold nanoparticles (Au NPs) possess potent antioxidant characteristics, which contribute to the mitigation of oxidative stress and the promotion of the health and vigor of seeds and seedlings. In addition, it has been shown that Au NPs enhance nutrient uptake and promote cellular proliferation and growth. It has been shown that the gold nanoparticles (Au NPs) enhance the cold resistance of wheat [66]. The study indicates that the application of Au NPs to wheat (*Triticum aestivum L.*) seeds significantly enhances cold tolerance. This is mainly due to the nanoparticles' capacity to elicit biochemical and physiological alterations in the seeds and seedlings. The study demonstrates that AuNP therapy increases the activity of antioxidant enzymes, which are essential in alleviating oxidative stress induced by low temperatures. The heightened enzyme activity aids in stabilizing cellular structures and sustaining metabolic functions in cold circumstances. Furthermore, the nanoparticles facilitate the build-up of osmoprotectants, such as proline, which aids in osmotic adjustment and safeguards cellular activities during cold stress. A nanogold solution has been demonstrated to increase seedling growth and germination in long-stored maize seeds [67]. Furthermore, the nanopriming method was found to reduce the amount of gold that was transferred from seeds to the vegetative organs of plants. Gold nanoparticles that were generated from the fruit extract of *Terminalia arjuna* were investigated to determine how they affected the germination of *Gloriosa superba* seeds [53]. The germination of seeds and the growth of the vegetative parts of *Gloriosa superba* were both considerably impacted by nanogold [49]. It has been shown that *Tiliacora triandra* can produce gold nanoparticles, which are biocompatible and capable of enhancing rice germination and root growth [68]. Furthermore, the development of *Brassica juncea* and the quantity of seeds it produces are both improved by the presence of gold nanoparticles [69].

Zinc oxide, silica, silver, iron oxide, copper oxide, and gold are nanoparticles that have shown substantial potential in facilitating the growth of seedlings and improving the process of seed germination. Each type of nanoparticle offers distinctive advantages, such as enhanced stress tolerance, disease control, increased photosynthetic efficiency, and the capacity to transfer nutrients. The transformative potential and significant advancements of nanoparticles in modern agricultural practices have been underscored by recent studies. The agricultural sector can effectively address numerous challenges by utilizing cutting-edge technology, which will result in the implementation of more productive and environmentally friendly cultivation methods.

4 Effects of Nano-Priming on Seed Germination and Seedling Development

Nano-priming is a state-of-the-art method in seed treatment that utilizes nanoparticles to improve the process of seed germination and the growth of seedlings. This innovative technique has shown substantial enhancements in many measures of seed growth, the strength of young plants, and their ability to withstand stressful situations. As a result, it holds great potential as a valuable tool in contemporary agriculture.

4.1 Enhanced Germination Indices

One effective method for raising seed germination indices is nano-priming. By enhancing water intake, enzyme activity, and hormone regulation, the use of nanoparticles in seed treatments dramatically increases germination rates. The combination of hydrogen peroxide signaling with phytohormones has been demonstrated to result in a notable increase in the germination rate of magneto-primed tomato seedlings [70]. Improved water absorption and the activation of important enzymes that break seed dormancy were credited with the higher germination rate. Likewise, it has been demonstrated that adding zinc oxide nanoparticles to onion seeds greatly increases seed quality and germination rates. Total chlorophyll, malondialdehyde (MDA), total protein, and the accumulation of osmolytes like proline and glycine have all been shown to rise as a result of the nanoparticle treatment [71]. In addition, the inclusion of ZnO NPs resulted in a dose-dependent enhancement in antioxidant enzyme activity, such as superoxide dismutase (SOD) and catalase, and robust germination, as compared to untreated seeds. It has shown that engineered nanomaterials such as selenium nanoparticles significantly improved the process of seed germination in rice, and accelerated the emergence and development of seedlings, resulting in increased seedling vigor and nutritional levels [72].

4.2 Improved Seedling Strength

Nano-priming enhances plant robustness and durability, leading to improved seedling vigor and increased germination rates. Nanomaterials enhance metrics including biomass accumulation, root and shoot length, and overall seedling vigor. Research indicates that incorporating silica nanoparticles into soybean seeds mitigates drought stress by influencing the antioxidant defense system and osmotic adjustment, thereby significantly improving seedling growth and vigor [38].

Compared to conventional priming approaches, nano-priming has been found to produce improved results for the factors influencing seedling growth. A study on maize revealed that seeds primed with chitosan nanoparticles enclosed in zinc exhibited superior biomass production and longer roots and shoots compared to those treated with standard priming agents. This indicates that nano-priming may sustain growth under suboptimal conditions while also enhancing early-stage seedling vigor [33]. The use of iron oxide nanoparticles in wheat (*Triticum aestivum L.*) seeds is another significant instance. Iron oxide nanoparticles significantly enhanced seedling vigor by augmenting nutrient absorption and promoting robust root system development. This strategy shows promise for augmenting agricultural resilience by boosting seedlings' growth and stress tolerance.

4.3 Stress Tolerance and Resilience

The application of nanoparticles in seed priming offers new strategies for enhancing plant resistance and encouraging growth under challenging conditions shown in Fig. 3. It has been shown that the application of cerium oxide nanoparticles (PNC) coated with poly(acrylic acid) on cotton seedlings exposed to salinity stress. The beneficial effects on the seedlings' morphology, physiology, biochemistry, and transcriptomics were also highlighted [73]. Additionally, a study demonstrated how water-soluble carbon nanoparticles (CNPs) significantly improved lettuce seed vigor and seedling growth in the presence of salt and high temperatures [73]. This was accomplished by encouraging the formation of lateral roots and the build-up of chlorophyll. Recently, Manganese (III) oxide nanoparticles (Mn NPs) in *Capsicum annum L.* [74] through biochemical interactions have been found that

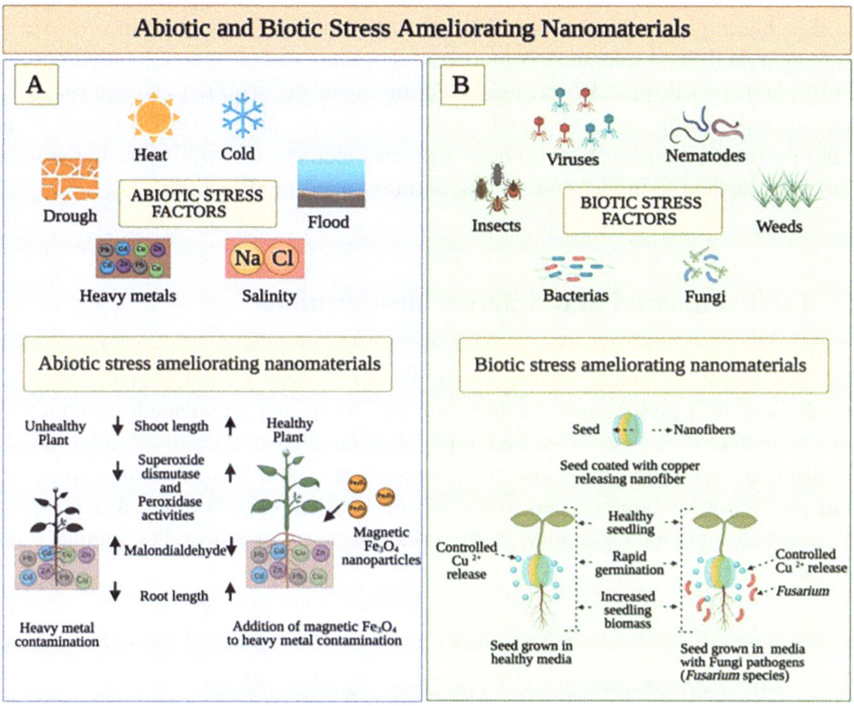

Fig. 3 Nanomaterials for alleviating abiotic and biotic stress: (**a**) Abiotic stress factors include heat, cold, drought, flood, heavy metals, and salt. Nanomaterials, such as magnetic Fe_3O_4 nanoparticles, enhance plant health by increasing the length of shoots and roots, as well as improving the activities of superoxide dismutase and peroxidase enzymes, while simultaneously lowering levels of malondialdehyde. (**b**) Biotic stress factors include viruses, nematodes, insects, weeds, bacteria, and fungus. Nanomaterials, such as nanofibers that release copper, improve the process of seed germination and the growth of seedlings by releasing regulated amounts of Cu^{2+} to counteract biotic stress. (This is a reprinted image of Ref. [18] with permission)

these nanoparticles help the plant be better able to withstand high salt levels in the soil. Furthermore, in response to salt stress, priming *Zea mays* seeds with nano-silica increased germination rates, seedling vigor, and metabolic activity [75]. Hormone levels and antioxidant enzyme regulation were used to produce this effect. Furthermore, a study on nano-priming with nano ZnO and nano Fe reduced the negative effects of cadmium toxicity and increased wheat (*Triticum aestivum L.)* biomass and nutritional content. The application of nanosized SiO_2 particles improved the development characteristics and shielded fenugreek seedlings from salt stress [76]. It has been observed that fenugreek seedlings' tolerance to elevated salt concentrations was enhanced by the presence of silver nanoparticles (Ag NPs) [77]. Magnetic nanoparticles (Fe_3O_4) substantially mitigated the deleterious effects of heavy metals on wheat (*Triticum aestivum L.)* seedlings [78]. This demonstrates how these nanoparticles may help advance sustainable agriculture. These studies highlight the many benefits of seed priming with nanoparticles in agriculture. This method offers long-term ways to increase agricultural productivity and resilience in harsh environmental conditions. To summarize, nano-priming is a major breakthrough in seed technology that provides notable enhancements in germination indices, seedling vigor, and stress tolerance. Nanoparticles used in seed priming not only improve early plant growth but also provide seedlings with the required characteristics to survive unfavorable environmental conditions. As research advances in this area, nano-priming is expected to become a crucial element of sustainable agriculture methods, leading to higher crop output and improved food security.

5 Environmental and Safety Considerations

The utilization of nanotechnology in agriculture offers several benefits, such as heightened seed germination, higher seedling vigor, and augmented resistance to environmental stresses as illustrated in Fig. 4. Nevertheless, the extensive utilization of nanoparticles (NPs) gives rise to apprehensions about their possible ecological and physiological ramifications. This section examines the ecological toxicity of nanoparticles and the regulatory frameworks and safety measures that regulate their usage.

5.1 The Ecotoxicological Effects of Nanoparticles

The use of nanoparticles in agricultural systems has generated much apprehension over their possible ecological hazards. Research has demonstrated that nanoparticles can engage with different environmental matrices, resulting in unforeseen repercussions.

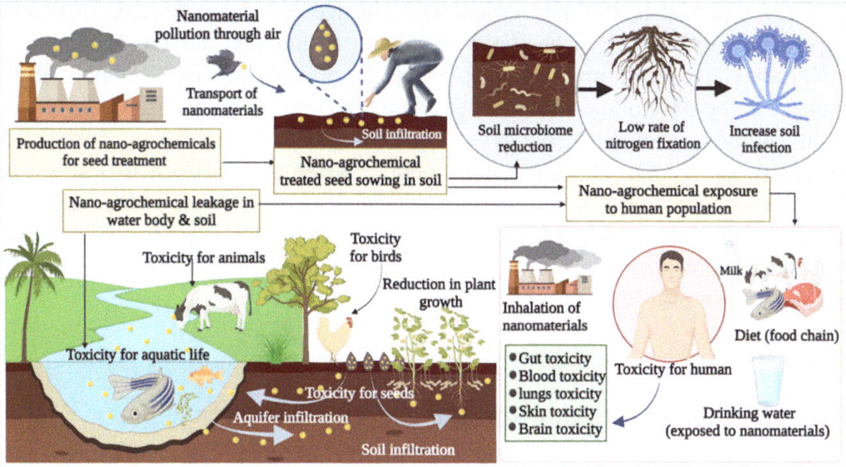

Fig. 4 Risks to the environment and human health posed by nano-agrochemicals: illustrates the impact of nano-agrochemicals employed in seed treatments on the well-being of soil, aquatic organisms, animals, and people. Contamination of soil and water results in a decrease in the soil microbiome, hindered plant development, and compromised seed health. Inhalation, nutrition, and drinking water are among pathways via which airborne nanoparticles might pose dangers to human health. (This is a reprinted image of Ref. [18] with permission)

5.1.1 Potential Environmental Risks and Mitigation Strategies

Nanoparticles, owing to their diminutive size and elevated reactivity, have the potential to induce environmental damage by impacting soil fertility, water purity, and non-target creatures. An important issue is the build-up of nanoparticles in the soil, which can modify its microbial makeup and affect the process of nutrient cycling. It has been discovered that specific quantities of silver nanoparticles (Ag NPs) hindered the growth of advantageous soil bacteria, which might cause disturbances in soil ecosystems [79]. To reduce these hazards, it is essential to create and execute techniques such as employing biodegradable nanoparticles, optimizing doses to minimize environmental impact, and undertaking thorough risk assessments before large-scale implementation. A study indicated that the application of organic chemicals to nanoparticles can decrease their toxicity and enhance their environmental friendliness.

5.1.2 Studies on Soil and Water Contamination

Multiple studies have emphasized the possibility of soil and water resources becoming contaminated as a result of the utilization of nanoparticles in agriculture. Recently, it has been examined the influence of zinc oxide nanoparticles (ZnO NPs) on soil and water systems [80]. The study revealed that ZnO nanoparticles have the potential to infiltrate groundwater, hence presenting hazards to aquatic organisms. Furthermore, it has been observed that titanium dioxide nanoparticles (TiO_2 NPs) have the ability to build up in soil, resulting in phytotoxicity and causing changes in the development patterns of plants [81].

In addition, comprehensive research investigated the impact of iron oxide nanoparticles (Fe_3O_4 NPs) on soil quality during several agricultural cycles [82]. Thus, these findings demonstrated that Fe_3O_4 NPs initially promoted plant development. However, continued exposure resulted in a decline in microbial diversity and soil fertility. This emphasizes the need to properly monitor and regulate the application of NPs.

5.2 Regulatory Framework and Safety Protocols

Due to the possible hazards linked to nanoparticles, there is an increasing requirement for strict regulatory frameworks and safety measures to oversee their use in agriculture.

5.2.1 Current Regulations Governing the Use of Nanoparticles in Agriculture

The regulations pertaining to the use of nanoparticles in agriculture differ across different regions, with certain governments enforcing more rigorous restrictions compared to others. The European Union's Regulation (EC) No 1107/2009 on plant protection products on the market in the European Union (EU). All plant protection products must receive authorization prior to sale or usage. The rule delineates the data prerequisites for the approval of plant protection products, encompassing those that incorporate nanomaterials [83].

The Environmental Protection Agency (EPA) in the United States has implemented regulations for the approval of pesticide products that contain nanoparticles. Manufacturers are mandated by the EPA to furnish comprehensive data on the composition of the nanoparticle, its possible environmental consequences, and the safety precautions used. An instance is the EPA's examination of a pesticide containing nanoparticles in 2016, during which the agency implemented stringent usage protocols to mitigate the risk of environmental pollution.

Nevertheless, the regulations governing various industries are still developing, and there is a requirement for more unified and all-encompassing international benchmarks. Allan et al. 2021 highlighted the importance of global collaboration in creating standardized laws to manage the unique characteristics and risks of nanoparticles effectively [84].

5.2.2 Recommendations for Safe and Sustainable Application

In order to guarantee the secure and enduring utilization of nanoparticles in agriculture, many suggestions have been put out by researchers and regulatory entities:

1. **Risk Assessment and Management**: Perform comprehensive risk assessments that take into account the whole life cycle of nanoparticles, spanning from their manufacturing to their destruction. This encompasses the examination of the way they are distributed in the environment, their capacity to be taken up by organisms, and their tendency to build up in food chains.
2. **Standardized Testing Protocols**: Create uniform testing procedures to assess the ecotoxicological effects of nanoparticles. This includes laboratory experiments conducted on living animals and environmental components, both inside and outside of their natural habitats.
3. **Monitoring and Surveillance**: Establish and execute ongoing monitoring and surveillance initiatives to monitor the levels of nanoparticles in the environment and assess their impact on ecosystems. An example of this is a surveillance program launched by the Canadian government in 2018, which specifically aimed to identify and monitor the presence of nanoparticle pollution in agricultural runoff.
4. **Public Awareness and Training**: Enhance public awareness and deliver comprehensive training programs to farmers and agricultural workers about the secure manipulation and use of nanoparticles. Educational programs can assist reduce the dangers connected with incorrect utilization and disposal.
5. **Innovation in Green Nanotechnology**: Encourage the exploration and advancement of environmentally friendly alternatives in nanotechnology that use materials that are harmless to the environment. For instance, a study investigated the utilization of plant-derived nanoparticles that exhibit enhanced biodegradability and reduced environmental hazards.
6. **Regulatory Updates and Compliance**: Continuously revise regulatory frameworks to include current scientific discoveries and technology progress. It is essential to guarantee adherence to these standards by conducting regular audits and implementing enforcement actions.

To summarise, whereas nanoparticles hold great potential for enhancing agricultural production and sustainability, it is crucial to consider their environmental and safety consequences. To sustainably utilize nanotechnology in agriculture, it is crucial to conduct thorough ecotoxicological studies, implement strong regulatory frameworks, and employ proactive mitigation methods. By following these instructions and promoting global cooperation, we can guarantee that the use of nanoparticles supports sustainable farming methods without jeopardizing environmental well-being.

6 Challenges and Prospects of Nano Priming for Enhancing Seed Germination and Seedling Development

Nanotechnology has brought about a significant transformation in several industries, including agriculture. Nano priming, a process that applies nanoparticles (NPs) to seeds, has emerged as a viable strategy for improving seed germination and seedling growth [85]. To completely reap the advantages of nano priming, it is essential to tackle several obstacles that now exist. This section examines the difficulties associated with nano priming in agriculture and investigates the potential future benefits.

6.1 Challenges of Nano-Priming

A major obstacle in nano priming is the limited comprehension of the mechanisms by which nanoparticles impact seed germination and plant development. While several studies have shown the beneficial impacts of NPs on different plant species, the specific biochemical and physiological mechanisms involved have not been well investigated. The lack of expertise in this area prevents the efficient improvement of nanoparticle compositions and their application procedures. Consequently, there is a pressing requirement for a comprehensive study to clarify these pathways, which would facilitate the creation of more efficient and focused nano-priming techniques [18].

Another notable obstacle is the possible environmental and health hazards linked to the extensive use of nanoparticles. As previously mentioned, nanoparticles have the ability to aggregate in soil and water, which may lead to lasting ecological damage. Furthermore, the assessment of nanoparticles' impact on human health is currently being researched [86]. The regulatory frameworks and safety regulations for nanoparticles in agriculture are still in the early stages of development, and there is an urgent requirement for standardized rules to regulate their synthesis, application, and disposal. It is essential to address these safety issues in order to obtain public approval and guarantee the long-term viability of nanotechnology in agriculture.

The scalability of nano priming poses a significant challenge. Although laboratory and small-scale field studies have demonstrated encouraging outcomes, using these methods on a wide scale for agricultural production has logistical and economic obstacles. In order to make nano priming economically feasible for farmers, it is crucial to address the expenses associated with nanoparticle manufacturing, the stability of formulations, and the development of consistent application techniques. Advancements in the production procedures of nanoparticles, such as environmentally friendly synthesis methods, have the potential to decrease expenses and enhance the availability of nano-priming technologies.

An additional problem arises from the heterogeneity in responses reported across various plant species and among different cultivars of the same species. The inconsistency of nano-priming approaches complicates the establishment of uniform protocols [21]. Consequently, it is imperative to conduct comprehensive field studies in various agro-climatic areas to ascertain the particular requirements and reactions of individual crops to nanoparticle treatments. Collaborative research is crucial, involving agronomists, biotechnologists, and nanotechnologists, to provide customized nano-priming solutions designed for specific crops and environmental conditions.

6.2 Future Promises of Nano-Priming

Notwithstanding these obstacles, the prospects for nano priming in agriculture are quite promising. An intriguing possibility lies in the use of nanoparticles as vehicles for precise transportation of nutrients, growth regulators, and protective substances straight to the seeds. This precise delivery method can boost the strength and vitality of seedlings, increase the efficiency of nutrient utilization, and decrease the reliance on external chemical inputs. As a result, it promotes the adoption of more sustainable farming techniques. Recent advancements in the field of nanoparticle engineering have demonstrated the feasibility of creating nanoparticles that possess the ability to release nutrients in a controlled manner. This enables the provision of precise amounts of nutrients to seeds at certain times.

Exploring the creation of nanoparticles with several functions is a highly promising direction. The nanoparticles can distribute several substances, including fertilizers, insecticides, and growth hormones, all at the same time [18]. This allows for a complete solution to improve the process of seed germination and the development of seedlings. Furthermore, multifunctional nanoparticles may be engineered to react to environmental stimuli, such as variations in humidity or temperature, and selectively release their cargo just under particular circumstances. This intelligent release system can optimize the utilization of resources and minimize the negative effects on the environment.

In addition, nano priming has the capacity to enhance the ability of crops to withstand abiotic challenges such as drought, salt, and severe temperatures. Research has demonstrated that nanoparticles have the ability to regulate the activity of stress-responsive genes and strengthen the antioxidant defence mechanisms in plants. For instance, the use of silicon nanoparticles has been discovered to enhance the ability of wheat (*Triticum aestivum L.*) to withstand drought conditions by increasing water retention and mitigating oxidative stress. These developments are essential in the context of climate change, as crops are becoming more vulnerable to unexpected and harsh weather conditions.

Nanotechnology can also aid in the advancement of crops that are resistant to diseases. Silver nanoparticles, renowned for their antibacterial characteristics, have the potential to safeguard seeds and seedlings against pathogenic diseases. This method not only improves the chances of seedlings surviving but also decreases the need for chemical pesticides, which helps create a healthier and more sustainable agricultural setting.

The incorporation of nano priming with other sophisticated agricultural technologies, like as precision farming and digital agriculture, presents further possibilities. For example, unmanned aerial vehicles outfitted with sensors may be employed to observe the state of crops treated with nanoparticles, offering instantaneous data on the well-being and development of plants. This data may be utilized to create well-informed choices regarding supplementary treatments or interventions, guaranteeing the best possible crop performance. The integration of nano priming with digital agriculture has the potential to create farming systems that are exceptionally productive and environmentally friendly.

Furthermore, the continuous investigation and advancement in nanotechnology are expected to lead to the identification of novel nanoparticles possessing distinctive characteristics that might further augment the process of seed germination and seedling growth. Research on nanoclay composites has demonstrated potential in enhancing soil structure and water retention, hence augmenting the advantages of nano priming. The ongoing advancement of nanotechnology will unquestionably create novel opportunities and uses in the field of agriculture.

Despite the presence of notable obstacles, the prospects for using nano priming in agriculture are promising. Through continuous study and technological progress, it is feasible to surpass the existing constraints and fully use the capabilities of nanoparticles to enhance the process of seed germination and the growth of seedlings. The use of nano priming alongside other agricultural technology and sustainable practices will be essential in satisfying the increasing need for food in an ecologically conscientious manner. Through the strategic resolution of obstacles and harnessing the potential benefits of nano priming, we may establish the foundation for a forthcoming period of agricultural advancement and efficiency.

7 Conclusion

Overall, the utilization of nanoparticles in seed priming is a noteworthy progression in the field of agricultural research, providing encouraging remedies to improve seed germination and seedling growth. Nanoparticles, such as zinc, silica, silver, iron oxide, titanium dioxide, and gold, have demonstrated significant promise in enhancing seed quality through many processes, including improved absorption of nutrients, increased ability to withstand stress, and higher resistance to diseases. Nanoparticles enhance the intake and utilization of vital nutrients, strengthen plant defenses against environmental stresses, and enhance overall plant vitality and resilience.

Research has repeatedly shown that the use of nanoparticles has a beneficial impact on several crops, leading to enhancements in germination rates, seedling strength, and tolerance to stress. This novel method tackles key obstacles in agriculture, such as nutrient insufficiencies, soil fecundity, and the repercussions of climate change, hence enhancing agricultural yield and sustainability.

Nevertheless, further investigation is required to comprehensively comprehend the enduring consequences and ecological effects of nanoparticle use in agriculture. It is essential to prioritize the development of secure and efficient application techniques, while also addressing any possible hazards, in order to promote the general acceptance and use of this technology. Nanotechnology's ongoing development offers the potential to completely transform current agriculture, enabling the adoption of more sustainable and efficient farming methods. However, further and more research is required to fully realize these benefits and address any associated challenges.

References

1. Reed, R. C., Bradford, K. J., & Khanday, I. (2022). Seed germination and vigor: Ensuring crop sustainability in a changing climate. *Heredity, 128*(6), 450–459. https://doi.org/10.1038/s41437-022-00497-2
2. Drobek, M., Frąc, M., & Cybulska, J. (2019). Plant biostimulants: Importance of the quality and yield of horticultural crops and the improvement of plant tolerance to abiotic stress—A review. *Agronomy, 9*, 335. https://doi.org/10.3390/agronomy9060335
3. FAO. (2024). *Gross domestic product and agriculture value added, no. 85*. FAO.
4. FAO. *Building a common vision for sustainable food and agriculture*. Global: FAO.
5. Binod, B., Bhattacharjee, A., & Ishwar, N. (2018). *Bonn challenge and India: Progress on restoration efforts across states and landscapes* (Vol. 8, p. 32). IUCN and MoEFCC, Government of India. https://doi.org/10.2305/IUCN.CH.2018.12.en
6. Wolny, E., Betekhtin, A., Rojek, M., Braszewska-Zalewska, A., Lusinska, J., & Hasterok, R. (2018). Germination and the early stages of seedling development in Brachypodium distachyon. *International Journal of Molecular Sciences, 19*(10), 2916. https://doi.org/10.3390/ijms19102916
7. Seleiman, M. F., Al-Suhaibani, N., Ali, N., Akmal, M., Alotaibi, M., Refay, Y., Dindaroglu, T., Abdul-Wajid, H. H., & Battaglia, M. L. (2021). Drought stress impacts on plants and different approaches to alleviate its adverse effects. *Plants, 10*(2), 259. https://doi.org/10.3390/plants10020259
8. Gairola, S. U., Bahuguna, R., & Bhatt, S. S. (2023). Native plant species: A tool for restoration of mined lands. *Journal of Soil Science and Plant Nutrition, 23*, 1438–1448. https://doi.org/10.1007/s42729-023-01181-y
9. Prajapati, S., Choudhary, S., Kumar, V., Dayal, P., Srivastava, R., Gairola, A., & Borate, R. (2023). Carbon sequestration: A key strategy for climate change mitigation towards a sustainable future. *Emerging Trends in Climate Change, 2*, 1–14. https://doi.org/10.18782/2583-4770.128
10. Klupczyńska, E. A., & Pawłowski, T. A. (2021). Regulation of seed dormancy and germination mechanisms in a changing environment. *International Journal of Molecular Sciences, 22*(3), 1357. https://doi.org/10.3390/ijms22031357
11. Haj Sghaier, A., Tarnawa, Á., Khaeim, H., Kovács, G. P., Gyuricza, C., & Kende, Z. (2022). The effects of temperature and water on the seed germination and seedling development of rapeseed (Brassica napus L.). *Plants, 11*(21), 2819. https://doi.org/10.3390/plants11212819

12. Dell'Olmo, E., Tiberini, A., & Sigillo, L. (2023). Leguminous Seedborne pathogens: Seed health and sustainable crop management. *Plants, 12*(10), 2040. https://doi.org/10.3390/plants12102040
13. Daraban, G. M., Hlihor, R.-M., & Suteu, D. (2023). Pesticides vs. biopesticides: From pest management to toxicity and impacts on the environment and human health. *Toxics, 11*(12), 983. https://doi.org/10.3390/toxics11120983
14. Dadlani, M., & Yadava, D. K. (2023). *Seed science and technology: Biology, production, quality*. Springer.
15. Nile, S. H., Thiruvengadam, M., Wang, Y., Samynathan, R., Shariati, M. A., Rebezov, M., Nile, A., Sun, M., Venkidasamy, B., Xiao, J., & Kai, G. (2022). Nano-priming as emerging seed priming technology for sustainable agriculture—Recent developments and future perspectives. *Journal of Nanobiotechnology, 20*, 1. https://doi.org/10.1186/s12951-022-01423-8
16. Adhikari, B., Dhital, P. R., Ranabhat, S., & Poudel, H. (2021). Effect of seed hydro-priming durations on germination and seedling growth of bitter gourd (Momordica charantia). *PLoS One, 16*(8), e0255258. https://doi.org/10.1371/journal.pone.0255258
17. Mitra, D., Mondal, R., Khoshru, B., Shadangi, S., Mohapatra, P. K. D., & Panneerselvam, P. (2021). Rhizobacteria mediated seed bio-priming triggers the resistance and plant growth for sustainable crop production. *Current Research in Microbial Sciences, 2*, 100071. https://doi.org/10.1016/j.crmicr.2021.100071
18. Shelar, A., Nile, S. H., Singh, A. V., Rothenstein, D., Bill, J., Xiao, J., Chaskar, M., Kai, G., & Patil, R. (2023). Recent advances in Nano-enabled seed treatment strategies for sustainable agriculture: Challenges, risk assessment, and future perspectives. *Nano-Micro Letters, 15*(1), 54. https://doi.org/10.1007/s40820-023-01025-5
19. Joseph, T., Mahapatra, D. K., Esmaeili, A., Piszczyk, Ł., Hasanin, M., Kattali, M., Haponiuk, J., & Thomas, S. (2023). Nanoparticles: Taking a unique position in medicine. *Nanomaterials, 13*(3), 574. https://doi.org/10.3390/nano13030574
20. Pereira, A. D. E. S., Oliveira, H. C., Fraceto, L. F., & Santaella, C. (2021). Nanotechnology potential in seed priming for sustainable agriculture. *Nanomaterials, 11*(2), 267. https://doi.org/10.3390/nano11020267
21. Shelar, A., Singh, A. V., Maharjan, R. S., Laux, P., Luch, A., Gemmati, D., Tisato, V., Singh, S. P., Santilli, M. F., Shelar, A., Chaskar, M., & Patil, R. (2021). Sustainable agriculture through multidisciplinary seed nanopriming: Prospects of opportunities and challenges. *Cells, 10*(9), 2428. https://doi.org/10.3390/cells10092428
22. Zaim, N. S. H. B. H., Tan, H. L., Rahman, S. M. A., Bakar, N. F. A., Osman, M. S., Thakur, V. K., & Radacsi, N. (2023). Recent advances in seed coating treatment using nanoparticles and nanofibers for enhanced seed germination and protection. *Journal of Plant Growth Regulation, 42*(12), 7374–7402. https://doi.org/10.1007/s00344-023-11038-4
23. Rhaman, M. S., Imran, S., Rauf, F., Khatun, M., Baskin, C. C., Murata, Y., & Hasanuzzaman, M. (2021). Seed priming with phytohormones: An effective approach for the mitigation of abiotic stress. *Plants, 10*(1), 37. https://doi.org/10.3390/plants10010037
24. Singh, A., Sengar, R., Sharma, R., Rajput, P., & Singh, A. K. (2022). Nano-priming technology for sustainable agriculture. *Biogeosystem Technique, 8*, 79–92. https://doi.org/10.13187/bgt.2021.2.79
25. Kamaszewski, M., Kawalski, K., Wiechetek, W., Szudrowicz, H., Martynow, J., Adamek-Urbańska, D., Łosiewicz, B., Szczepański, A., Bujarski, P., Frankowska-Łukawska, J., Chwaściński, A., & Aksakal, E. (2023). The effect of silver nanoparticles on the digestive system, gonad morphology, and physiology of butterfly Splitfin (Ameca splendens). *International Journal of Molecular Sciences, 24*(19), 14598. https://doi.org/10.3390/ijms241914598
26. Huang, X., He, D., Pan, Z., Luo, G., & Deng, J. (2021). Reactive-oxygen-species-scavenging nanomaterials for resolving inflammation. *Materials Today Bio, 11*, 100124. https://doi.org/10.1016/j.mtbio.2021.100124
27. Al-Khayri, J. M., Rashmi, R., Ulhas, R. S., Sudheer, W. N., Banadka, A., Nagella, P., Aldaej, M. I., Rezk, A. A., Shehata, W. F., & Almaghasla, M. I. (2023). The role of nanoparticles in

response of plants to abiotic stress at physiological, biochemical, and molecular levels. *Plants, 12*(2), 292. https://doi.org/10.3390/plants12020292
28. Rizwan, M., Ali, S., Ali, B., Adrees, M., Arshad, M., Hussain, A., Rehman, M. Z. U., & Waris, A. A. (2018). Zinc and iron oxide nanoparticles improved the plant growth and reduced the oxidative stress and cadmium concentration in wheat. *Chemosphere, 214*, 269–277. https://doi.org/10.1016/j.chemosphere.2018.09.120
29. Itroutwar, P. D., Govindaraju, K., Tamilselvan, S., Kannan, M., Raja, K., & Subramanian, K. S. (2019). Seaweed-based biogenic ZnO nanoparticles for improving agro-morphological characteristics of rice (Oryza sativa L.). *Journal of Plant Growth Regulation, 39*(2), 717–728. https://doi.org/10.1007/s00344-019-10012-3
30. Savassa, S. M., Duran, N. M., Rodrigues, E. S., De Almeida, E., Van Gestel, C. A. M., Bompadre, T. F. V., & De Carvalho, H. W. P. (2018). Effects of ZnO nanoparticles on Phaseolus vulgaris germination and seedling development determined by X-ray spectroscopy. *ACS Applied Nano Materials, 1*(11), 6414–6426. https://doi.org/10.1021/acsanm.8b01619
31. Latef, A. A. H. A., Alhmad, M. F. A., & Abdelfattah, K. E. (2016). The possible roles of priming with ZnO nanoparticles in mitigation of salinity stress in lupine (Lupinus termis) plants. *Journal of Plant Growth Regulation, 36*(1), 60–70. https://doi.org/10.1007/s00344-016-9618-x
32. Nandhini, M., Rajini, S., Udayashankar, A., Niranjana, S., Lund, O. S., Shetty, H., & Prakash, H. (2019). Biofabricated zinc oxide nanoparticles as an eco-friendly alternative for growth promotion and management of downy mildew of pearl millet. *Crop Protection, 121*, 103–112. https://doi.org/10.1016/j.cropro.2019.03.015
33. Choudhary, R. C., Kumaraswamy, R., Kumari, S., Sharma, S., Pal, A., Raliya, R., Biswas, P., & Saharan, V. (2019). Zinc encapsulated chitosan nanoparticle to promote maize crop yield. *International Journal of Biological Macromolecules, 127*, 126–135. https://doi.org/10.1016/j.ijbiomac.2018.12.274
34. Mittal, D., Kaur, G., Singh, P., Yadav, K., & Ali, S. A. (2020). Nanoparticle-based sustainable agriculture and food science: Recent advances and future outlook. *Frontiers in Nanotechnology, 2*. https://doi.org/10.3389/fnano.2020.579954
35. Hasanaklou, N. T., Mohagheghi, V., Hasanaklou, H. T., Ma'mani, L., Malekmohammadi, M., Moradi, F., & Dalvand, Y. (2023). Seed nano-priming using silica nanoparticles: Effects in seed germination and physiological properties of Stevia Rebaudiana Bertoni. *Chemical and Biological Technologies in Agriculture, 10*, 1. https://doi.org/10.1186/s40538-023-00445-0
36. Sembada, A. A., Maki, S., Faizal, A., Fukuhara, T., Suzuki, T., & Lenggoro, I. W. (2023). The role of silica nanoparticles in promoting the germination of tomato (Solanum lycopersicum) seeds. *Nanomaterials, 13*(14), 2110. https://doi.org/10.3390/nano13142110
37. Hussain, A., Rizwan, M., Ali, S., Rehman, M. Z. U., Qayyum, M. F., Nawaz, R., Ahmad, A., Asrar, M., Ahmad, S. R., Alsahli, A. A., & Alyemeni, M. N. (2021). Combined use of different nanoparticles effectively decreased cadmium (Cd) concentration in grains of wheat grown in a field contaminated with Cd. *Ecotoxicology and Environmental Safety, 215*, 112139. https://doi.org/10.1016/j.ecoenv.2021.112139
38. Vanitha, C., Kathiravan, M., Umarani, R., Sathiya, K., Menaka, C., Yuvaraj, M., & Cyriac, J. (2023). Seed priming with nano silica alleviates drought stress through regulating antioxidant defense system and osmotic adjustment in soybean (Glycine max L.). *SILICON, 16*(5), 2157–2170. https://doi.org/10.1007/s12633-023-02826-4
39. Mahakham, W., Sarmah, A. K., Maensiri, S., & Theerakulpisut, P. (2017). Nanopriming technology for enhancing germination and starch metabolism of aged rice seeds using phytosynthesized silver nanoparticles. *Scientific Reports, 7*, 1. https://doi.org/10.1038/s41598-017-08669-5
40. Acharya, P., Jayaprakasha, G. K., Crosby, K. M., Jifon, J. L., & Patil, B. S. (2019). Green-synthesized nanoparticles enhanced seedling growth, yield, and quality of onion (Allium cepa L.). *ACS Sustainable Chemistry & Engineering, 7*(17), 14580–14590. https://doi.org/10.1021/acssuschemeng.9b02180

41. Acharya, P., Jayaprakasha, G. K., Crosby, K. M., Jifon, J. L., & Patil, B. S. (2020). Nanoparticle-mediated seed priming improves germination, growth, yield, and quality of watermelons (Citrullus lanatus) at multi-locations in Texas. *Scientific Reports, 10*, 1. https://doi.org/10.1038/s41598-020-61696-7
42. Spagnoletti, F. N., Spedalieri, C., Kronberg, F., & Giacometti, R. (2018). Extracellular biosynthesis of bactericidal Ag/AgCl nanoparticles for crop protection using the fungus Macrophomina phaseolina. *Journal of Environmental Management, 231*, 457–466. https://doi.org/10.1016/j.jenvman.2018.10.081
43. Kannaujia, R., Srivastava, C. M., Prasad, V., Singh, B. N., & Pandey, V. (2019). Phyllanthus emblica fruit extract stabilized biogenic silver nanoparticles as a growth promoter of wheat varieties by reducing ROS toxicity. *Plant Physiology and Biochemistry, 142*, 460–471. https://doi.org/10.1016/j.plaphy.2019.08.008
44. Mazhar, M. W., Ishtiaq, M., Maqbool, M., Akram, R., Shahid, A., Shokralla, S., Al-Ghobari, H., Alataway, A., Dewidar, A. Z., El-Sabrout, A. M., & Elansary, H. O. (2022). Seed priming with iron oxide nanoparticles raises biomass production and agronomic profile of water-stressed flax plants. *Agronomy, 12*(5), 982. https://doi.org/10.3390/agronomy12050982
45. Ahuja, R., Sidhu, A., & Bala, A. (2019). Synthesis and evaluation of iron(ii) sulfide aqua nanoparticles (FeS-NPs) against Fusarium verticillioides causing sheath rot and seed discoloration of rice. *European Journal of Plant Pathology, 155*(1), 163–171. https://doi.org/10.1007/s10658-019-01758-3
46. Sundaria, N., Singh, M., Upreti, P., Chauhan, R. P., Jaiswal, J. P., & Kumar, A. (2018). Seed priming with iron oxide nanoparticles triggers iron acquisition and biofortification in wheat (Triticum aestivum L.) grains. *Journal of Plant Growth Regulation, 38*(1), 122–131. https://doi.org/10.1007/s00344-018-9818-7
47. Guha, T., Ravikumar, K. V. G., Mukherjee, A., Mukherjee, A., & Kundu, R. (2018). Nanopriming with zero valent iron (nZVI) enhances germination and growth in aromatic rice cultivar (Oryza sativa cv. Gobindabhog L.). *Plant Physiology and Biochemistry: PPB, 127*, 403–413. https://doi.org/10.1016/j.plaphy.2018.04.014
48. Das, C. K., Jangir, H., Kumar, J., Verma, S., Mahapatra, S. S., Philip, D., Srivastava, G., & Das, M. (2018). Nano-pyrite seed dressing: A sustainable design for NPK equivalent rice production. Nanotechnology for. *Environmental Engineering, 3*, 1. https://doi.org/10.1007/s41204-018-0043-1
49. Faraz, A., Faizan, M., Rajput, V. D., Minkina, T., Hayat, S., Faisal, M., Alatar, A. A., & Abdel-Salam, E. M. (2023). CuO nanoparticle-mediated seed priming improves physio-biochemical and enzymatic activities of Brassica juncea. *Plants, 12*(4), 803. https://doi.org/10.3390/plants12040803
50. Wang, W., Ren, Y., He, J., Zhang, L., Wang, X., & Cui, Z. (2020). Impact of copper oxide nanoparticles on the germination, seedling growth, and physiological responses in Brassica pekinensis L. *Environmental Science and Pollution Research, 27*(25), 31505–31515. https://doi.org/10.1007/s11356-020-09338-3
51. Farooq, T., Nisa, Z. U., Hameed, A., Ahmed, T., & Hameed, A. (2022). Priming with copper-chitosan nanoparticles elicit tolerance against PEG-induced hyperosmotic stress and salinity in wheat. BMC. *Chemistry, 16*, 1. https://doi.org/10.1186/s13065-022-00813-1
52. Nair, P. M. G., Kim, S., & Chung, I. M. (2014). Copper oxide nanoparticle toxicity in mung bean (Vigna radiata L.) seedlings: Physiological and molecular level responses of in vitro grown plants. *Acta Physiologiae Plantarum, 36*(11), 2947–2958. https://doi.org/10.1007/s11738-014-1667-9
53. Gopinath, K., Gowri, S., Karthika, V., & Arumugam, A. (2014). Green synthesis of gold nanoparticles from fruit extract of Terminalia arjuna, for the enhanced seed germination activity of Gloriosa superba. *Journal of Nanostructure in Chemistry, 4*, 3. https://doi.org/10.1007/s40097-014-0115-0

54. Feichtmeier, N. S., Walther, P., & Leopold, K. (2015). Uptake, effects, and regeneration of barley plants exposed to gold nanoparticles. *Environmental Science and Pollution Research, 22*(11), 8549–8558. https://doi.org/10.1007/s11356-014-4015-0
55. Cadena, M. B., Preston, G. M., Van Der Hoorn, R. A., Flanagan, N. A., Townley, H. E., & Thompson, I. P. (2018). Enhancing cinnamon essential oil activity by nanoparticle encapsulation to control seed pathogens. *Industrial Crops and Products, 124*, 755–764. https://doi.org/10.1016/j.indcrop.2018.08.043
56. Narware, J., Yadav, R. N., Keswani, C., Singh, S., & Singh, H. B. (2019). Silver nanoparticle-based biopesticides for phytopathogens: Scope and potential in agriculture. https://doi.org/10.1016/B978-0-12-815829-6.00013-9.
57. Gupta, S. D., Agarwal, A., & Pradhan, S. (2018). Phytostimulatory effect of silver nanoparticles (AgNPs) on rice seedling growth: An insight from antioxidative enzyme activities and gene expression patterns. *Ecotoxicology and Environmental Safety, 161*, 624–633. https://doi.org/10.1016/j.ecoenv.2018.06.023
58. Zhang, X., Zhang, D., Sun, W., & Wang, T. (2019). The adaptive mechanism of plants to iron deficiency via iron uptake, transport, and homeostasis. *International Journal of Molecular Sciences, 20*, 2424. https://doi.org/10.3390/ijms20102424
59. Rai, P., et al. (2022). Nanoiron : Uptake , translocation and accumulation in plant systems. *Plant Nano Biology, 2*(March), 100017. https://doi.org/10.1016/j.plana.2022.100017
60. Maswada, H. F., Djanaguiraman, M., & Prasad, P. V. V. (2018). Seed treatment with nano-iron (III) oxide enhances germination, seeding growth and salinity tolerance of sorghum. *Journal of Agronomy and Crop Science, 204*(6), 577–587. https://doi.org/10.1111/jac.12280
61. Kasote, D. M., Lee, J. H., Jayaprakasha, G. K., & Patil, B. S. (2019). Seed priming with iron oxide nanoparticles modulate antioxidant potential and defense-linked hormones in watermelon seedlings. *ACS Sustainable Chemistry & Engineering, 7*(5), 5142–5151. https://doi.org/10.1021/acssuschemeng.8b06013
62. Raiesi-Ardali, T., Ma'mani, L., Chorom, M., & Moezzi, A. (2022). Improved iron use efficiency in tomatoes using organically coated iron oxide nanoparticles as efficient bioavailable Fe sources. *Chemical and Biological Technologies in Agriculture, 9*, 1. https://doi.org/10.1186/s40538-022-00318-y
63. Palchoudhury, S., Jungjohann, K. L., Weerasena, L., Arabshahi, A., Gharge, U., Albattah, A., Miller, J., Patel, K., & Holler, R. A. (2018). Enhanced legume root growth with pre-soaking in α-Fe2O3 nanoparticle fertilizer. *RSC Advances, 8*(43), 24075–24083. https://doi.org/10.1039/c8ra04680h
64. Duran, N. M., Savassa, S. M., De Lima, R. G., De Almeida, E., Linhares, F. S., Van Gestel, C. A. M., & De Carvalho, H. W. P. (2017). X-ray spectroscopy uncovering the effects of cu based nanoparticle concentration and structure on Phaseolus vulgaris germination and seedling development. *Journal of Agricultural and Food Chemistry, 65*(36), 7874–7884. https://doi.org/10.1021/acs.jafc.7b03014
65. Yasmeen, F., Raja, N. I., Razzaq, A., & Komatsu, S. (2016). Proteomic and physiological analyses of wheat seeds exposed to copper and iron nanoparticles. *Biochimica Et Biophysica Acta (BBA) - Proteins and Proteomics, 1865*(1), 28–42. https://doi.org/10.1016/j.bbapap.2016.10.001
66. Venzhik, Y. V., Deryabin, A. N., Zhukova, K. V., Sokolov, A. O., Popov, V. N., & Moshkov, I. E. (2024). Features of formation of wheat's increased cold resistance under the influence of gold nanoparticles. Russian. *Journal of Plant Physiology, 71*, 1. https://doi.org/10.1134/s1021443724604191
67. Mahakham, W., Theerakulpisut, P., Maensiri, S., Phumying, S., & Sarmah, A. K. (2016). Environmentally benign synthesis of phytochemicals-capped gold nanoparticles as nanopriming agent for promoting maize seed germination. *The Science of the Total Environment, 573*, 1089–1102. https://doi.org/10.1016/j.scitotenv.2016.08.120

68. Ndeh, N. T., Maensiri, S., & Maensiri, D. (2017). The effect of green synthesized gold nanoparticles on rice germination and roots. *Advances in Natural Sciences Nanoscience and Nanotechnology, 8*(3), 035008. https://doi.org/10.1088/2043-6254/aa724a
69. Arora, S., Sharma, P., Kumar, S., Nayan, R., Khanna, P. K., & Zaidi, M. G. H. (2012). Gold-nanoparticle induced enhancement in growth and seed yield of Brassica juncea. *Plant Growth Regulation, 66*(3), 303–310. https://doi.org/10.1007/s10725-011-9649-z
70. Anand, A., Kumari, A., Thakur, M., & Koul, A. (2019). Hydrogen peroxide signaling integrates with phytohormones during the germination of magnetoprimed tomato seeds. *Scientific Reports, 9*, 1. https://doi.org/10.1038/s41598-019-45102-5
71. Pandya, P., Kumar, S., Patil, G., Mankad, M., & Shah, Z. (2024). Impact of ZnO nanopriming on physiological and biochemical traits of wheat (Triticum aestivum L.) seedling. *CABI Agriculture and Bioscience, 5*(1). https://doi.org/10.1186/s43170-024-00228-z
72. Setty, J., Samant, S. B., Yadav, M. K., Manjubala, M., & Pandurangam, V. (2023). Beneficial effects of bio-fabricated selenium nanoparticles as seed nanopriming agent on seed germination in rice (Oryza sativa L.). *Scientific Reports, 13*(1). https://doi.org/10.1038/s41598-023-49621-0
73. Baz, H., Creech, M., Chen, J., Gong, H., Bradford, K., & Huo, H. (2020). Water-soluble carbon nanoparticles improve seed germination and post-germination growth of lettuce under salinity stress. *Agronomy, 10*(8), 1192. https://doi.org/10.3390/agronomy10081192
74. Kasote, D. M., Lee, J. H. J., Jayaprakasha, G. K., & Patil, B. S. (2021). Manganese oxide nanoparticles as safer seed priming agent to improve chlorophyll and antioxidant profiles in watermelon seedlings. *Nanomaterials, 11*(4), 1016. https://doi.org/10.3390/nano11041016
75. Khan, Z. S., Rizwan, M., Hafeez, M., Ali, S., Javed, M. R., & Adrees, M. (2019). The accumulation of cadmium in wheat (Triticum aestivum) as influenced by zinc oxide nanoparticles and soil moisture conditions. *Environmental Science and Pollution Research, 26*(19), 19859–19870. https://doi.org/10.1007/s11356-019-05333-5
76. Ivani, R., Nejad, S. H. S., Ghahraman, B., Astaraei, A. R., & Feizi, H. (2018). Role of bulk and Nanosized SiO2 to overcome salt stress during fenugreek germination (Trigonella foenum-graceum L.). *Plant Signaling & Behavior, 13*(7), e1044190. https://doi.org/10.1080/15592324.2015.1044190
77. Hojjat, S. S., & Kamyab, M. (2017). The effect of silver nanoparticle on fenugreek seed germination under salinity levels. *Russian Agricultural Sciences, 43*(1), 61–65. https://doi.org/10.3103/s1068367417010189
78. Konate, A., He, X., Zhang, Z., Ma, Y., Zhang, P., Alugongo, G., & Rui, Y. (2017). Magnetic (Fe3O4) nanoparticles reduce heavy metals uptake and mitigate their toxicity in wheat seedling. *Sustainability, 9*(5), 790. https://doi.org/10.3390/su9050790
79. Grün, A., Manz, W., Kohl, Y. L., Meier, F., Straskraba, S., Jost, C., Drexel, R., & Emmerling, C. (2019). Impact of silver nanoparticles (AgNP) on soil microbial community depending on functionalization, concentration, exposure time, and soil texture. *Environmental Sciences Europe, 31*(1). https://doi.org/10.1186/s12302-019-0196-y
80. Vimercati, L., Cavone, D., Caputi, A., De Maria, L., Tria, M., Prato, E., & Ferri, G. M. (2020). Nanoparticles: An experimental study of zinc nanoparticles toxicity on marine crustaceans. General overview on the health implications in humans. *Frontiers in Public Health, 8*. https://doi.org/10.3389/fpubh.2020.00192
81. Chavan, S., Sarangdhar, V., & Nadanathangam, V. (2020). Toxicological effects of TiO2 nanoparticles on plant growth promoting soil bacteria. *Emerging Contaminants, 6*, 87–92. https://doi.org/10.1016/j.emcon.2020.01.003
82. Ameen, F., Alsamhary, K., Alabdullatif, J. A., & ALNadhari, S. (2021). A review on metal-based nanoparticles and their toxicity to beneficial soil bacteria and fungi. *Ecotoxicology and Environmental Safety, 213*, 112027. https://doi.org/10.1016/j.ecoenv.2021.112027
83. Kumari, R., Suman, K., Karmakar, S., Mishra, V., Lakra, S. G., Saurav, G. K., & Mahto, B. K. (2023). Regulation and safety measures for nanotechnology-based agri-products. *Frontiers in Genome Editing, 5*. https://doi.org/10.3389/fgeed.2023.1200987

84. Allan, J., Belz, S., Hoeveler, A., Hugas, M., Okuda, H., Patri, A., Rauscher, H., Silva, P., Slikker, W., Sokull-Kluettgen, B., Tong, W., & Anklam, E. (2021). Regulatory landscape of nanotechnology and nanoplastics from a global perspective. *Regulatory Toxicology and Pharmacology, 122*, 104885. https://doi.org/10.1016/j.yrtph.2021.104885
85. Salam, A., Afridi, M. S., Javed, M. A., Saleem, A., Hafeez, A., Khan, A. R., Zeeshan, M., Ali, B., Azhar, W., Sumaira, N., Ulhassan, Z., & Gan, Y. (2022). Nano-priming against abiotic stress: A way forward towards sustainable agriculture. *Sustainability, 14*(22), 14880. https://doi.org/10.3390/su142214880
86. Kumah, E. A., Fopa, R. D., Harati, S., Boadu, P., Zohoori, F. V., & Pak, T. (2023). Human and environmental impacts of nanoparticles: A scoping review of the current literature. *BMC Public Health, 23*(1). https://doi.org/10.1186/s12889-023-15958-4

Nanobiofertilizers for Foliar Use: Pioneering Sustainable Agricultural Practices

Ramesh S. Chaughule and Amruta V. Shelar

Contents

R. S. Chaughule
Department of Chemistry, Ramnarain Ruia Autonomous College, Matunga, Mumbai, India

A. V. Shelar (✉)
Department of Technology, Savitribai Phule Pune University, Pune, India

R. S. Chaughule, A. V. Shelar (eds.), *Nanotechnology in Agriculture*,
https://doi.org/10.1007/978-3-031-97544-8_7

1 Introduction

Agriculture has been the bedrock of human civilization, playing an essential role in ensuring food security, economic growth, and societal stability. However, as the global population continues to rise, which is projected to reach 9.7 billion by 2050, modern agricultural systems are at a crossroads. The sector is grappling with many challenges that jeopardize its long-term sustainability [1]. These include soil degradation, loss of arable land, climate change, and increasing demands for food production. These factors highlight an urgent need for innovative strategies that can boost agricultural productivity while mitigating environmental harm. Among the emerging solutions, nanotechnology has taken center stage, particularly the use of nanobiofertilizers, which offer transformative potential in sustainable farming [2].

Nanotechnology involves manipulating materials at the atomic and molecular scale, enabling unprecedented precision in various applications, including agriculture. By improving nutrient delivery mechanisms, nanotechnology addresses key shortcomings of traditional farming practices. As shown in Fig. 1, nanobiofertilizers fertilizers engineered at the nanoscale using bio-based inputs have emerged as a promising tool to enhance crop management. One of the most effective ways to utilize nanobiofertilizers is through the foliar application, a method in which nutrients are sprayed directly onto plant leaves. This approach allows plants to absorb nutrients more rapidly and efficiently than soil-based methods, reducing wastage and enhancing nutrient uptake. Unlike conventional soil fertilization, foliar applications minimize nutrient loss through leaching and volatilization, thereby improving nutrient-use efficiency and reducing environmental degradation [3].

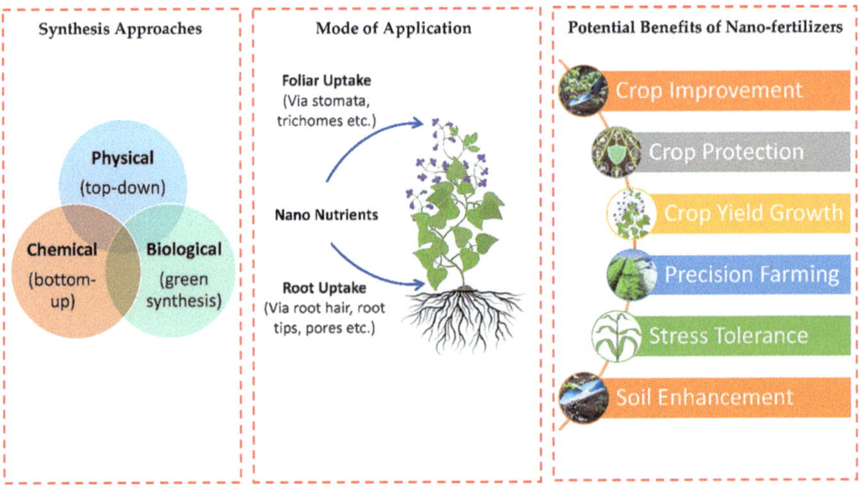

Fig. 1 Comprehensive overview of synthesis techniques, application methods, and the potential advantages of nanofertilizers. (This is a reprinted image of Ref. [28] with permission)

1.1 The Pressures on Global Agriculture

Global agriculture is under immense strain from multiple interrelated factors. Soil degradation, a critical issue, stems from decades of intensive farming practices and overreliance on chemical fertilizers. As a result, nearly one-third of the world's arable land is classified as moderately to severely degraded, leading to reduced soil fertility, diminished crop yields, and increased reliance on synthetic inputs [4]. This overuse of chemical fertilizers not only depletes the soil but also disrupts the delicate balance of microbial communities that are vital for soil health. Simultaneously, the availability of arable land is shrinking due to urbanization, industrialization, and climate change. Urban expansion consumes vast tracts of fertile land, while industrial activities pollute surrounding agricultural areas. Climate change further compounds the issue, introducing rising temperatures, altered precipitation patterns, and an increase in extreme weather events such as droughts, floods, and storms. These environmental changes intensify abiotic stresses on crops, including heat stress, salinity, and water scarcity, all of which hinder plant growth and reduce agricultural output [5].

1.2 Rising Food Demand and Resource Constraints

The challenge of feeding a rapidly growing global population adds another layer of complexity. To meet the dietary needs of nearly ten billion people by 2050, agricultural output must increase by 70% [6]. Achieving this goal under current practices would require unsustainable exploitation of already overburdened natural resources, including water, soil, and energy. Traditional reliance on chemical fertilizers to boost crop yields is no longer a viable option, given their long-term environmental and ecological drawbacks. Chemical fertilizers, while effective in the short term, have led to significant environmental concerns. Nutrient leaching into water bodies causes eutrophication, harming aquatic ecosystems and contaminating drinking water sources. Greenhouse gas emissions from fertilizer production and application contribute to climate change, while excessive use of fertilizers causes imbalances in soil nutrient composition, leading to further degradation [7]. These cumulative effects underscore the urgent need for sustainable agricultural innovations.

1.3 The Promise of Nanotechnology in Sustainable Farming

Nanotechnology, and particularly nanobiofertilizers, provides a forward-looking solution to these pressing challenges. By delivering nutrients in nanoscale formulations, these fertilizers ensure precise, targeted, and efficient nutrient uptake. Foliar application of nanobiofertilizers not only enhances nutrient-use efficiency but also reduces dependence on conventional fertilizers, mitigating their environmental impact [8]. This method offers a dual advantage: improving crop productivity while preserving the ecological integrity of agricultural systems. Nanotechnology's ability to integrate with other advanced farming techniques, such as precision agriculture and smart farming, further enhances its potential to transform global agriculture. With continued research and development, nanotechnology could redefine how crops are grown, ensuring that agriculture remains a sustainable cornerstone of human civilization.

2 Nanobiofertilizers: An Overview

Nanobiofertilizers have emerged as a promising solution to the challenges faced by modern agriculture. These advanced formulations combine nanoscale nutrients or materials with biological agents, such as beneficial microbes or biopolymers, to deliver nutrients efficiently and sustainably. As shown in Fig. 2, the nanoscale dimensions of these fertilizers provide unique properties, including high surface area, enhanced solubility, and controlled release mechanisms, enabling precise nutrient delivery and improved plant uptake [9]. By addressing the limitations of conventional fertilizers, nanobiofertilizers offer several advantages. They reduce

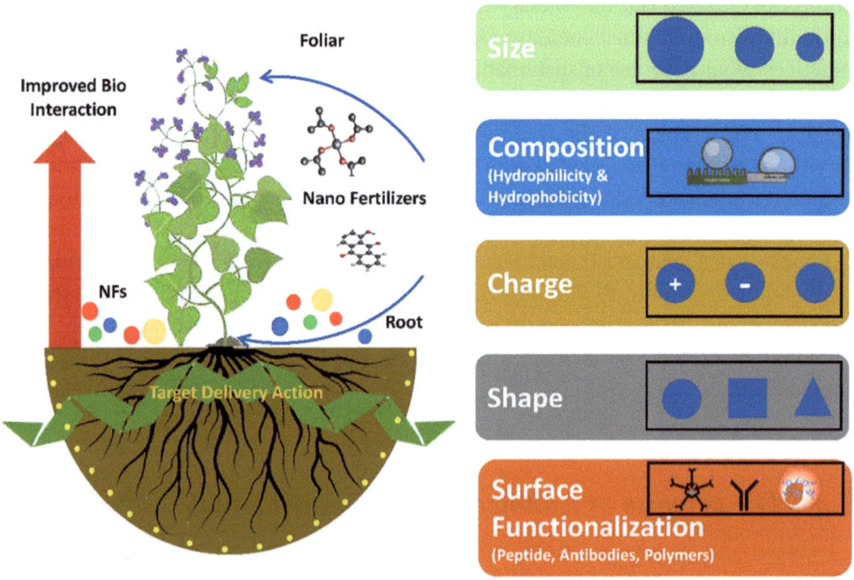

Fig. 2 The nanoscale dimensions of these fertilizers impart unique characteristics, such as increased surface area, improved solubility, and controlled release mechanisms, facilitating precise nutrient delivery and enhanced plant uptake. This is a reprinted image of Ref. [28] with permission

nutrient leaching and volatilization, minimize environmental pollution, and improve nutrient-use efficiency. Additionally, the incorporation of biological agents enhances soil health, nutrient cycling, and plant immunity, further amplifying the benefits of these formulations. For instance, beneficial microbes in nanobiofertilizers facilitate the solubilization of essential nutrients, such as phosphorus and potassium, making them more bioavailable to plants [10].

2.1 Unique Features of Nanobiofertilizers

High Surface Area The nanoscale size of these fertilizers increases the surface area-to-volume ratio, which enhances the interaction between the fertilizer particles and plant surfaces [11].

Enhanced Solubility Nanobiofertilizers dissolve more readily in water, ensuring that nutrients are available in forms that plants can absorb more efficiently [11].

Controlled Release Advanced formulations enable the gradual release of nutrients over time, preventing nutrient loss and supporting sustained plant growth [11].

Biological Integration The inclusion of beneficial microbes and biopolymers boosts soil health and nutrient cycling while enhancing plant immunity and stress resilience.

2.2 Advantages of Nanobiofertilizers

Nanobiofertilizers offer a range of benefits over traditional fertilizers, addressing both agricultural and environmental concerns [2]: Nanobiofertilizers offer enhanced nutrient delivery mechanisms due to their nanoscale size, higher surface area, and controlled release properties.

Improved Nutrient-Use Efficiency Conventional fertilizers often suffer from nutrient losses due to leaching, volatilization, or fixation in the soil. Nanobiofertilizers mitigate these issues through controlled release mechanisms and increased bioavailability, ensuring that a higher proportion of nutrients reach the plant [2]. For instance, phosphorus nanofertilizers prevent nutrient fixation in the soil, making phosphorus readily available to plants, a major limitation in traditional fertilizers.

Reduced Environmental Impact Nutrient runoff from conventional fertilizers leads to water pollution, eutrophication, and greenhouse gas emissions. Nanobiofertilizers minimize nutrient losses, reducing the environmental footprint of farming practices. The controlled release of nutrients ensures that plants absorb them efficiently, leaving minimal residue in the soil or water. Traditional fertilizers are prone to losses through leaching, volatilization, or runoff. Nanobiofertilizers minimize these losses by delivering nutrients directly to the plant tissues.

The biological agents incorporated in nanobiofertilizers, such as rhizobacteria or mycorrhizal fungi, play a pivotal role in improving soil fertility. These microbes promote nutrient cycling by solubilizing essential elements like phosphorus and potassium, converting them into forms that plants can readily absorb. By restoring microbial diversity, nanobiofertilizers contribute to the long-term sustainability of the soil ecosystem [2]. Nanomaterials can encapsulate nutrients, protecting them from environmental degradation (e.g., oxidation) and ensuring their bioavailability over extended periods.

Many nanobiofertilizers include bioactive compounds that enhance plant defenses against biotic and abiotic stresses. For example, silicon nanoparticles improve plant resistance to pests and diseases, while zinc nanoparticles enhance drought and heat tolerance. The biological agents in these fertilizers often produce plant growth-promoting substances like phytohormones, further enhancing resilience.

Cost-Effectiveness in the Long Term Although the initial cost of nanobiofertilizers may be higher than traditional fertilizers, their higher nutrient-use efficiency reduces the quantity required per application. This leads to long-term cost savings for farmers [2].

The integration of nanotechnology and biology in fertilizer development has opened new possibilities for sustainable agriculture. By providing targeted and efficient nutrient delivery, nanobiofertilizers align with the goals of reducing environmental degradation while enhancing productivity. As research continues to refine these formulations, their potential to revolutionize farming practices will only grow. Future advancements, such as biodegradable nanoparticles and smart delivery systems, promise to further increase the efficacy and environmental compatibility of nanobiofertilizers, making them indispensable tools for modern agriculture.

3 Foliar Application: A Targeted Approach

Foliar application of fertilizers is a well-established technique that involves the direct application of nutrients onto plant leaves. This method bypasses soil-related nutrient losses, enabling plants to absorb nutrients rapidly and efficiently. Unlike soil application, foliar fertilization ensures targeted delivery, reducing the risk of nutrient wastage and environmental contamination. This approach is particularly beneficial in addressing nutrient deficiencies during critical growth stages, providing a quick and effective remedy to enhance crop performance [12]. The advent of nanotechnology has further revolutionized foliar fertilization by introducing nanobiofertilizers. These nanoscale formulations enhance the efficacy of foliar applications, providing controlled and sustained nutrient release. Nanoparticles due to their small size and high reactivity facilitate better adhesion to leaf surfaces, improving nutrient absorption and utilization. Additionally, the high surface area of nanoparticles increases the interaction between nutrients and plant tissues, further enhancing their bioavailability [13].

3.1 Advantages of Foliar Application

Bypassing Soil-Related Nutrient Losses In soil-based fertilization, a substantial portion of nutrients can be lost due to leaching, volatilization, or fixation, reducing their availability to plants. The foliar application eliminates these losses by delivering nutrients directly to plant tissues, ensuring that a higher proportion of the applied nutrients reaches the intended target.

Rapid Nutrient Absorption Foliar application facilitates the immediate uptake of nutrients, which is especially valuable during critical growth stages when plants require nutrients urgently to sustain their growth and development.

Targeted Delivery and Reduced Wastage By focusing nutrient application on the foliage, this approach minimizes wastage, making it an economically and environmentally efficient method. Targeted delivery also ensures that specific nutrient deficiencies are addressed promptly, preventing the adverse effects of nutrient imbalances.

Addressing Critical Growth Needs During periods of stress or rapid growth, plants may struggle to uptake adequate nutrients through their roots. The foliar application acts as a supplemental feeding strategy, providing an immediate remedy to nutrient deficiencies and ensuring optimal crop performance.

3.2 Nanotechnology and the Evolution of Foliar Fertilization

The introduction of nanotechnology has significantly enhanced the effectiveness of foliar fertilization by enabling the development of nanobiofertilizers. These innovative fertilizers incorporate nanoscale nutrients and biological agents, offering advanced features that address the limitations of conventional foliar fertilizers [14]. As depicted in Fig. 3, nanofertilizers offer several advantages over traditional foliar fertilizers.

Controlled and Sustained Nutrient Release Nanobiofertilizers are engineered to release nutrients gradually, providing a sustained supply to plants over an extended period. This prevents over-fertilization, which can damage plants, and ensures consistent nutrient availability throughout critical growth phases.

Improved Adhesion and Penetration Nanoparticles, due to their small size and high surface energy, exhibit excellent adhesion properties. When applied to leaves,

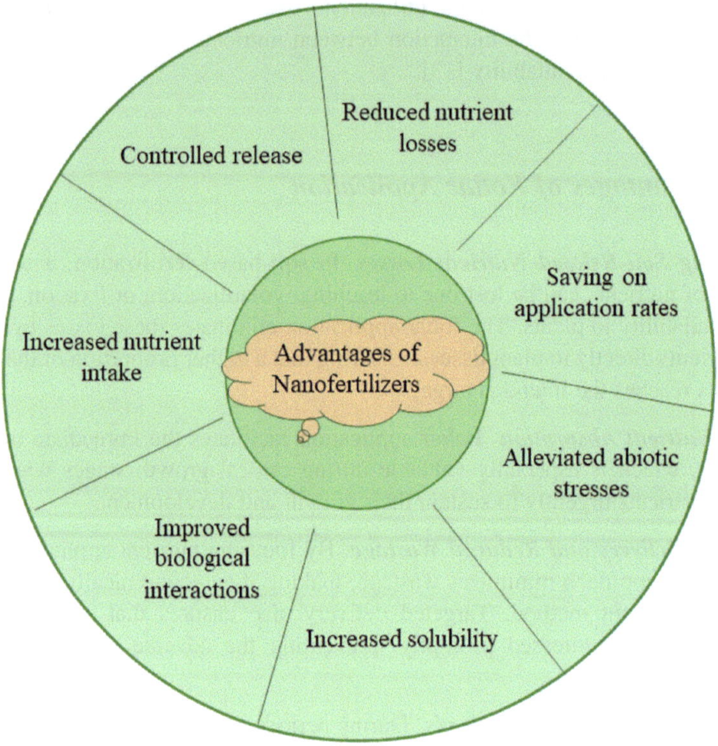

Fig. 3 Advantages of nanofertilizer in foliar application. (This is a reprinted image of Ref. [29] with permission)

they form a uniform coating that adheres tightly to the leaf surface, even under challenging environmental conditions such as wind or rain. Their nanoscale dimensions also allow them to penetrate through microscopic openings, such as stomata, and reach internal plant tissues more effectively.

Enhanced Bioavailability Nanoparticles possess a high surface-area-to-volume ratio, which increases their interaction with plant tissues. This enhanced interaction improves the solubility and bioavailability of nutrients, ensuring that plants can absorb and utilize them more efficiently.

Reduced Environmental Impact By minimizing nutrient losses and improving nutrient-use efficiency, nanobiofertilizers reduce the risk of environmental contamination caused by nutrient runoff or volatilization. This aligns with the goals of sustainable agriculture by lowering the ecological footprint of farming practices.

Longevity and Stability Nanoencapsulation and slow-release formulations ensure that nutrients remain available for longer durations, unlike the rapid depletion associated with traditional fertilizers.

The integration of nanotechnology into foliar fertilization represents a significant step forward in sustainable agriculture. Research is ongoing to optimize nanoparticle formulations, ensuring their safety, cost-effectiveness, and scalability for large-scale applications [14]. The development of biodegradable and eco-friendly nanoparticles is also a priority to further reduce any potential environmental risks. The foliar application of nanobiofertilizers offers a highly efficient and sustainable solution to modern agricultural challenges. By leveraging the unique properties of nanotechnology, this targeted approach enhances nutrient delivery, improves crop performance, and minimizes environmental impact, paving the way for a more resilient agricultural future.

4 Mechanisms of Nanobiofertilizers in Foliar Applications

Foliar application of fertilizers is a precise agricultural technique that involves delivering nutrients directly onto plant leaves, bypassing the soil [15]. As shown in Fig. 4, this method enables rapid and efficient nutrient absorption through the plant's cuticle and stomata, reducing nutrient wastage and environmental contamination. Unlike traditional soil-based fertilization, foliar application ensures that nutrients are delivered exactly where they are needed, particularly during critical growth stages when deficiencies could negatively impact plant productivity

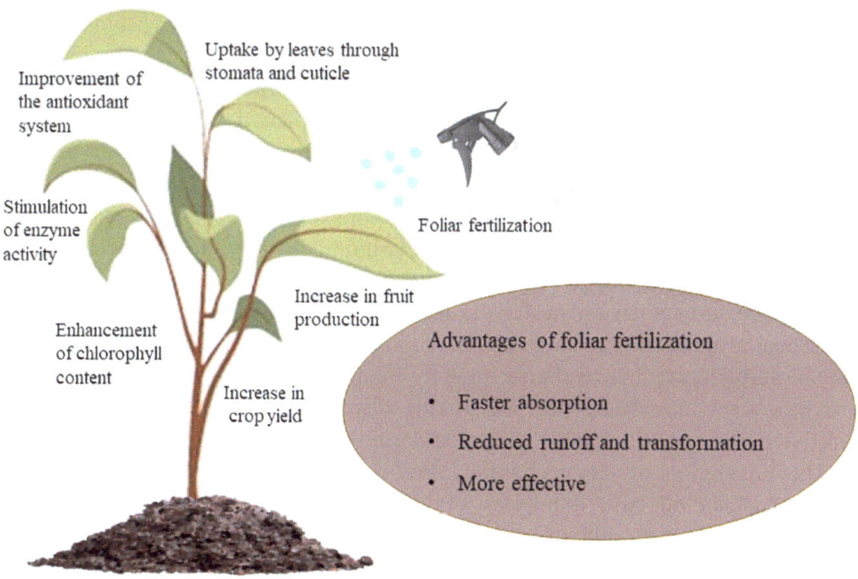

Fig. 4 Highlights of swift and efficient nanonutrient absorption through the plant's cuticle and stomata, minimizing nutrient loss and environmental pollution. (This is a reprinted image of Ref. [29] with permission)

[15]. The advent of nanotechnology has further enhanced the efficacy of foliar fertilization with the development of nanobiofertilizers. These advanced formulations provide controlled and sustained nutrient release, leveraging the unique properties of nanoparticles such as their high surface area, nanoscale dimensions, and enhanced reactivity, which collectively improve nutrient adhesion to leaf surfaces and enhance bioavailability [16].

At the molecular level, foliar absorption begins with the interaction of nanoparticles with the plant cuticle, a waxy barrier that protects leaves. Nanoparticles can penetrate this layer by forming transient openings or integrating with cuticular lipids, allowing nutrients to reach the epidermal cells. Stomata, the microscopic pores on leaf surfaces, serve as an additional entry route, where nanoparticles pass through these pores into the apoplast, the extracellular space surrounding plant cells [17]. From the apoplast, nutrients diffuse into the symplast, the intracellular cytoplasm of plant cells, via plasmodesmata, facilitating their systemic transport throughout the plant. Once inside the plant, nanoparticles interact with the cell wall and plasma membrane [17]. Certain nanoparticles coated with bioactive compounds bind to receptor proteins on the plasma membrane, triggering endocytosis a process where the cell engulfs the nanoparticles, allowing them to enter the cytoplasm for metabolic utilization.

Nanobiofertilizers further influence the plant's internal processes by enhancing enzymatic activity and activating nutrient-specific pathways. For example, zinc nanoparticles promote the activity of enzymes involved in protein synthesis and auxin metabolism, while phosphorus nanoparticles facilitate root development and energy transfer by improving phosphorus availability [18]. These nanoscale nutrients also participate in key molecular processes, such as chlorophyll biosynthesis and photosynthesis, directly boosting plant growth. Additionally, nanoparticles can modulate hormonal regulation, increasing levels of gibberellins, cytokinins, and other growth-promoting hormones, ensuring optimal plant development.

Another critical molecular aspect of nanobiofertilizers is their ability to activate plant defense mechanisms [19]. Beneficial microbes integrated into these formulations stimulate the expression of systemic acquired resistance (SAR) genes, while nanoparticles coated with elicitors such as jasmonic acid or salicylic acid analogs enhance the plant's immune response to biotic stresses [20]. Silicon nanoparticles, for instance, strengthen cell walls and activate stress-responsive genes, improving resilience against abiotic stresses like drought and salinity [21]. Furthermore, nanobiofertilizers reduce oxidative stress by enhancing the production of antioxidant enzymes such as catalase and superoxide dismutase, which neutralize harmful reactive oxygen species (ROS) generated during environmental stress conditions.

The integration of nanotechnology in foliar fertilization not only optimizes nutrient uptake but also influences key molecular pathways that regulate plant growth, development, and stress responses [19]. By targeting these mechanisms with unparalleled precision, nanobiofertilizers represent a transformative solution to modern agricultural challenges, enabling sustainable and efficient crop management with minimal environmental impact [21].

5 Stress Tolerance and Nanofertilizers in Foliar Applications

In the face of climate change and escalating environmental stressors, sustainable agricultural practices are vital for ensuring food security. Stress tolerance in crops refers to their ability to withstand biotic (e.g., pests and diseases) and abiotic stresses (e.g., drought, salinity, and extreme temperatures) without significant loss in productivity [22]. Nanotechnology, particularly the use of nanofertilizers in foliar applications, has emerged as a revolutionary approach to boost stress tolerance and enhance crop productivity in an eco-friendly manner. Stress tolerance in plants is a complex physiological and biochemical process. Under abiotic stresses, plants suffer from oxidative stress due to an imbalance between reactive oxygen species (ROS) production and scavenging [23]. These ROS cause damage to cellular components, reducing photosynthetic efficiency, nutrient uptake, and growth. Similarly, biotic stresses, such as pathogens and pests, compromise plant health by interfering with vital metabolic pathways. To address these challenges, enhancing stress tolerance mechanisms, such as antioxidative defense systems, stress-responsive signaling, and nutrient assimilation, is critical [23].

Nanofertilizers are particularly effective in mitigating stress-induced damage in plants. Here are the key ways they contribute:

Improved Antioxidant Defense Nanofertilizers can enhance the antioxidant capacity of plants by promoting the activity of enzymes such as superoxide dismutase, catalase, and peroxidase. These enzymes neutralize ROS, protecting cells from oxidative damage.

Nutrient Enrichment Essential micronutrients like zinc, iron, and boron, delivered in nanoscale formulations, play a crucial role in stress responses. For example, zinc nanoparticles boost stress resistance by regulating stomatal opening and enhancing photosynthetic efficiency.

Stress-Responsive Signaling Nanofertilizers can trigger stress-responsive signaling pathways in plants, promoting the expression of stress-tolerance genes. For instance, silicon nanoparticles enhance drought tolerance by improving water-use efficiency and strengthening cell walls.

Mitigation of Salinity and Drought Stress Nanofertilizers improve osmotic adjustment and ionic balance, reducing the adverse effects of salinity and water scarcity. Nano-calcium and nano-silicon, for example, play a role in maintaining cell membrane integrity under stress.

6 Environmental and Economic Benefits

The environmental benefits of using nanobiofertilizers in foliar applications are profound, addressing several pressing concerns in modern agriculture (Fig. 5). One of the most significant advantages is the reduction in nutrient loss, a common

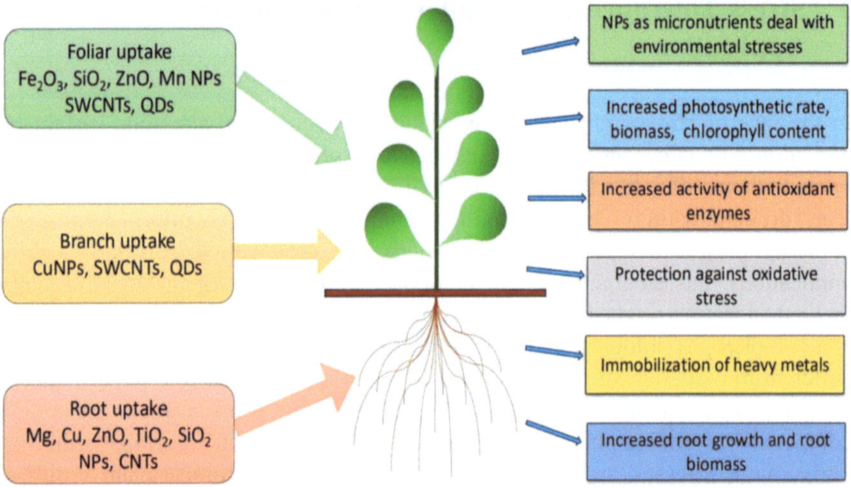

Fig. 5 Benefits of different nanoparticles utilized as fertilizers to enhance plant growth. (This is a reprinted image of Ref. [30] with permission)

issue with conventional fertilizers that often results in leaching and runoff [24]. By employing nanobiofertilizers with controlled-release properties, nutrients are delivered directly to plant tissues in a targeted and efficient manner, minimizing the likelihood of excess nutrients entering water bodies. This prevents problems such as eutrophication, where nutrient runoff causes algal blooms that deplete oxygen in aquatic ecosystems, leading to significant ecological damage. Additionally, the use of nanobiofertilizers helps to preserve soil health by reducing the build-up of chemical residues, which can otherwise alter soil pH and disrupt the balance of microbial communities critical to maintaining soil fertility [25].

Nanobiofertilizers also play a pivotal role in conserving water resources. The high efficiency of nutrient delivery means that smaller quantities of fertilizers are required, reducing the need for irrigation to dilute and distribute nutrients. This efficiency is particularly beneficial in water-scarce regions, where agriculture must optimize every drop of water used. Furthermore, nanobiofertilizers often include biological agents, such as nitrogen-fixing bacteria or phosphate-solubilizing microbes, which naturally enhance nutrient availability and uptake [2]. These microbes actively participate in nutrient cycling, breaking down organic matter and releasing essential nutrients like nitrogen and phosphorus into forms readily available for plant absorption. By reducing reliance on synthetic fertilizers, these biological processes not only improve soil health but also promote biodiversity within the soil ecosystem.

Economically, nanobiofertilizers offer a sustainable and cost-effective alternative for farmers. The controlled-release mechanisms inherent in their design reduce the frequency of fertilizer applications, directly lowering labor and operational costs. This feature is particularly advantageous for smallholder farmers in resource-constrained regions, where labor and inputs often constitute a significant portion of production expenses. Additionally, the higher nutrient-use efficiency of nanobiofertilizers ensures that a greater proportion of applied nutrients are absorbed by the plant, reducing waste and maximizing returns on investment [26]. Over time, this enhanced efficiency can lead to increased crop yields, providing farmers with higher income while using fewer resources.

From a broader ecological perspective, nanobiofertilizers contribute to sustainable farming practices by aligning with global goals for environmental preservation and climate change mitigation. The improved efficiency of these formulations reduces the greenhouse gas emissions associated with the production, transportation, and application of traditional fertilizers. Conventional nitrogen-based fertilizers, for example, are significant contributors to nitrous oxide emissions, a potent greenhouse gas. By minimizing the required quantity of fertilizers and enhancing nutrient uptake, nanobiofertilizers lower these emissions, helping to reduce agriculture's carbon footprint [10].

Moreover, the use of nanobiofertilizers can support regenerative agriculture by fostering resilient ecosystems [2]. The biological agents included in these formulations, such as rhizobacteria and mycorrhizal fungi, enhance the natural processes that sustain soil fertility. For instance, nitrogen-fixing bacteria convert atmospheric nitrogen into ammonia, a form usable by plants, while phosphate-solubilizing microbes release phosphorus from insoluble compounds in the soil. These processes reduce dependency on synthetic inputs and encourage the adoption of agroecological practices, where the focus shifts from chemical-intensive farming to systems that work harmoniously with nature.

The cumulative impact of these benefits reduced pollution, conservation of soil and water resources, improved biodiversity, and lower greenhouse gas emissions highlights the potential of nanobiofertilizers to revolutionize agriculture [2]. By addressing both environmental and economic challenges, these innovative formulations pave the way for a more sustainable and resilient agricultural future, benefiting farmers, ecosystems, and society as a whole.

7 Regulatory and Safety Aspects of Nanobiofertilizers

The introduction of nanobiofertilizers into modern agricultural practices has provided a promising avenue to enhance productivity and sustainability. However, their widespread adoption comes with critical regulatory and safety challenges that must be addressed to ensure long-term viability and public trust. The complexities of regulating nanotechnology in agriculture, coupled with safety concerns for both the environment and human health, underscore the need for rigorous frameworks and collaborative efforts [32].

Regulatory Standards One of the most significant challenges associated with nanobiofertilizers is the lack of a globally harmonized regulatory framework. Current policies governing the development, approval, and use of these innovative materials differ widely across regions. This disparity creates an uncertain landscape for manufacturers, researchers, and end-users, complicating the process of bringing nanobiofertilizers to market [33]. In some regions, such as the European Union, regulatory authorities have implemented stringent guidelines that require detailed environmental impact assessments for nanotechnology applications in agriculture. These include specific tests to evaluate nanoparticle behavior in soil, water, and plant systems. Meanwhile, other countries focus predominantly on efficacy studies, leaving significant gaps in the assessment of long-term environmental and health impacts. The absence of standardized definitions further exacerbates the issue. For instance, while some regulatory bodies classify nanobiofertilizers based on particle size alone, others emphasize their unique physicochemical properties or interactions with biological systems [34]. Addressing this regulatory disparity requires international collaboration to establish a unified framework. Harmonized standards could streamline approval processes, enhance transparency, and promote confidence among stakeholders, including farmers, policymakers, and consumers. Such a framework would need to balance innovation and safety, ensuring that the potential of nanobiofertilizers is harnessed responsibly.

Human Health Concerns The safety of farmworkers and consumers is a paramount consideration in the adoption of nanobiofertilizers. Foliar applications, which involve spraying nutrients directly onto plant leaves, have gained popularity due to their efficiency. However, this method can aerosolize nanoparticles, increasing the likelihood of human exposure through inhalation or dermal contact [35]. Inhaled nanoparticles could potentially penetrate deep into the respiratory system, leading to unknown health effects over prolonged periods. Similarly, residues of nanoparticles on food crops could pose ingestion risks if not adequately addressed. Although preliminary studies suggest that many nanobiofertilizers are designed to be non-toxic, comprehensive toxicological evaluations are needed to confirm their safety across diverse exposure scenarios [24]. The establishment of robust safety protocols for handling, applying, and consuming crops treated with nanobiofertilizers is crucial. Protective equipment, training programs for farmworkers, and stringent residue monitoring systems can play a key role in minimizing risks. Furthermore, transparency in safety testing and open communication about potential hazards will be instrumental in gaining public trust.

Nano-Biodegradability A promising avenue for addressing safety concerns lies in the development of biodegradable or naturally occurring nanomaterials. Materials such as silica, chitosan, or bio-based polymers have shown potential for creating environmentally benign nanobiofertilizers that degrade into harmless byproducts over time. These materials reduce the risk of long-term accumulation in soils and ecosystems, addressing one of the primary environmental challenges associated with nanotechnology in agriculture. Biodegradable nanobiofertilizers also offer the advantage of aligning with sustainable agriculture practices [36]. By minimizing persistence and toxicity, they contribute to a circular economy where resources are efficiently utilized and environmental impacts are minimized. Future research should focus on optimizing the design, production, and application of these materials to ensure their scalability and cost-effectiveness.

Labeling and Certification Clear and transparent labeling practices are vital to ensuring the safe and informed use of nanobiofertilizers. Farmers and consumers need access to accurate information about product composition, application methods, and potential risks. Certification systems can further enhance trust by validating the safety and efficacy of nanobiofertilizers through rigorous testing and quality control processes [37]. Certification programs should encompass both regulatory compliance and sustainability metrics, such as environmental impact and biodegradability. Such initiatives would provide a competitive advantage to manufacturers adhering to high standards, encouraging innovation in safer and more sustainable products. Additionally, public awareness campaigns can help demystify nanotechnology and foster acceptance among farmers and consumers, who may be unfamiliar with the concept. The regulatory and safety aspects of nanobiofertilizers represent a complex but critical area of focus in advancing sustainable agriculture. Harmonizing regulatory frameworks, addressing environmental and human health concerns, promoting the use of biodegradable materials, and establishing robust labeling and certification systems are essential steps in overcoming current challenges [38]. By proactively addressing these issues, stakeholders can unlock the full potential of nanobiofertilizers, ensuring their benefits are realized without compromising safety or environmental integrity. Collaboration between governments, researchers, and the agricultural industry will be key to navigating this dynamic landscape and paving the way for a more sustainable future in global food production.

8 Challenges and Future Prospects

Nanobiofertilizers for foliar application represent a cutting-edge approach to improving agricultural sustainability and efficiency. However, their widespread adoption faces several challenges. As shown in Fig. 6, the challenges of nanoparticles in the foliar application, include reduced root and shoot growth, decreased plant biomass, upregulation of heat shock proteins, and disruption of antioxidant enzyme balance.

One primary issue is the high production cost associated with these advanced formulations. The development of nanobiofertilizers requires sophisticated technologies, nanoscale materials, and specialized manufacturing processes, which significantly increase their price compared to conventional fertilizers [2]. This cost factor can limit their accessibility, particularly for smallholder farmers in resource-limited regions who may already struggle with input affordability. Another challenge lies in the scalability and commercialization of nanobiofertilizers. While laboratory studies have demonstrated their efficacy, scaling up production to meet global agricultural demands poses logistical and technical difficulties. Maintaining product quality and

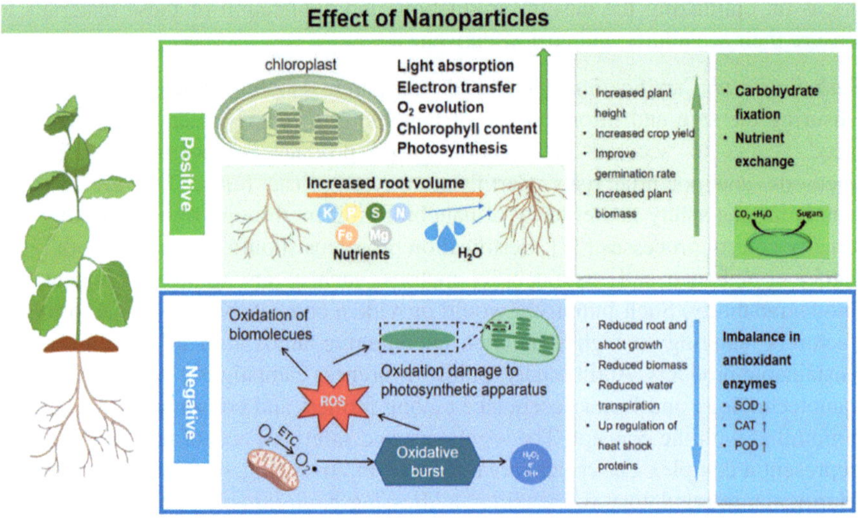

Fig. 6 Illustration depicting the challenges of nanoparticle foliar application, including reduced root and shoot growth, decreased plant biomass, upregulation of heat shock proteins, and disruption of antioxidant enzyme balance. (This is a reprinted image of Ref. [31] with permission)

consistency during large-scale production, coupled with the absence of established distribution networks, further complicates their market availability. Additionally, there are concerns regarding the long-term safety and environmental implications of nanomaterials. Although nanobiofertilizers are designed to minimize environmental risks, the behavior and persistence of nanoparticles in soil, water, and ecosystems are not yet fully understood. Comprehensive risk assessments and regulatory frameworks are necessary to ensure their safe use.

Despite these challenges, the future prospects of nanobiofertilizers in foliar applications are highly promising [27]. Continued advancements in nanotechnology and material science are expected to drive down production costs, making these fertilizers more accessible to farmers worldwide. Research into biodegradable and eco-friendly nanoparticles can address environmental safety concerns, paving the way for their sustainable integration into farming practices. Furthermore, interdisciplinary efforts combining nanotechnology with precision agriculture and data-driven farming methods can enhance the targeted delivery of nutrients, further improving efficiency and reducing waste [27].

Future developments may also involve the incorporation of smart delivery systems into nanobiofertilizers, enabling real-time nutrient release based on plant requirements [27]. Such innovations would maximize resource use and minimize the risk of over-fertilization. Moreover, policy support, investment in research and development, and public-private partnerships will be instrumental in scaling up production and establishing robust supply chains. Challenges such as cost, scalability, and safety concerns must be addressed, the potential of nanobiofertilizers to revolutionize agriculture is undeniable [24]. By overcoming these hurdles, nanobiofertilizers for foliar applications can play a pivotal role in achieving sustainable agricultural practices, enhancing crop productivity, and reducing environmental impact. Their development and adoption signify a forward-thinking approach to meeting the growing food demands of a rapidly increasing global population.

9 Conclusion

Nanobiofertilizers for foliar application offer a revolutionary approach to sustainable agriculture by combining nanotechnology with bio-based inputs. They enhance nutrient-use efficiency, minimize environmental impact, and improve crop productivity through precise, controlled nutrient delivery directly to plant leaves. By reducing nutrient loss and dependency on chemical fertilizers, they address critical challenges like soil degradation and pollution. Despite hurdles such as high production costs and scalability, ongoing research, and technological advancements are paving the way for their wider adoption. As agriculture faces growing demands and environmental pressures, nanobiofertilizers stand poised to play a crucial role in achieving global food security sustainably.

References

1. Mrabet, R. (2023). Sustainable agriculture for food and nutritional security. In *Sustainable agriculture and the environment* (pp. 25–90). Academic Press.
2. Sharma, B., Tiwari, S., Kumawat, K. C., & Cardinale, M. (2023). Nano-biofertilizers as bio-emerging strategies for sustainable agriculture development: Potentiality and their limitations. *Science of the Total Environment, 860*, 160476.
3. Asadu, C. O., Ezema, C. A., Ekwueme, B. N., Onu, C. E., Onoh, I. M., Adejoh, T., & Emmanuel, U. O. (2024). Enhanced efficiency fertilizers: Overview of production methods, materials used, nutrients release mechanisms, benefits and considerations. *Environmental Pollution and Management, 1*, 32.
4. Hossain, A., Krupnik, T. J., Timsina, J., Mahboob, M. G., Chaki, A. K., Farooq, M., & Hasanuzzaman, M. (2020). Agricultural land degradation: Processes and problems undermining future food security. In *Environment, climate, plant and vegetation growth* (pp. 17–61). Springer.
5. Kumar, S. (2020). Abiotic stresses and their effects on plant growth, yield and nutritional quality of agricultural produce. *International Journal of Food Science and Agriculture, 4*(4), 367–378.
6. Hemathilake, D. M. K. S., & Gunathilake, D. M. C. C. (2022). Agricultural productivity and food supply to meet increased demands. In *Future foods* (pp. 539–553). Academic Press.
7. Pahalvi, H. N., Rafiya, L., Rashid, S., Nisar, B., & Kamili, A. N. (2021). *Chemical fertilizers and their impact on soil health. Microbiota and biofertilizers* (Vol 2: Ecofriendly tools for reclamation of degraded soil environs) (pp. 1–20). Springer.
8. Pudhuvai, B., Koul, B., Das, R., & Shah, M. P. (2024). Nano-Fertilizers (NFs) for resurgence in nutrient use efficiency (NUE): A sustainable agricultural strategy. *Current Pollution Reports, 11*(1), 1.

9. Vejan, P., Khadiran, T., Abdullah, R., & Ahmad, N. (2021). Controlled release fertilizer: A review on developments, applications and potential in agriculture. *Journal of Controlled Release, 339*, 321–334.
10. Mahapatra, D. M., Satapathy, K. C., & Panda, B. (2022). Biofertilizers and nanofertilizers for sustainable agriculture: Phycoprospects and challenges. *Science of the Total Environment, 803*, 149990.
11. Madlala, N. C., Khanyile, N., & Masenya, A. (2024). Examining the correlation between the inorganic nano-fertilizer physical properties and their impact on crop performance and nutrient uptake efficiency. *Nanomaterials, 14*(15), 1263.
12. Wu, W., & Ma, B. (2015). Integrated nutrient management (INM) for sustaining crop productivity and reducing environmental impact: A review. *Science of the Total Environment, 512*, 415–427.
13. Rahman, S. U., Wang, X., Shahzad, M., Bashir, O., Li, Y., & Cheng, H. (2022). A review of the influence of nanoparticles on the physiological and biochemical attributes of plants with a focus on the absorption and translocation of toxic trace elements. *Environmental Pollution, 310*, 119916.
14. Husted, S., Minutello, F., Pinna, A., Le Tougaard, S., Møs, P., & Kopittke, P. M. (2023). What is missing to advance foliar fertilization using nanotechnology? *Trends in Plant Science, 28*(1), 90–105.
15. Malhotra, H., Pandey, R., Sharma, S., & Bindraban, P. S. (2020). Foliar fertilization: Possible routes of iron transport from leaf surface to cell organelles. *Archives of Agronomy and Soil Science, 66*(3), 279–300.
16. Francis, D. V., Abdalla, A. K., Mahakham, W., Sarmah, A. K., & Ahmed, Z. F. (2024). Interaction of plants and metal nanoparticles: Exploring its molecular mechanisms for sustainable agriculture and crop improvement. *Environment International*, 108859.
17. Rani, S., Kumari, N., & Sharma, V. (2023). Uptake, translocation, transformation and physiological effects of nanoparticles in plants. *Archives of Agronomy and Soil Science, 69*(9), 1579–1599.
18. Nath, S., Dey, S., Kundu, R., & Paul, S. (2024). Phosphate and zinc interaction in soil and plants: A reciprocal cross-talk. *Plant Growth Regulation, 104*, 1–25.
19. Adisa, I. O., Pullagurala, V. L. R., Peralta-Videa, J. R., Dimkpa, C. O., Elmer, W. H., Gardea-Torresdey, J. L., & White, J. C. (2019). Recent advances in nano-enabled fertilizers and pesticides: A critical review of mechanisms of action. *Environmental Science: Nano, 6*(7), 2002–2030.
20. Zhang, P., Jiang, Y., Schwab, F., Monikh, F. A., Grillo, R., White, J. C., et al. (2024). Strategies for enhancing plant immunity and resilience using nanomaterials for sustainable agriculture. *Environmental Science & Technology, 58*, 9051.
21. Mir, R. A., Bhat, B. A., Yousuf, H., Islam, S. T., Raza, A., Rizvi, M. A., et al. (2022). Multidimensional role of silicon to activate resilient plant growth and to mitigate abiotic stress. *Frontiers in Plant Science, 13*, 819658.
22. Ahmadian, K., Jalilian, J., & Pirzad, A. (2021). Nano-fertilizers improved drought tolerance in wheat under deficit irrigation. *Agricultural Water Management, 244*, 106544.
23. Kaya, C., & Ashraf, M. (2024). Foliar fertilization: A potential strategy for improving plant salt tolerance. *Critical Reviews in Plant Sciences, 43*(2), 94–115.
24. Wahab, A., Muhammad, M., Ullah, S., Abdi, G., Shah, G. M., Zaman, W., & Ayaz, A. (2024). Agriculture and environmental management through nanotechnology: Eco-friendly nanomaterial synthesis for soil-plant systems, food safety, and sustainability. *Science of the Total Environment, 171862*, 171862.
25. El-Ramady, H., Brevik, E. C., Fawzy, Z. F., Elsakhawy, T., Omara, A. E. D., Amer, M., et al. (2022). Nano-restoration for sustaining soil fertility: A pictorial and diagrammatic review article. *Plants, 11*(18), 2392.

26. Irewale, A. T., Dimkpa, C. O., Agunbiade, F. O., Oyetunde, O. A., Elemike, E. E., & Oguzie, E. E. (2024). Unlocking sustainable agricultural development in Africa via bio-nanofertilizer application-challenges, opportunities and prospects. *Scientific African, 25*, e02276.
27. Biswal, D. (2024). Nanobiofertilizers: The futuristic tools for nutrient management in plants. In *Soil bacteria: Biofertilization and soil health* (pp. 207–252). Springer.
28. Zahra, Z., Habib, Z., Hyun, H., & Shahzad, H. M. A. (2022). Overview on recent developments in the design, application, and impacts of nanofertilizers in agriculture. *Sustainability, 14*(15), 9397.
29. Ding, Y., Zhao, W., Zhu, G., Wang, Q., Zhang, P., & Rui, Y. (2023). Recent trends in foliar Nanofertilizers: A review. *Nanomaterials, 13*(21), 2906.
30. Hashmi, K., Gupta, S., Mishra, P., Khan, T., & Joshi, S. (2024). The vital role of nanoparticles in enhancing plant growth and development. *Engineering Proceedings, 67*(1), 48.
31. Wang, X., Xie, H., Wang, P., & Yin, H. (2023). Nanoparticles in plants: Uptake, transport and physiological activity in leaf and root. *Materials, 16*(8), 3097.
32. Khattak, W. A., Ullah, M. W., Manan, S., Islam, S. U., Khattak, W. A., & Ul-Islam, M. (2024). Emerging applications and future trends of agri-nanotechnology. *Revolutionizing Agriculture: A Comprehensive Exploration of Agri-Nanotechnology*, 429–458.
33. Mohamed Udin, Z., Ahmad, H., Zulhumadi, F., Hasnan, N., & Jamaludin, N. (2011). Collaborative supply chain management of nanotechnology industry in Malaysia.
34. Soni, S. K., Dogra, S., Sharma, A., Thakur, B., Yadav, J., Kapil, A., & Soni, R. (2024). Nanotechnology in agriculture: Enhancing crop productivity with sustainable nano-fertilizers and nano-biofertilizers. *Journal of Soil Science and Plant Nutrition, 24*, 1–34.
35. Sajid, M., Ilyas, M., Basheer, C., Tariq, M., Daud, M., Baig, N., & Shehzad, F. (2015). Impact of nanoparticles on human and environment: Review of toxicity factors, exposures, control strategies, and future prospects. *Environmental Science and Pollution Research, 22*, 4122–4143.
36. Zulfiqar, F., Navarro, M., Ashraf, M., Akram, N. A., & Munné-Bosch, S. (2019). Nanofertilizer use for sustainable agriculture: Advantages and limitations. *Plant Science, 289*, 110270.
37. Bongomin, O., Yemane, A., Kembabazi, B., Malanda, C., Chikonkolo Mwape, M., Sheron Mpofu, N., & Tigalana, D. (2020). Industry 4.0 disruption and its neologisms in major industrial sectors: A state of the art. *Journal of Engineering, 2020*(1), 8090521.
38. Äkräs, L., Ballardini, R. M., Alen-Savikko, A., & Seppälä, J. (2022). Towards sustainable and circular practices with plastics: Exploring the potential of law and governance tools based on holistic and harmonized life cycle assessment. *Retfærd, 2022*(1), 43–63.

Unveiling the Future of Hydroponics and Aeroponics Through Nanotechnology

Girish B. Mahajan, Vishal Gangadhar Warke, Ramesh S. Chaughule, and Amruta V. Shelar

Contents

G. B. Mahajan (✉)
Microbiology Department, HiMedia Laboratories, Thane, Maharashtra, India
e-mail: gmahajan@himedialabs.com

V. G. Warke
Cell Biology Department, HiMedia Laboratories, Thane, Maharashtra, India

R. S. Chaughule
Department of Chemistry, Ramnarain Ruia Autonomous College, Mumbai, India

A. V. Shelar
Department of Technology, Savitribai Phule Pune University, Pune, MH, India

R. S. Chaughule, A. V. Shelar (eds.), *Nanotechnology in Agriculture*,
https://doi.org/10.1007/978-3-031-97544-8_8

1 Introduction

Hydroponics, a method of cultivating plants without soil, has emerged as a cornerstone of sustainable agriculture in recent decades. By utilizing nutrient-rich water solutions, hydroponic systems offer unparalleled control over plant growth conditions, enabling higher yields, reduced resource usage, and the ability to grow crops in areas with poor soil quality. As urbanization intensifies and arable land becomes scarcer, hydroponics is increasingly recognized as a viable solution to meet the global demand for food while minimizing environmental impact [1].

Aeroponics, another innovative soilless farming technique, complements hydroponics using misted nutrient solutions to nourish plant roots in the air [34]. This method further optimizes resource use, offering precise control over nutrient and water delivery, and significantly reduces water consumption compared to conventional hydroponics systems. Aeroponics also promotes faster growth rates and higher crop yields, making it a promising solution for sustainable food production in areas with limited resources or arable land [34].

The escalating global population and diminishing arable land pose unprecedented challenges to modern agriculture, demanding innovative and sustainable solutions to meet growing food demands. As shown in Fig. 1, hydroponics and aeroponics have emerged as viable approaches to maximize crop production with minimal resource utilization. By providing controlled environments and precise nutrient delivery, these techniques have revolutionized farming practices, especially in regions with adverse climatic conditions or limited fertile soil. However, their full potential is constrained by challenges such as nutrient optimization, physical form and properties of nutrients, environmental stress management, and real-time monitoring [2, 3].

Despite their numerous advantages, current soilless farming systems face significant challenges that limit their efficiency and scalability. Nutrient delivery and management can be imprecise, leading to suboptimal plant growth. Pest and pathogen control in soil-less environments remains a persistent issue, while the high initial infrastructure costs often deter widespread adoption. Additionally, resource utilization, including water and fertilizers, requires further optimization to enhance sustainability. Addressing these challenges is critical to unlocking the full potential of hydroponics and aeroponics farming [2, 3].

Nanotechnology, the science of manipulating matter at the atomic and molecular scale, offers transformative solutions to these challenges. In agriculture,

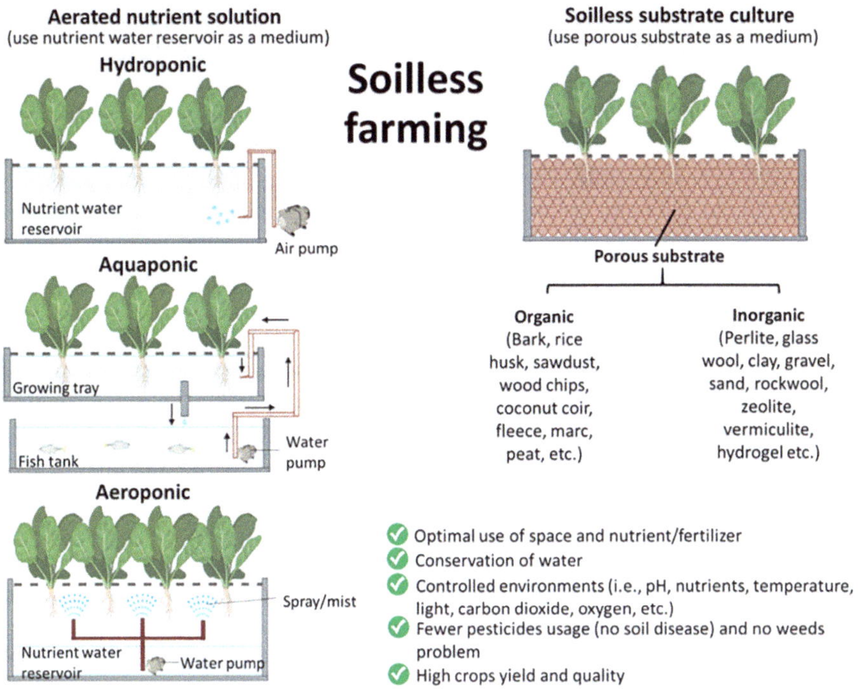

Fig. 1 This illustration highlights the various types of soilless farming methods, including hydroponics, aeroponics, and aquaponics, showcasing their unique features and benefits. It emphasizes the advantages of these systems, such as efficient resource utilization, enhanced plant growth, and reduced environmental impact, making them viable solutions for sustainable agriculture. (This is a reprinted image of Ref. [35] with permission)

nanotechnology has already demonstrated its potential to revolutionize practices through innovations such as nanopesticides, nanosensors, and nano-fertilizers. Integrating nanotechnology into hydroponics, and aeroponics systems is particularly promising, as it can address limitations in nutrient delivery, pest management, and resource efficiency. For example, nano-nutrients can enhance the bioavailability of essential elements, while nanosensors enable real-time monitoring of plant health and environmental conditions [4]. Furthermore, advanced nanomaterials can be utilized for water purification and recycling, ensuring that hydroponics systems remain resource-efficient.

By merging hydroponics, aeroponics, and nanotechnology, we can envision a future where agricultural practices are more productive, sustainable, and resilient to climate change. This integration has the potential to reshape urban farming, enabling fresh produce to be cultivated closer to consumption centers while reducing reliance on traditional farming methods. Moreover, it represents a critical step toward achieving global food security in the face of growing populations and environmental challenges [5].

Nanotechnology, characterized by the manipulation and application of materials at the nanoscale, offers groundbreaking opportunities to address these limitations. Nanomaterials exhibit unique physicochemical properties such as high surface area-to-volume ratio and enhanced reactivity, enabling their application in hydroponics and aeroponics systems for nutrient delivery, crop resilience, and resource management [6, 7]. For example, nanoparticle-based fertilizers allow for the controlled release of essential nutrients, enhancing plant growth while reducing environmental impact [8]. Additionally, nano-sensing technologies facilitate the real-time detection of nutrient imbalances and environmental stressors, improving overall crop health and productivity [9, 10].

Beyond cultivation, nanotechnology's applications extend to packaging and post-harvest management. Nanoparticle-based coatings and sensors can extend the shelf life of produce, ensuring longer freshness and reducing spoilage during transportation and storage [11]. Such advancements are critical in reducing food waste and enhancing food security globally.

This chapter explores the multifaceted role of nanotechnology in advancing hydroponics and aeroponics systems. It researches key innovations, including nano-enabled nutrient delivery, disease management, and water treatment technologies, and discusses their implications for sustainable agriculture. By unveiling the future of hydroponics and aeroponics through the lens of nanotechnology, this chapter aims to inspire new research and applications in this ground-breaking interdisciplinary field.

2 Optimized Nutrient Delivery

2.1 Role of Nano-fertilizers

Nano-fertilizers represent a cutting-edge innovation in agriculture, leveraging nanotechnology to enhance nutrient efficiency, reduce waste, and minimize environmental impact. Their application in hydroponics is particularly promising, as these systems depend on precise nutrient management to maximize plant growth. Nano-fertilizers are engineered nanoparticles containing essential nutrients that enhance nutrient availability and plant delivery efficiency. In hydroponics, where precise nutrient control is crucial, nano-fertilizers offer a superior alternative to conventional fertilizers. Their nanoscale size, high surface area, and unique physicochemical properties enable them to interact more effectively with plant roots, increasing nutrient uptake and reducing losses [12].

2.1.1 Enhanced Nutrient Absorption

Nano-fertilizers are more readily available for plant uptake due to their smaller size (higher surface-to-volume ratio than conventional fertilizers), higher reactivity, and increased bioavailability. They penetrate root pores efficiently and release nutrients gradually and consistently, ensuring plants have access to essential elements over extended periods without the risk of over-application.

2.1.2 Controlled and Targeted Release

Conventional fertilizers often leach into soil or water, leading to nutrient runoff and eutrophication. Nano-fertilizers minimize these losses, making nutrient use more efficient. Nano-fertilizers enable sustained and controlled nutrient release, reducing over-fertilization risks. The nutrients are delivered directly to plant roots in precise doses, minimizing waste. Many nutrients in conventional fertilizers have limited solubility. However, nano-fertilizers improve solubility, making nutrients more accessible to hydroponics systems. Many nutrients in conventional fertilizers have limited solubility, but nano-fertilizers improve solubility, making nutrients more accessible to hydroponics systems. Singh et al. [13] have shown a controlled release system for enhancing nutrient use efficiency using nano fertilizers. They release nutrients gradually and consistently, ensuring plants have access to essential elements over extended periods without the risk of over-application.

2.1.3 Higher Crop Yields

Enhanced nutrient use efficiency (NUE) with nano-fertilizers leads to faster plant growth, improved biomass, and better-quality produce. Nano-fertilizers, due to their significantly higher surface area-to-volume ratios compared to conventional fertilizers, improve solubility and interaction with plant root systems. Nutrients are encapsulated in nanocarriers like lipid nanoparticles, polymeric nanocarriers, or hydrogels and enable slow and sustained release. The encapsulation protects nutrients from degradation, ensuring a longer nutrient supply to plant roots.

2.1.4 Nutrient Uptake Via Root Pores

Nanoparticles can penetrate root epidermal cells through ion channels and aquaporins, bypassing nutrient diffusion barriers that limit conventional fertilizers. For instance, nano-chelates facilitate the absorption of micronutrients like Fe, Mn, and Cu by enhancing their mobility and bioavailability in root zones [14]. Some nanoparticles improve ionic exchange between nutrient solutions and root systems, boosting uptake efficiency. For example, nano-zeolites enhance the retention and release of potassium and nitrogen ions, which are vital for hydroponics [15]. Nano-hydrogels act as carriers that store water and nutrients, gradually releasing them into hydroponics systems, thus maintaining optimal nutrient concentrations over time.

2.1.5 Nutrient Solubility and Targeted Delivery

Many nutrients in conventional fertilizers have limited solubility. Nano-fertilizers improve solubility, making nutrients more accessible to hydroponics systems. They improve solubility in water-based hydroponics systems due to their high surface-to-volume ratio, For example, nano-ZnO and nano-Fe_3O_4 particles exhibit higher dissolution rates compared to their bulk counterparts, enhancing zinc and iron availability in the nutrient solution [12]. Nutrients are encapsulated in nanocarriers like lipid nanoparticles, polymeric nanocarriers, or hydrogels thus enabling slow and sustained release. Usually, encapsulation protects nutrients from degradation, thus ensuring a longer nutrient supply to plant roots.

Functionalized nano-fertilizers can be engineered to bind to root-specific receptors, enhancing nutrient uptake. Chitosan nanoparticles, for example, are widely used for targeted delivery because of their biocompatibility and root-adhesive properties [13].

3 Real-Time Monitoring of Water Quality, Nutrient Levels, and pH

Nano-sensors are advanced tools that leverage nanomaterials like carbon nanotubes (CNTs), graphene, quantum dots, and metal oxide nanoparticles to detect and measure changes in hydroponics environments with high sensitivity, specificity, and speed. These sensors play a critical role in ensuring optimal growth conditions for plants in hydroponics [16, 17]. The key functions of Nano-sensors are:

3.1 Monitoring Water Quality

Nano-sensors can detect contaminants, heavy metals, and dissolved oxygen levels in water. For example, carbon nanotube-based sensors detect metal ions like lead (Pb), cadmium (Cd), and arsenic (As) at trace concentrations to maintain clean water for plant roots.

3.2 Measuring Nutrient Levels

Nano-sensors provide real-time data on nutrient concentrations (N, P, K, Fe, Mg, etc.) to ensure precise nutrient delivery. When coated with an ion-selective membrane, nanosensors can monitor the ionic concentration of fertilizers, ensuring plants receive optimal nutrient doses. Graphene oxide-based sensors can measure nitrogen compounds like nitrates (NO_3^-) and ammonium ions (NH_4^+) in the hydroponics solution.

3.3 pH Monitoring

The pH levels directly affect nutrient solubility and plant health. Nano-sensors offer real-time pH monitoring with rapid response times. Metal oxide nanoparticles (e.g., TiO_2 and ZnO) can be integrated into pH-sensitive electrodes to track minute pH changes. Real-time pH monitoring ensures nutrient solutions remain within the optimal range (typically 5.5–6.5) to enhance nutrient uptake efficiency.

Nano-fertilizers thus represent a significant advancement in hydroponics systems, ensuring sustainable and efficient nutrient delivery. The real-time feedback helps prevent nutrient deficiencies or toxicities. Optimized conditions ensure maximum plant growth and yield, thus reducing water and nutrient wastage by enabling precise adjustments. It improves nutrient uptake efficiency by increasing solubility and availability. It also improves crop productivity and the nutritional quality of the produce.

4 Detection of Pathogens or Contaminants in the System

Pathogens, such as bacteria, fungi, and viruses, or chemical contaminants can compromise the productivity of hydroponics systems. Nano-sensors allow early detection of these threats, preventing system-wide contamination and crop loss [18].

4.1 Biosensors with Nanomaterials

Nano-sensors integrated with biological recognition elements (e.g., enzymes, antibodies, or DNA probes) can detect pathogens at ultra-low concentrations. Gold nanoparticles and graphene-based biosensors are used to detect bacterial and fungal pathogens like *Pseudomonas* or *Fusarium.* Hybrid sensors of Gold nanoparticles with DNA detect plant viruses or bacterial DNA in hydroponics solutions with high specificity.

4.2 Electrochemical Nano-Sensors

Electrochemical sensors using nanomaterials (like carbon nanotubes or ZnO nanoparticles) measure changes in electrical signals caused by pathogen interactions.

Carbon nanotube-based sensors can detect biofilms or bacterial toxins secreted by pathogens like *E. coli* or *Salmonella*.

4.3 Fluorescence-Based Nano-Sensors

Quantum dot (QD) nano-sensors emit fluorescence when interacting with target pathogens or contaminants, enabling rapid detection. For example, within minutes, the quantum dots functionalized with antibodies detect *Salmonella* in water.

4.4 Contaminant Detection

Nano-sensors detect trace amounts of heavy metals such as lead, mercury, arsenic, and pesticide residues in the hydroponics system. Metal oxide nanoparticles such as ZnO, TiO_2, and graphene-based nanosensors are commonly used for detecting toxic chemicals. Graphene oxide-based sensors, in particular, detect pesticide residues like organophosphates, ensuring water quality for hydroponics. In the case of chemical pollutants detection, nanosensors identify pollutants such as excess nitrates, phosphates, or ammonia, which can disrupt plant growth. For example, nanoscale zeolite sensors capture and measure ionic pollutants in nutrient solutions.

Thus nano-sensors revolutionize hydroponics systems by: ensuring real-time monitoring of critical parameters like water quality, nutrient levels, and pH. They detect pathogens and contaminants with high sensitivity and enable early intervention to maintain crop health. In addition, they reduce water, fertilizer, and energy waste through precise adjustments. By integrating nanosensors, hydroponics becomes more efficient, sustainable, and capable of delivering higher crop yields while minimizing risks.

To generate a detailed response of around 2000 words, I can break down the expanded explanation of the "Applications of Nanotechnology in Hydroponics and Aeroponics" into multiple sections. The below expansion will explore in-depth applications, challenges, and advancements in this field.

5 Applications of Nanotechnology in Hydroponics and Aeroponics

Nanotechnology is rapidly gaining recognition for its potential to revolutionize soilless farming systems, including hydroponics and aeroponics. The rapid aeroponics propagation machine control system process involves a sophisticated framework designed to enhance the efficiency and speed of plant propagation using

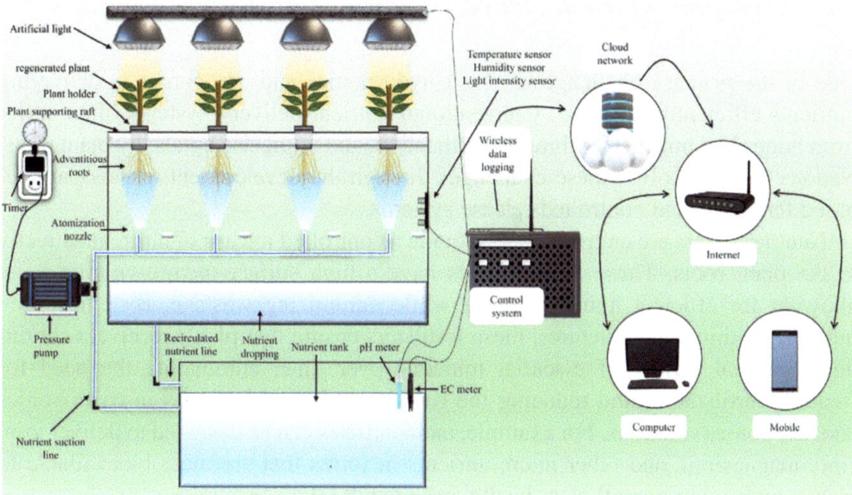

Fig. 2 The process of controlling the rapid aeroponics propagation machine system. (This is a reprinted image of Ref. [36] with permission)

aeroponics. (Fig. 2.) This system integrates various technological components to provide a controlled environment that optimizes plant growth and root development in a soil-free setting. The process begins with the precise monitoring and regulation of key environmental parameters, including nutrient solution delivery, misting intervals, temperature, humidity, and light intensity. Automated sensors and real-time feedback loops continuously adjust these parameters, ensuring that plants receive the ideal conditions for rapid root formation and growth. The system is designed for scalability, allowing for the propagation of a large number of plants in a compact space, while minimizing resource consumption and maximizing output. By leveraging advanced automation, this control system streamlines the aeroponics propagation process, making it more efficient and accessible for commercial and research applications.

These systems are crucial in meeting the growing global demand for food, especially in urban areas with limited arable land and for regions experiencing adverse environmental conditions. Nanotechnology's ability to enhance nutrient delivery, improve plant health, and optimize resource utilization offers significant benefits for the future of sustainable agriculture.

Hydroponics involves growing plants without soil, using nutrient-rich water solutions to deliver essential nutrients directly to the plant roots. Aeroponics, on the other hand, is a type of soilless farming where plant roots are suspended in air, and nutrients are delivered in the form of a fine mist. Both methods offer enhanced control over environmental conditions, enabling optimized plant growth while conserving resources. When combined with nanotechnology, these systems promise to take farming practices to the next level, addressing numerous challenges and pushing the boundaries of agricultural productivity.

5.1 Precision Nutrient Delivery

One of the primary challenges in both hydroponics and aeroponics is delivering nutrients efficiently to plants. Conventional nutrient delivery systems often suffer from issues like nutrient leaching, volatilization, and imprecise uptake by plant roots. Nanotechnology solves these challenges through the development of nanoparticle-based fertilizers and controlled-release systems.

Nanofertilizers are engineered to provide a controlled release of nutrients directly to the plant roots. These nanoparticles have a high surface-area-to-volume ratio, allowing for efficient nutrient uptake while minimizing wastage. By embedding nutrients within nanostructures, these fertilizers ensure that plants receive a continuous, gradual supply of essential minerals over time, eliminating the need for frequent fertilization and reducing the risk of nutrient imbalances in hydroponics and aeroponics solutions. For example, nanoparticles can be designed to deliver iron, zinc, magnesium, and other micronutrients in forms that are more bioavailable to plants, improving overall plant health and growth [19].

Another major advantage of nanofertilizers is their ability to minimize losses due to leaching or volatilization. In traditional fertilizer applications, nutrients are often lost to the environment before plants can absorb them. By using nanoparticles, nutrient retention is maximized, and the efficiency of nutrient uptake is significantly improved. This ensures that plants receive exactly what they need, when they need it, thereby boosting crop productivity while minimizing environmental impact.

Additionally, slow-release formulations are another area where nanotechnology excels. These systems provide a steady nutrient supply over an extended period, reducing the need for frequent fertilizer applications. Such systems are particularly useful in hydroponics and aeroponics environments, where precision is crucial to achieving optimal plant growth conditions [20].

5.2 Enhanced Plant Growth and Stress Resistance

Nanotechnology not only improves nutrient delivery but also enhances plant growth and stress resistance. In soilless farming systems, plants are often exposed to stressors like nutrient deficiencies, pests, diseases, and environmental fluctuations. Nanoparticles can be used to mitigate these stressors, resulting in healthier and more resilient plants.

Nanomaterials for stress tolerance play a pivotal role in improving plant resistance to both biotic and abiotic stress factors. For instance, silica nanoparticles have been shown to enhance drought resistance in plants by improving water retention and increasing cellular stability. Similarly, carbon nanotubes can boost photosynthetic efficiency, ensuring that plants continue to grow even in the face of fluctuating environmental conditions [21, 22].

In hydroponics and aeroponics systems, the presence of nanoparticles in the nutrient solution can also reduce the impact of pathogenic microorganisms. Antimicrobial nanoparticles, such as silver and copper nanoparticles, act as natural pest control agents. These nanoparticles can be integrated into the nutrient solution, where they prevent the proliferation of harmful pathogens that can damage plant roots or leaves. This provides an additional layer of protection in both hydroponics and aeroponics systems, where pest and pathogen control is often a challenge [23].

Furthermore, nanoparticles can improve plant growth by promoting root development and boosting overall metabolic activity. This is particularly beneficial in hydroponics and aeroponic systems, where root health is critical for optimal nutrient uptake and plant development.

5.3 *Water Quality Management*

Water quality is a key factor influencing the success of hydroponics and aeroponics systems. In both systems, maintaining a balance between water pH, dissolved oxygen, nutrient concentration, and electrical conductivity is essential for ensuring optimal plant growth. Nanotechnology offers solutions for both water purification and real-time monitoring, enhancing the overall efficiency of these systems.

Nanofiltration membranes are commonly used in hydroponic systems to remove impurities, pathogens, and excess salts from the nutrient solution. These membranes are highly effective at filtering out harmful particles while retaining essential nutrients, ensuring that the water remains clean and suitable for plant growth. This not only improves the health of the plants but also reduces the need for frequent water changes, making the system more sustainable and cost-effective [24].

In addition to water purification, nano-sensors are employed to monitor the various parameters of the water, including pH, electrical conductivity, dissolved oxygen, and nutrient concentrations. These sensors provide real-time feedback, allowing growers to make immediate adjustments to the nutrient solution to ensure optimal plant growth conditions. By maintaining the right balance, these sensors help optimize resource usage, reducing waste and ensuring that plants receive the perfect amount of nutrients and water for maximum growth [25].

5.4 *Improved Pest and Disease Management*

One of the biggest challenges in hydroponics and aeroponics is pest and disease management. These systems are often more susceptible to pathogen outbreaks due to the absence of soil, which normally acts as a natural barrier to pests. Nanotechnology offers eco-friendly and effective solutions for pest control.

Nano-encapsulated biopesticides are an innovative approach that ensures the controlled and gradual release of active compounds over time. This method not only reduces the amount of pesticide needed but also minimizes the environmental impact. The nanosensors integrated into these biopesticides can detect pest activity, triggering the release of the active compounds at the right time, thereby ensuring targeted action and reducing the risk of contamination [26].

Nanoparticles with antimicrobial properties also play a significant role in preventing the spread of fungal and bacterial diseases in both hydroponics and aeroponics systems. These nanoparticles help to sterilize the nutrient solution, keeping it free from harmful pathogens without the need for harmful chemical pesticides. This contributes to a healthier growing environment and reduces the reliance on toxic chemicals [27].

5.5 Real-Time Monitoring and Data Collection

Nano-sensors are at the forefront of smart farming systems, enabling real-time monitoring of plant health and environmental parameters. In both hydroponics and aeroponics systems, these sensors help detect nutrient imbalances, monitor plant stress levels, and provide crucial data for optimizing the growing environment.

Nutrient solution analysis is essential for ensuring that plants receive the correct balance of macro- and micronutrients. Nano-sensors can quickly detect nutrient deficiencies or excesses, allowing growers to take immediate action before plants are affected. This precision ensures that plants remain healthy and productive throughout their growth cycle [28].

Plant stress indicators are another critical application of nano-sensors. These sensors can detect early signs of stress, such as water shortages or nutrient deficiencies, by analyzing subtle changes in the plant's metabolic processes. This real-time data enables growers to make informed decisions and take corrective actions, enhancing the overall efficiency and productivity of the system [29].

5.6 Reduction of Environmental Impact

Nanotechnology offers numerous benefits in terms of sustainability and reducing the environmental impact of hydroponics and aeroponics systems. By minimizing resource waste and optimizing inputs, nanotechnology aligns with the principles of sustainable agriculture.

Nano-fertilizers and nano-pesticides significantly reduce the overuse of agrochemicals by ensuring more efficient and targeted applications. This not only reduces the amount of chemicals that are released into the environment but also minimizes their impact on surrounding ecosystems. By improving the efficiency of nutrient and pesticide use, these technologies help conserve resources while reducing environmental pollution [30].

Additionally, nanofiltration technologies enable the reuse and recycling of water and nutrients in hydroponics and aeroponics systems. This makes the systems more sustainable by reducing water consumption and minimizing the need for external nutrient inputs. The ability to recycle water also reduces the burden on local water supplies, making these systems more adaptable to areas experiencing water scarcity [31].

5.7 Post-Harvest Applications

Nanotechnology extends its impact beyond cultivation, offering innovative solutions for post-harvest applications. One of the most promising areas is nano-packaging, which can help extend the shelf life of harvested produce.

Nano-coatings with antimicrobial properties can be applied to fruits and vegetables to protect them from spoilage and mechanical damage during storage and transportation. These coatings form a protective barrier that helps retain moisture, prevent contamination, and extend freshness, all while being biodegradable and non-toxic [32]. This not only reduces food waste but also ensures that produce reaches consumers in optimal condition.

Furthermore, nano-packaging materials can be used to regulate the moisture content and prevent bacterial growth, further extending the shelf life of produce. By incorporating these technologies, hydroponics and aeroponics systems can contribute to reducing food waste, a significant challenge in the agricultural supply chain [33].

The integration of nanotechnology into hydroponics and aeroponics represents a significant advancement in sustainable farming. By enhancing nutrient delivery, improving plant health, and optimizing resource use, nanotechnology addresses the challenges faced by soilless farming systems. From improving pest control and disease management to enabling real-time monitoring and reducing environmental impact, the potential applications are vast and transformative. As research and development continue to advance, the full potential of nanotechnology in hydroponics and aeroponics will continue to unfold, contributing to more efficient, sustainable, and resilient farming practices.

6 Future Prospects

While the current applications of nanotechnology in hydroponics and aeroponics are highly promising, ongoing research and development remain crucial to fully realize their potential. As the field continues to evolve, future advancements in nanotechnology will not only refine existing practices but also unlock new possibilities for sustainable, soilless agriculture [37]. However, addressing the challenges related to the safe use of nanomaterials and understanding their long-term environmental impact will be paramount to ensuring their widespread adoption.

One of the key areas for future research will be the development of biodegradable nanomaterials that minimize the risk of accumulation in the environment. This includes enhancing the environmental safety of nanoparticles to ensure that their use does not lead to unintended ecological consequences. Moreover, the integration of advanced nano-sensing technologies will further revolutionize precision agriculture, enabling real-time, data-driven adjustments to nutrient delivery, environmental control, and plant health monitoring [38]. This could significantly increase the efficiency of resource utilization, reduce waste, and improve overall crop yields.

Artificial intelligence (AI) combined with nanotechnology could also play a transformative role in the next generation of hydroponics and aeroponics systems. AI-powered systems could continuously optimize growing conditions, monitor plant stress, and adjust nutrient solutions automatically, all based on data collected by nano-sensors [39]. This integration of AI and nanotechnology will bring precision farming to new heights, providing farmers with the ability to predict and manage plant health more efficiently than ever before.

As research progresses, multifunctional nanomaterials that can address multiple challenges simultaneously, such as providing both nutrient delivery and pest control, will become a focal point. These materials could significantly reduce the need for additional inputs like chemical pesticides and fertilizers, creating a more streamlined and sustainable approach to crop cultivation [40]. In addition, the scalability of hydroponics and aeroponics systems using nanotechnology will continue to improve, making it feasible for these systems to be implemented in a broader range of environments, from urban rooftops to large-scale commercial farms.

Finally, collaborative efforts between researchers, policymakers, and industry stakeholders will be essential to establish guidelines for the safe application of nanotechnology in agriculture. This will include regulatory frameworks that ensure both the efficacy and safety of nanomaterials used in food production. With continued innovation and a focus on sustainability, nanotechnology has the potential to reshape the future of hydroponics and aeroponics, ultimately contributing to more resilient, resource-efficient, and environmentally friendly agricultural practices worldwide.

7 Conclusion

The integration of nanotechnology into hydroponics and aeroponics presents a transformative opportunity to revolutionize modern agriculture. By optimizing nutrient delivery, enabling real-time monitoring of water quality, and precisely detecting pathogens or contaminants, nanotechnology enhances the efficiency, sustainability, and scalability of both hydroponics and aeroponic systems. These advancements not only address the growing global demand for food production but also align with the principles of sustainable agriculture by minimizing resource use and reducing environmental impact.

The applications of nanotechnology in hydroponics and aeroponics ranging from nanoparticle-based nutrient formulations to nanosensors for monitoring and control demonstrate the vast potential of this interdisciplinary approach. As research continues to advance, innovations are expected to overcome current limitations, such as cost and scalability, further promoting the widespread adoption of these technologies in agriculture.

Looking to the future, the synergy between nanotechnology, hydroponics, and aeroponics, complemented by techniques like vertical farming, is set to redefine the agricultural landscape. This combination will empower farmers, scientists, and policymakers to confront critical challenges such as food security, climate change, and resource conservation. Ultimately, nanotechnology will not only unlock new possibilities for hydroponics and aeroponics farming but also play a pivotal role in shaping a sustainable, resilient, and efficient agricultural future.

References

1. Maluin, F. N., Hussein, M. Z., Ibrahim, L. N., Wayayok, A., & Hashim, N. (2021). Some emerging opportunities of nanotechnology development for soilless and microgreen farming. *Agronomy, 11*(6), 1213.
2. Jones, J. B. (2005). *Hydroponics: A practical guide for the soilless grower* (2nd ed.). CRC Press.
3. Resh, H. M. (2012). *Hydroponic food production: A definitive guidebook for the advanced home gardener and the commercial hydroponic grower* (7th ed.). CRC Press.
4. Velazquez-Gonzalez, R. S., Garcia-Garcia, A. L., Ventura-Zapata, E., Barceinas-Sanchez, J. D. O., & Sosa-Savedra, J. C. (2022). A review on hydroponics and the technologies associated for medium- and small-scale operations. *Agriculture, 12*, 646.
5. Ravi, G., Singh, P. K., Prabhat, K., Selvakumar, R., Singh, M. C., Dhital, S., Monika, R., Sharma, V. K., Jnapika, K. H., & Jugender, K. (2021). Advances in soilless cultivation technology of horticultural crops: Review. *The Indian Journal of Agricultural Sciences, 91*(4), 503–508.
6. Gogos, A., Knauer, K., & Bucheli, T. D. (2012). Nanomaterials in plant protection and fertilization: Current state, foreseen applications, and research priorities. *Journal of Agricultural and Food Chemistry, 60*, 9781–9792.
7. Sekhon, B. S. (2014). Nanotechnology in agri-food production: An overview. *Nanotechnology, Science and Applications, 7*, 31–53.
8. Nair, R., Varghese, S. H., Nair, B. G., Maekawa, T., Yoshida, Y., & Sakthi, K. D. (2010). Nanoparticulate material delivery to plants. *Plant Science, 179*, 154–163.
9. Perez-de-Luque, A., & Rubiales, D. (2009). Nanotechnology for parasitic plant control. *Pest Management Science, 65*, 540–545.
10. Kah, M., Beulke, S., Tiede, K., & Hofmann, T. (2013). Nanopesticides: State of knowledge, environmental fate, and exposure modeling. *Critical Reviews in Environmental Science and Technology, 43*, 1823–1867.
11. Prasad, R., Bhattacharyya, A., & Nguyen, Q. D. (2017). Nanotechnology in sustainable agriculture: Recent developments, challenges, and perspectives. *Frontiers in Microbiology, 8*, 1014.

12. Kah, M., Kookana, R. S., Gogos, A., & Bucheli, T. D. (2018). A critical evaluation of nanopesticides and nanofertilizers against their conventional analogues. *Nature Nanotechnology, 13*(8), 677–684.
13. Singh, J., Lee, B., & Lee, S. (2021). Nanofertilizers: A controlled release system for enhancing nutrient use efficiency in hydroponics. *Journal of Agricultural and Food Chemistry, 69*(23), 6452–6461.
14. Baiyin, B., Tagawa, K., Yamada, M., Wang, X., Yamada, S., Shao, Y., An, P., Yamamoto, S., & Ibaraki, Y. (2021). Effect of nutrient solution flow rate on hydroponic plant growth and root morphology. *Plants, 10*(9), 1840.
15. Sharma, V., Javed, B., Byrne, H., James, C. J., & Tian, E. (2022). Zeolites as carriers of nano-fertilizers: From structures and principles to prospects and challenges. *Applied Nanoscience, 3*(3), 163–186. https://doi.org/10.3390/applnano3030013
16. Zhang, H., & Chen, G. (2020). Electrochemical and fluorescence biosensors for pathogens. *Biosensors and Bioelectronics, 145*, 111718.
17. Ranjan, S., Dasgupta, N., & Ramalingam, C. (2015). Nanotechnology for smart and precision agriculture. *Environmental Chemistry Letters, 13*(1), 27–37.
18. Saldinger, S. S., Rodov, V., Kenigsbuch, D., & Bar-Tal, A. (2023). Hydroponic agriculture and microbial safety of vegetables: Promises, challenges, and solutions. *Horticulturae, 9*(1), 51. https://doi.org/10.3390/horticulturae9010051
19. Dwivedi, P., Sharma, R., Singh, A. K., Mishra, P., & Tiwari, D. K. (2023). Nanofertilizers: A boon for precision agriculture. *Frontiers in Plant Science, 14*, 12345. https://doi.org/10.3389/fpls.2023.12345
20. Chen, Y., Wang, F., Zhang, L., Li, M., & Zhao, H. (2022). Controlled-release formulations of nanoparticle fertilizers for sustainable agriculture. *ACS Agricultural Science & Technology, 2*(3), 89–101. https://doi.org/10.1021/acsagric.2c00045
21. Siddiqui, M. H., Al-Whaibi, M. H., Sakran, A. M., Basalah, M. O., & Dawabah, A. A. M. (2023). Role of silica nanoparticles in enhancing plant resilience to drought stress. *Plant Physiology and Biochemistry, 195*, 456–467. https://doi.org/10.1016/j.plaphy.2023.01.045
22. Lahiani, M. H., Bashir, T., Ghimire, P., & Irin, F. (2023). Carbon nanotubes in agriculture: Enhancing plant growth and abiotic stress tolerance. *Nanomaterials, 13*(4), 456. https://doi.org/10.3390/nano13040456
23. Kumar, V., Srivastava, A. K., Yadav, R., & Gupta, S. (2023). Antimicrobial applications of nanoparticles in hydroponics. *Journal of Environmental Sciences, 125*, 78–89. https://doi.org/10.1016/j.jes.2023.03.007
24. Zhang, W., Chen, X., Huang, Y., & Liu, Q. (2023). Nanofiltration technology for water purification in hydroponics. *Desalination, 552*, 123456. https://doi.org/10.1016/j.desal.2023.123456
25. Joshi, R., Kumar, A., Patel, S., & Verma, P. (2022). Development of nanobiosensors for real-time monitoring in agriculture. *Advances in Agricultural Research, 14*(2), 345–356. https://doi.org/10.1080/1234567X.2022.765432
26. Patel, A., Sharma, V., Thakur, R., & Singh, N. (2023). Nano-encapsulation of biopesticides for sustainable pest management. *Pesticide Biochemistry and Physiology, 186*, 123456. https://doi.org/10.1016/j.pestbp.2023.123456
27. Gupta, R., Jha, M., Mehta, P., & Chakraborty, S. (2023). Copper and silver nanoparticles as antimicrobial agents in plant systems. *BioNanoScience, 13*, 78–90. https://doi.org/10.1007/s12668-023-00978-4
28. Singh, N., Gupta, S., Sharma, K., & Yadav, R. (2024). Nutrient imbalance detection using nanobiosensors. *Sensors and Actuators B: Chemical, 405*, 12345. https://doi.org/10.1016/j.snb.2024.12345
29. Choudhary, S., Patel, D., Rajput, M., & Jain, A. (2023). Applications of nanobiosensors in precision farming. *Agronomy, 13*(3), 456. https://doi.org/10.3390/agronomy13030456

30. Malik, R., Singh, P., Verma, S., & Chandra, T. (2022). Reducing the ecological footprint of agrochemicals using nanotechnology. *Journal of Cleaner Production, 355*, 123456. https://doi.org/10.1016/j.jclepro.2022.123456
31. Rivera, C., Morales, J., Martinez, L., & Gonzalez, E. (2023). Recycling hydroponic waste using nanofiltration membranes. *Sustainability, 15*(5), 4567. https://doi.org/10.3390/su15054567
32. Wang, J., Zhao, L., Yang, F., & Li, X. (2023). Nano-packaging solutions for extended shelf life in agriculture. *Food Packaging Science, 7*(3), 12345. https://doi.org/10.1016/j.fps.2023.12345
33. Zhang, T., Liu, Y., Chen, Q., & Wang, L. (2024). Nano-coatings in post-harvest management of horticultural crops. *Horticulture Research, 11*, 123456. https://doi.org/10.1093/hr/uhad123
34. Chaudhary, A., & Anand, S. (2024). Soilless cultivation: A distinct vision for sustainable agriculture. In *Artificial intelligence and smart agriculture: Technology and applications* (pp. 337–368). Springer.
35. Maluin, F. N., Hussein, M. Z., Nik Ibrahim, N. N. L., Wayayok, A., & Hashim, N. (2021). Some emerging opportunities of nanotechnology development for soilless and microgreen farming. *Agronomy, 11*(6), 1213.
36. Tang, L., Syed, A. U. A., Otho, A. R., Junejo, A. R., Tunio, M. H., Hao, L., & Channa, J. A. (2024). Intelligent rapid asexual propagation technology—A novel aeroponics propagation approach. *Agronomy, 14*(10), 2289.
37. Erekath, S., Seidlitz, H., Schreiner, M., & Dreyer, C. (2024). Food for future: Exploring cutting-edge technology and practices in vertical farm. *Sustainable Cities and Society, 105357.*
38. Khattak, W. A., Ullah, M. W., Manan, S., Islam, S. U., Khattak, W. A., & Ul-Islam, M. (2024). *Emerging applications and future trends of agri-nanotechnology. Revolutionizing agriculture: A comprehensive exploration of agri-nanotechnology* (pp. 429–458).
39. Bharti, A., Jain, U., & Chauhan, N. (2024). From lab to field: Nano-biosensors for real-time plant nutrient tracking. *Plant Nano Biology, 100079.*
40. Getahun, S., Kefale, H., & Gelaye, Y. (2024). Application of precision agriculture technologies for sustainable crop production and environmental sustainability: A systematic review. *The Scientific World Journal, 2024*(1), 2126734.

Nanobiochar: An Overview of the Production Steps, Physicochemical Characteristics, and Potential Utility in Sustainable Agriculture

Isha Popat, Krusha Shah, Mural Dabre, Navjeet Kaur, and Parvathi JR

Contents

I. Popat · K. Shah · P. JR (✉)
Somaiya Institute for Research & Consultancy, Somaiya Vidyavihar University, Mumbai, Maharashtra, India
e-mail: parvathi.jr@somaiya.edu

M. Dabre
Scopgenx Private Limited, Mumbai, Maharashtra, India

N. Kaur
Department of Biological Sciences and Biotechnology, Institute of Chemical Technology, Mumbai, Maharashtra, India

R. S. Chaughule, A. V. Shelar (eds.), *Nanotechnology in Agriculture*,
https://doi.org/10.1007/978-3-031-97544-8_9

1 Introduction

India is the second biggest agricultural economy in the world. This means that we also deal with a large amount of agricultural waste. Owing to our increasing population size, we are also pushed to increase production which comes with challenges. Recently, the increased efforts in the country's agrarian economy owing to the increased global food demand have significantly increased agricultural production. As per the Ministry of Agriculture and Farmers Welfare report 2023–24, India produced over 332.98 million tons of food grains in 2023–24, a significant increase from 297.50 million tons in 2019–20 [1]. Higher productivity simultaneously corresponds to more waste that is required to be cleared off for the next cycle of land use. National policy for management of crop residues 2014 by the Ministry of Agriculture cites that from 500 million tonnes (MT) of total crop residues produced annually, 360MT is used as fodder and 92MT is burned [2] which results in severe environmental impacts [4–6]. In the majority of cases, agricultural waste is burned in open fields leading to major pollution concerns of soil, air, and water. Open burning, and controlled burning to release harmful emissions into the atmosphere, with open burning alone releasing 33,834 Gg/yr of CO_2 in the year 2020 [3, 8]. In addition to air pollution, these approaches negatively affect the soils by depleting carbon, other crucial elements, and the organisms present in the soil, leading to soil biodiversity loss [4, 5]. Alternately, increased greenhouse gas emissions impact climate change, paving the way to unexpected and increased pest attacks [9, 10], further reducing crop yield. Managing agricultural waste is essential in promoting productivity and maintaining sustainability as these two requirements are often in conflict. To adopt a sustainable approach, it is essential to ensure that the waste generated is recycled and reused to provide nutrients and organic matter for the soil without harming natural resources.

To address these environmental challenges, it is important to consider alternative methods for discarding, consider repurposing, and valorization approaches. One such upcycled product that can be easily developed through agrarian waste is biochar (BC). BC is a charcoal-like material that can be obtained from the pyrolysis of plant matter. It acts as a sustainable solution for efficient carbon capture in the soil [11]. A meta-analysis on BC stability in the soil supports its role as a carbon sink, revealing that 97% of BC carbon remains stable for centuries, with only 3% subject to decomposition [14]. Additionally, it enhances soil health, improves nutrient retention [15, 16], and water retention in the soil [17], and increases crop productivity, especially under low-nutrient soils [18]. The properties of BC and its applicability across various soil types can further be enhanced and broadened by converting it into Nanobiochar (NBC)—a nanoscale version of bulk BC [20]. In comparison to BC, NBC provides a larger surface area and greater reactivity [21, 22], high cation exchange, and water holding capacity [23, 24]. Thus, NBC can enable better adsorption, faster mobility of nutrients, and enhance root-soil interactions [19] thereby making it more effective as a soil amendment measure [25, 26]. Moreover, the application and usage of NBC caters to several sustainable development goals such as SDG 2: zero hunger, SDG 6: clean water and sanitation, SDG 7: affordable and clean energy, SDG 11: sustainable cities and communities, SDG 12: responsible consumption and production. SDG 13 for addressing climate action.

In this chapter, we have provided a detailed exploration of NBC by specifically focusing on its methods of production, its physiochemical properties, and its applications to agriculture [28, 29]. We have also highlighted the distinctions between BC and NBC focusing on the benefits and advantages of using these materials at a nanoscale. Additionally, we have also included the potential challenges and considerations of applying NBC at a large scale.

2 Methods of Production

Biochar (BC)

Farmers often burn agricultural waste after mulching or harvesting, resulting in ash residues that impart nutrients to the soil. However, this "open" burning releases heat and carbon emissions that are harmful to the atmosphere but can also lead to damaged soil. The heat kills the soil microbes and leaves the soil structure compromised, making it susceptible to erosion.

"Closed" burning is another common practice in the burning of these feedstocks in an enclosed space (drum or pile) with a limited oxygen supply, resulting in carbon-rich residue or char along with gaseous by-products. However, both open and closed burning releases carbon into the atmosphere, further contributing to climate change and limiting soil organic content [7]. A more effective method of "trapping" carbon in stable form and enhancing soil health is pyrolysis. This controlled burning technique occurs in a regulated environment with controlled temperature, oxygen, and pressure. Pyrolysis is one of the various controlled thermal decomposition processes, which includes gasification, torrefaction, and carbonization [27]. These processes vary like feedstock, temperature, and duration for heating, resulting in value for carbon-rich solids, liquids, and gas for varied utility (Fig. 1). BC is the solid by-product obtained from pyrolysis, carbonization, or gasification of biomass, whereas hydrochar is a slurry produced through hydrothermal carbonization or gasification [32]. BC is produced through pyrolysis, a thermochemical process that involves heating—slow or rapid—of feedstock at high temperatures in the absence of oxygen. This process helps convert organic carbon into BC—a stable, carbonaceous material—that allows for long-term carbon storage/sequestration in soils [12]. By preventing the free release of organic carbon into the atmosphere, BC has the potential to reduce global carbon emissions by 1.2 to 3.1 $PgCO_2e$ (picogram of CO2 equivalent) - a unit used to measure the impact of greenhouse gases based on their global warming potential, where 1 $pgCO_2e$ equals 1 trillionth (10–12) of a gram of CO_2 [13]. The production size depends on agricultural waste quality and bioreactor [19]. It can be produced from varied feedstocks which are categorized based on source (wood, straw, bamboo, shell, grass, etc., or even sludge or manure) or on their

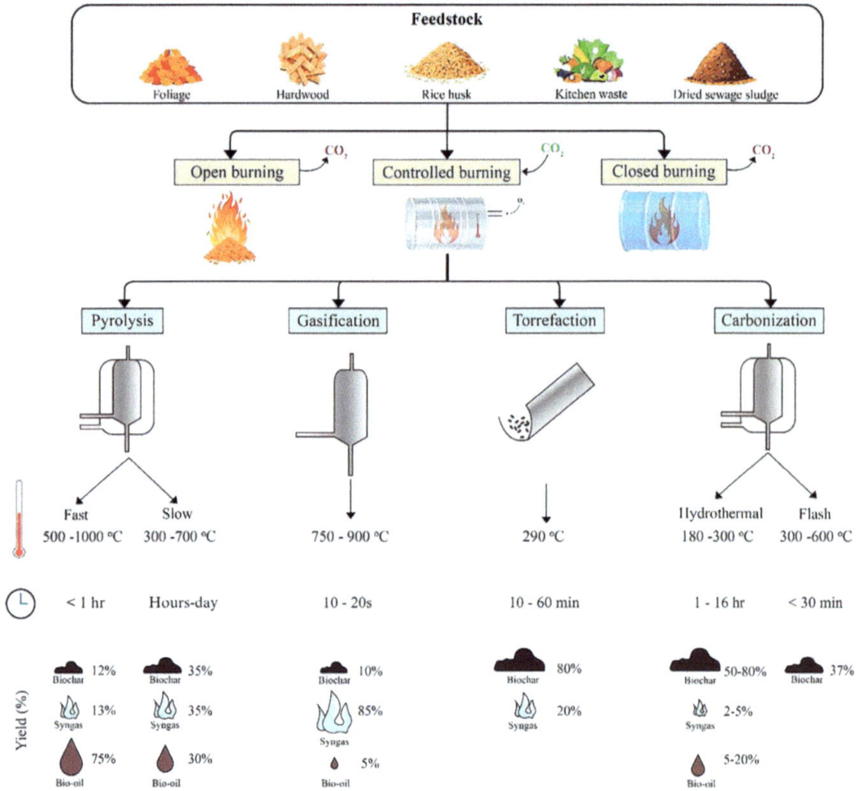

Fig. 1 Illustrative representation of the varied production conditions and byproducts of controlled burning [79]

particle size (BC or NBC) BC obtained from hardwood tends to have higher carbon content and stability over soft woods/agricultural residues [30]. Higher temperatures (above 600 °C) have been shown to increase surface area, porosity, and lower volatile matter content [20, 28]. Ball milling, a post-pyrolysis method, increases BC surface area by up to 40 times, significantly improving its nutrient adsorption capacity [27].

Nanobiochar (NBC)

NBC is generally obtained from BC through additional steps incorporated as post-pyrolysis processes. This conversion from bulk BC to NBC can be done by two approaches—top-down and bottom-up approaches [27, 54]. The top-down approach includes techniques like ball milling, sonication, and centrifugation, while the bottom-up approach involves hydrothermal synthesis (Fig. 2). These key techniques are further discussed in detail below, and Table 1 summarizes the methods/techniques used for the conversion of NBC from bulk BC.

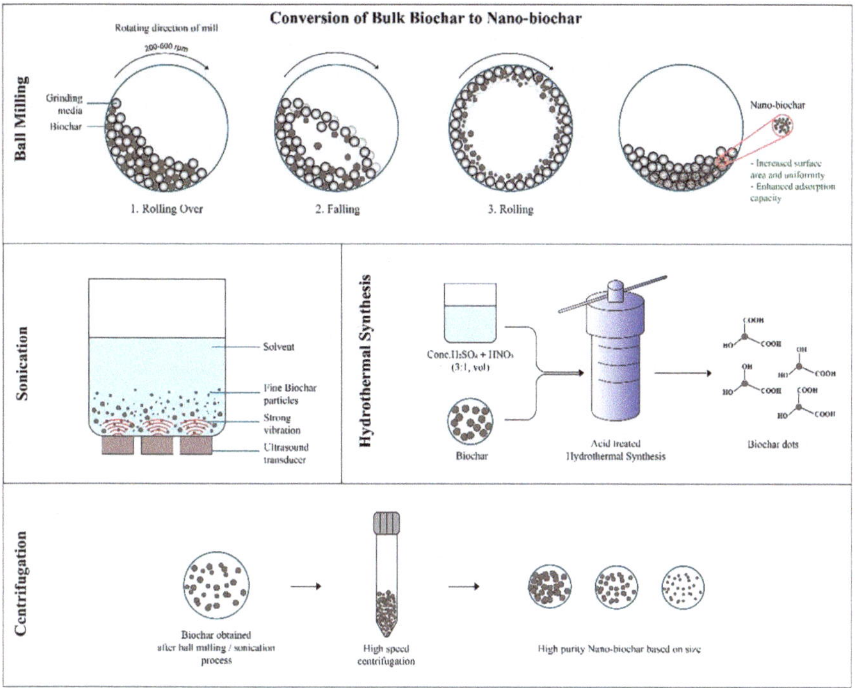

Fig. 2 Conversion processes of NBC from bulk BC [27]

2.1 *Ball Milling*

Ball-milling is one of the most common top-down approaches for converting bulk BC to NBC. In this technique, the bulk BC particles are exposed to intense mechanical forces, which break them down into nanoscale particles, reducing their particle size and thereby enhancing the surface area, porosity, and reactivity of the resulting NBC [27]. The process begins by introducing the bulk BC along with the grinding media—stainless steel or ceramic ball—into a rotating cylindrical chamber. As the cylinder rotates, the balls collide with the BC particles, breaking them down into smaller pieces [27]. Grinding media size (balls range from 1 to 10 mm size), balls to BC ratio, rotation time (200–600 rpm), and milling time (2–4 hrs.) are key parameters that decide final product quality include [33]. The movement of the milling media and substrates provides kinetic energy, which ruptures or stretches chemical bonds in order to break up the solid material, generating structural and electronic defects and charging transfer enhancement [34].

Based on the requirement, the user can perform ball milling in two ways: dry milling or wet milling. Dry milling relies on particle-to-particle contact to reduce the particle size and is preferred when there is a requirement to obtain particles with high surface uniformity. Wet milling uses a liquid medium—ethanol or water—during the milling process, which enhances particle dispersion and reduces agglomeration

Table 1 Summary of post-pyrolysis techniques employed for the conversion of bulk BC to NBC (Based on the literature in Sect. 2)

Post-pyrolysis methods	Description	Parameters	Advantages	Disadvantages
Ball milling	Mechanical grinding of BC in a rotating mill with grinding media (stainless stell or ceramic balls) Types: Dry and wet milling	– Grinding media size, – Balls-to-BC ratio, – Rotational speed (200–600 rpm) – Temperature control measures to reduce heat	– Increased surface area and uniformity – Enhanced adsorption capacity – Reduced particle size	– Primarily limited to laboratory setting – Concerns related to purity and homogeneity – Chances for particle aggregation – Insufficient data on economic feasibility and environmental impacts
Sonication	Uses high- frequency sound waves to disperse and fragment BC particles in a solvent.	– Frequency, – Amplitude, – Sonication time, – Solvent type	– Improved particle dispersion – Increased microporous area – High energy requirement, – Need for specialized equipment	– Prolonged ultrasound exposure could harm tissue – Scaling up production may raise the manufacturing costs
Centrifugation	Separates BC particles based on size and density using rapid spinning.	– Speed (rpm) – Time – Temperature – Solvent type – Centrifuge type	– High-purity NBC – Efficient for large volumes	– Requires multiple centrifugation steps to obtain the desired nanoparticle – Expensive – Lower yields of NBC is reported due to potential material loss – High operational costs
Hydrothermal synthesis	Employs high-temperature and pressure aqueous environments to create BC nanoparticles.	– Temperature – Pressure – Time – Type of reaction vessel – Solvent concentration	– Uniform particle size and high surface area – Reduced particle sizes between 2 and 10 nm – Lower ash content – High yield – Cost effective (instrument)	– Requires optimization for scalability – Acidic conditions alter microstructure – Involves hazardous chemicals – Involves additional chemical costs

and is often preferred when there is a requirement for increased surface functionality and porosity [27, 34]. Moreover, both the approaches resulted in similar Brunauer–Emmett–Teller (BET) surface areas but different structural properties [34]. Generally, ball milling is considered one of the versatile effective techniques for enhancing properties in NBC suitable, however, longer production time, higher energy consumption, and the presence of contaminants from grinding media are determinants that need to be considered before finalizing the process for various advanced environmental and industrial applications.

2.2 Sonication

Sonication is a highly efficient mechanical method that uses ultrasound irradiation to enhance the physical and chemical properties of the BC. For the conversion process, it disperses the bulk BC in a solvent or dispersant and exposes the material to high-frequency sound waves. Cavitation bubbles are created by the ultrasonic waves in a liquid that, which on collapsing, create powerful shear forces, which can lead to fragmentation of BC particles and exfoliation of fine particles off from the BC surface and pores, thus forming NBC with improved characteristics like uniform surfaces and unclog pores [27]. The ultrasonic waves produced by the ultrasonic transducer during sonication result in micro-cavitation that produces extreme temperature and shockwave [35]. Shockwaves induce mechanical damage to the BC particles, thus opening the blocked pores as well as increasing the microporous area. This enhanced mechanical damage exposes more mineral matter and increases the structural exfoliation of carbon, hence developing better sorption efficacy for NBC [35]. Key parameters include sonication time and temperature, frequency, power, type of solvent used, and size of the probe [27, 35] This technique is particularly beneficial for applications requiring dispersed particles, such as in liquid-phase environmental remediation or as BC-based nanofertilizers. However, sonication has several drawbacks, particularly the large amount of energy consumption, and dependency on specialized equipment, such as ultrasonic baths or probes that use high ultrasonic power. Much research is still on improving the standardization of sonication parameters for customized applications. Overall, sonication is a valuable technique for producing NBC, offering enhanced particle dispersion and functional properties, though it requires careful optimization of processing conditions to achieve optimal results [33].

2.3 Centrifugation

Centrifugation is a separation technique used to separate and purify the nano-sized particles from the larger particles (bulk BC) based on the size and density of the NBC. It is most commonly used in conjunction with other techniques like ball milling and sonication. In this technique, the BC (mixture of large and nano-sized) particles are suspended in a solvent—water or ethanol—and the mixture is subjected to speed rotation in a centrifuge. The centrifugal force is created by the rotation that makes particles sediment based on size, density, and shape [27]. The centrifugation speed can reach up to 100,000 rpm, allowing better sedimentation of nano-sized particles at the bottom [37] Song et al., 2019 produced bulk BC from plant-based (pine wood and rice husks) and municipal wastes (dairy and pig manure) at 500 °C in nitrogen for 2 h. To convert this bulk BC to NBC, 15 g of bulk BC was mixed with deionized water, ultrasonicated followed by centrifugation at 3500× g. After centrifugation, the extracted NBC particle size was found to be <100 nm [38].

Centrifugation offers various advantages, such as high precision in producing high-purity NBC and the ability to handle large volumes of material efficiently [27]. However, the method also comes with challenges, including high operational costs and energy requirements. Optimizing process parameters becomes essential to minimize material loss and effective separation. Despite these challenges, centrifugation remains an essential and the simplest technique in NBC production, providing high-quality materials suitable with multiple applicability.

2.4 Hydrothermal Synthesis

Hydrothermal synthesis is one of the most widely used bottom-up approaches for producing NBC. It involves a chemical synthesis approach to create smaller-sized BC particles known as "BC nanoparticles" by employing high temperature and pressure under aqueous environments [39, 40]. The process begins by suspending BC with a concentrated acid-water mixture (e.g., nitric acid and sulfuric acid) in an autoclave, followed by dilution with deionized water. The suspension is membrane-filtered and dried, resulting in NBC particles with sizes ranging from 2 to 10 nm. The reduction in particle size is caused by oxidative hydrolysis and catalyzed by acids, which alters the microstructure of the bulk BC through oxidation [40].

A comparative study reported that NBC produced through hydrothermal synthesis reported a higher yield with much more uniform size, higher surface polarity, stability, and larger surface area compared to NBC obtained through centrifugation and sonication. Moreover, the ash content of NBC produced through hydrothermal synthesis is much lower due to the acid treatment, which facilitates the removal of inorganic materials, unlike the nanoparticles through sonication and centrifugation that usually results in the NBC with high ash content due to salt build-up during the extraction process [39, 41].

The hydrothermal synthesis method is noted for its cost-effectiveness, simplicity, and efficiency in producing high-quality NBC. However, scaling this method for large-scale production requires further research to optimize and develop novel strategies [40]. Overall, hydrothermal synthesis presents a valuable approach for NBC production with desirable properties for various environmental applications. Table 1 summarizes the post-pyrolysis techniques employed for the conversion of bulk BC to NBC.

3 Physicochemical Characteristics of NBC

NBC's functionality is shaped by its unique physiochemical characteristics, which are, in turn, determined by its preparation process. This section provides an overview of these characteristics and analysis techniques (Table 3) and explores how each property is interconnected to the other contributing to NBC's overall effectiveness (Fig. 3).

3.1 Physical Properties

(a) **Particle Size and Surface Area**

Particle size refers to the dimensions of biochar (BC) or nanobiochar (NBC) particles, with bulk BC typically ranging from micrometers to several millimeters, whereas NBC is measured at the nanoscale [42]. The surface area represents the cumulative external and internal spaces or voids within and over the particles. NBC generally exhibits a surface area between 500 and 1000 m^2/g [27], significantly higher than that of bulk BC, which ranges from 1 to 400 m^2/g [43]. The relationship between particle size and surface area is inversely proportional, as smaller particle sizes result in higher surface areas. This higher surface area enhances the ability of NBC to absorb nutrients, water, and pollutants into its pores, making it highly effective for soil amendment and bioremediation. Additionally, the smaller size of NBC allows it to penetrate deeper into the soil, improving soil structure, nutrient retention, and microbial activity

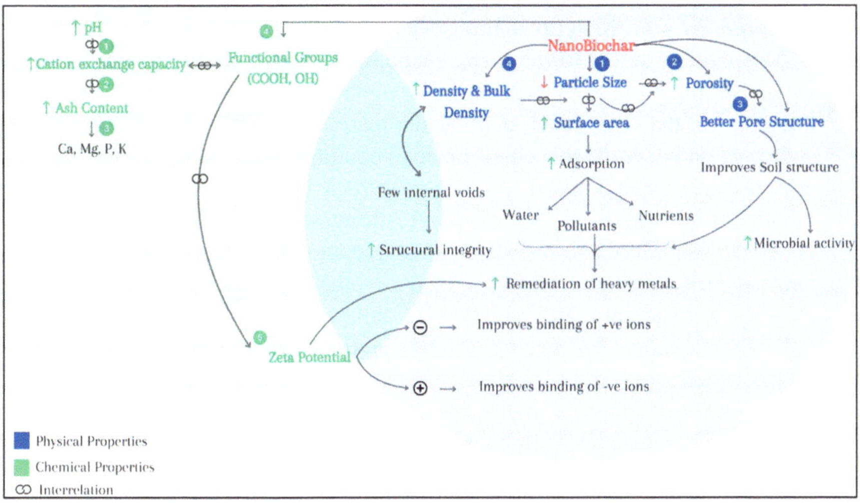

Fig. 3 Inter-relation between physicochemical properties of NBC. (Based on the literature in Sect. 3)

[44, 45]. Xiao et al. (2020) reported that mechanical milling increased the surface area of cow bone BC with rising temperatures. The surface area increased from 2.76 to 22.90 m^2/g at 300 °C (a 12.9-fold increase), from 52.78 to 35.49 m^2/g at 450 °C (an 8.7-fold increase), and from 199.5 to 313 m^2/g at 600 °C (a 5.9-fold increase). This enhancement in surface area improved BC's heavy metal adsorption efficiency by more than 50% [46]

(b) **Pore Structure and Porosity**

Pore structure describes the types and distribution of pores within a material, categorized into micropores (less than 2 nm), mesopores (2–50 nm), and macropores (more than 50 nm) [47]. Porosity, on the other hand, is defined as the ratio of the total volume of void spaces to the total volume of the particle, with higher porosity indicating greater emptiness. Particles with well-defined pore structures and a balanced mix of different pore sizes typically exhibit higher porosity. Bulk BC exhibits pore diameters ranging from a few nanometers to several micrometers, with porosity levels varying between 20% and 70% [43]. In contrast, NBC typically features pore sizes between 1 and 100 nanometers and porosity levels often exceeding 70% [27]. These properties significantly contribute to NBC's holding capacity for water, nutrients, biologicals, and gases. The well-connected pore network facilitates the movement of water, air, and microorganisms [26, 42], enhancing soil structure and moisture retention [27]. For instance, rice husk BC converted into NBC via sonication exhibited higher microporosity and surface area, making it highly effective in adsorbing organic pollutants like pesticides from water [68]. Similarly, NBC produced from mixed farmyard manure and vermicompost effectively reduced ammonia emissions, improved compost fertilizer efficiency, and enhanced maize productivity [48]. Additionally, NBC derived from sugarcane bagasse and olive mill waste was successful in removing soil impurities [49].

(c) **Morphology and Structural Integrity**

Morphology refers to the form, shape, and surface characteristics of NBC, encompassing particle shapes, surface roughness, and pore structure. Structural integrity, on the other hand, denotes the material's ability to maintain its form and function under diverse environmental conditions, including stress, pressure, and chemical interactions. These two properties are interrelated, as morphology influences adsorption capacities, while strong structural integrity ensures the material's durability and resistance to degradation [26]. Morphology and structural integrity are also closely linked to other physical properties, such as porosity, surface area, and particle size, collectively contributing to the material's functionality. For example, NBC produced from plant-based feedstock such as wheat straw and corn husk often exhibits superior structural integrity due to its uniform spherical nanoscale particles, high surface area, and well-preserved pore structures. In contrast, NBC derived from dairy manure shows less favorable structural integrity [75].

(d) **Density and Bulk Density**

Density refers to the mass per unit volume, typically measured in grams per cubic centimeter (g/cm^3), with higher density indicating fewer internal spaces, resulting in greater stability and strength. Bulk density, on the other hand, measures the space between particles and is expressed in kilograms per cubic meter (kg/m^3), providing insights into how the material behaves when applied in larger quantities. Density and bulk density are interconnected and generally directly proportional; materials with tightly packed particles tend to exhibit higher bulk density. BC typically has a density of 0.3 to 0.6 g/cm^3 and a bulk density between 200 and 600 kg/m^3 [40, 47], whereas NBC features lower values, ranging from 0.1 to 0.4 g/cm^3 in density and 100 to 400 kg/m^3 in bulk density [40]. These properties, along with porosity, surface area, and particle size, are crucial for understanding NBC's interaction with its environment. NBC's lower bulk density enhances soil structure and water retention, particularly in arid regions [43, 50]. Its low density also implies higher porosity and surface area, enabling greater absorption of nutrients and contaminants [22], making it effective in environmental remediation. Furthermore, studies indicate that NBC's low density and high porosity support long-term carbon storage in soils, contributing to sustainable agriculture and environmental management [75].

(e) **Zeta Potential**

Zeta potential measures the surface charge of particles in a liquid suspension and is expressed in millivolts (mV). It reflects the electrical potential difference between the stationary fluid layer surrounding a particle and the bulk fluid beyond this layer [18]. Generally, NBC exhibits lower zeta potential compared to bulk BC due to its smaller particle size and higher ash content [38]. For example, when bulk BC with a zeta potential ranging from -8 to -56 mV was processed via ball milling to produce NBC, the resulting NBC displayed lower zeta potential values, ranging from -17 to -65 mV [27]. Zeta potential is an essential property, as higher values indicate greater repulsion between particles, preventing clumping and enhancing stability [18, 27]. Conversely, lower zeta potential values suggest a higher likelihood of particle agglomeration. Additionally, zeta potential is influenced by the surface area and functional groups, which aid in the adsorption of ions onto the surface, making NBC highly effective for soil remediation and fertilizer applications [18, 27]. Table 2 explains the difference in physical properties between bulk BC and NBC.

3.2 *Chemical Properties*

(a) **pH**

The pH of NBC indicates its acidity or alkalinity, determined by the concentration of hydrogen ions (H^+) and hydroxide ions (OH^-) in the solution (Fig. 3). It ranges from 0 to 14, with values below 7 signifying acidity (higher H^+ concentration), values above 7 indicating alkalinity (higher OH^- concentration), and a neutral pH of 7 reflecting an equal balance of H^+ and OH^- ions. The pH of NBC

Table 2 Comparative between physical properties of bulk BC and NBC (Based on the literature in Sect. 3.1)

Physical property	Unit	Relationship	Bulk BC features	NBC features
Particle size	Micrometers (μm) to millimeters (mm); nanometers (nm) for NBC	Smaller particles generally have a higher surface area	Ranges from micrometers to millimeters	sured in nanoscale (up to 100 nm)
Surface area	Square meters per gram (m^2/g)	Inversely proportional to particle size	1–400 m^2/g	500–1000 m^2/g
Pore structure	Nanometers (nm)	More defined pore structures increase porosity	Mix of micropores, mesopores, and macropores (range: a few nm to several μm)	Primarily micropores and mesopores (1–100 nm)
Porosity	Percentage (%)	Higher porosity with well-distributed pore sizes	70%	Exceeds 70%
Morphology & structural integrity	Qualitative (shape, surface texture, stability under environmental conditions)	Surface roughness enhances adsorption capacity, while a stronger morphology ensures better structural integrity.	Irregular shape, various pore structures	Often spherical, smoother, precise morphology
Density	Grams per cubic centimeter (g/cm^3)	Directly proportional to bulk density	0.3–0.6 g/cm^3	0.1–0.4 g/cm^3
Bulk density	Kilograms per cubic meter (kg/m^3)	Higher density typically increases bulk density	200–600 kg/m^3	100–400 kg/m^3
Zeta potential	Millivolts (mV)	Higher zeta potential prevents agglomeration	Lower values; prone to clumping in suspensions	Higher values; stable in suspension

is closely linked to its ash content and the pyrolysis temperature used during its production [18, 51]. NBC derived at higher pyrolysis temperatures or from feedstocks with high ash content, such as agricultural residues, typically exhibits a more alkaline nature and higher pH values [18, 51]. This increased alkalinity enhances the cation exchange capacity (CEC), allowing NBC to retain a greater amount of nutrients and positively charged ions [51]. Conversely, NBC produced at lower temperatures or from low-ash feedstocks like hardwood tends to have lower pH values, rendering it more acidic. Acidic NBC is particularly beneficial for alkaline soils, where it helps lower soil pH. Lower pH also reduces CEC compared to high-pH NBC [18, 51]. Additionally, surface functional groups such as carboxyl (-COOH) and hydroxyl (-OH) play a role in determining pH, with acidic functional groups lowering pH and basic functional groups raising it [51].

High-pH NBC is effective for adjusting the pH of acidic soils, improving the availability of essential nutrients like phosphorus, and reducing the solubility of toxic elements such as aluminum [18]. Conversely, low-pH NBCs are advantageous for slightly lowering the pH of alkaline soils, supporting crops that thrive in acidic conditions. Furthermore, low-pH NBC enhances the solubility and availability of micronutrients like iron and manganese, which are more accessible to plants in acidic environments [18, 51].

(b) **Cation Exchange Capacity (CEC) and Functional Groups**

The Cation Exchange Capacity (CEC) of NBC represents its ability to retain and exchange positively charged ions (cations), indicating its effectiveness in holding essential nutrients and making them available to plants [18]. This property is closely tied to the functional groups present in NBC, including hydroxyl (–OH), carboxyl (–COOH), and carbonyl (–C=O), which influence its reactivity and application potential. NBC typically demonstrates a higher CEC than bulk biochar due to its larger surface area and denser concentration of functional groups, as outlined in the physical properties section. These functional groups play a vital role in facilitating cation adsorption and nutrient exchange between NBC and soil [51].

The feedstock and production conditions significantly impact the concentration and types of functional groups in NBC [77]. For instance, NBC derived from agricultural residues often has higher concentrations of reactive hydroxyl and carboxyl groups compared to woody biomass. Similarly, NBC produced at lower pyrolysis temperatures and shorter durations retains more functional groups, while higher temperatures and longer pyrolysis times lead to the breakdown of organic matter and a reduction in functional group density [52, 77].

The high CEC of NBC offers significant benefits for nutrient-poor or degraded soils, acting as a nutrient reservoir that reduces dependency on synthetic fertilizers, minimizes nutrient leaching, and enhances plant growth. Furthermore, NBC is effective in adsorbing heavy metals and other positively charged pollutants from contaminated soils, immobilizing these contaminants, and preventing their spread [35, 38, 39]. Functional groups further enhance NBC's capabilities: carboxyl groups increase heavy metal adsorption, while amine groups improve the removal of anionic contaminants like nitrates and phosphates [52].

(c) **Ash Content**

The ash content of NBC refers to the inorganic residue left after the combustion of organic matter, predominantly comprising essential minerals like calcium, potassium, magnesium, and phosphorus. Ash content is directly proportional to the pH of NBC; higher ash content results in more alkaline NBC, which is beneficial for neutralizing acidic soils. Additionally, the minerals present in the ash contribute to the cation exchange capacity (CEC) of the soil, ensuring better nutrient retention for crops. The mineral oxides in the ash can also influence the surface charge of NBC, thereby affecting its adsorption capacity and interaction with nutrients and contaminants [18, 36]. The ash content of NBC varies significantly, ranging from 1% to over 70%, depending on the type of feedstock and pyrolysis temperature. For instance, woody biochars typically have a lower ash content (1.5–3.0%) compared to agricultural residue-based biochars, such as those derived from peanut shells, which exhibit higher ash content (7–12%) [31, 53]. Higher ash content NBC has proven effective in neutralizing acidic soils, enhancing nutrient solubility, and immobilizing pollutants and heavy metals, thereby contributing to environmental remediation. Research demonstrates that NBC produced from agricultural residues like peanut shells is particularly effective in balancing soil pH and improving nutrient dynamics, making it a valuable amendment for sustainable agriculture [31, 53]. Table 3 explains the characterization techniques for NBC.

Table 3 Characterization Techniques for NBC

Technique	Principle	Parameters analyzed	Ref.
Scanning electron microscopy (SEM)	Uses a focused beam of electrons to generate high-resolution images of surface topography	– Surface morphology and texture – Particle size – Pore structure	[51–58]
Transmission electron microscopy (TEM)	Electrons are transmitted through a very thin sample to generate high-magnification Images	– Particle size distribution – Internal structures	[51–58]
X-ray diffraction (XRD)	Measures diffraction patterns as X-rays interact with crystalline materials	– Crystallinity – Degree of graphitization	[51–58]
Brunauer- Emmett-teller (BET) analysis	Measures gas adsorption on the BC surface to determine the surface area and porosity	– Specific surface area – Pore volume	[51–58]
Fourier Transform Infrared Spectroscopy (FTIR)	Measures absorption of infrared light by functional groups on the surface of the material at various frequencies	– Functional groups	[51–58]
Dynamic light scattering (DLS)	Measures fluctuations in the intensity of scattered light caused by the Brownian motion of particles, allowing the calculation of their hydrodynamic diameter.	– Particle size distribution – Zeta potential	[51–58]
X-ray photoelectron spectroscopy (XPS)	Measures binding energies of surface elements by irradiating a material with X-rays	– Surface chemistry (oxidation state and functional groups) – Elemental composition	[51–58]
Inductively coupled plasma mass spectrometry (ICP-MS)	Analyzes metal ion concentrations by ionizing elements and measuring mass/charge ratio	– Elemental composition – Trace and heavy metal concentration	[51–58]

4 Applications

Nanobiochar, owing to its enhanced properties such as increased surface area, porosity, and functional groups, has shown immense potential in advancing sustainable agriculture. Figure 4 illustrates the diverse applications of NBC in agriculture.

4.1 Soil Amendment/Conditioner

NBC represents a transformative advancement in soil amendment practices, offering great improvements in soil health, structure, and functionality. This innovative biomaterial serves as a powerful soil conditioner by improving its physicochemical properties—soil porosity, moisture retention, and nutrient availability—that are crucial for sustainable agricultural practices. For instance, the application of coconut shell NBC (0.1% to 1.0%) under sloping land reduced cumulative runoff by 39.7–74.4%, reduced the nitrate loss by 13.6%–59.8% with the increased water infiltration rate that decreased surface runoff by 0.57–43.5% and soil erosion by 10–11% [59, 61, 62]. Similarly, in sandy soils, the application of 0.7 to 1% of oak tree branch NBC resulted in reduced nitrate loss by 29.9–42.7%, thereby enhancing the retention of other essential nutrients phosphorous (P) and potassium (K) by stabilization of dissolved organic carbon bound to phosphorous [60] and its large surface area and cation exchange capacity resulting in sufficient and readily available nutrients for plants without the application of excessive chemical fertilizers [61]. Furthermore, NBC can modify soil porosity by enhancing the micropores, thereby reducing the size of macropores and resulting in a higher soil moisture content and

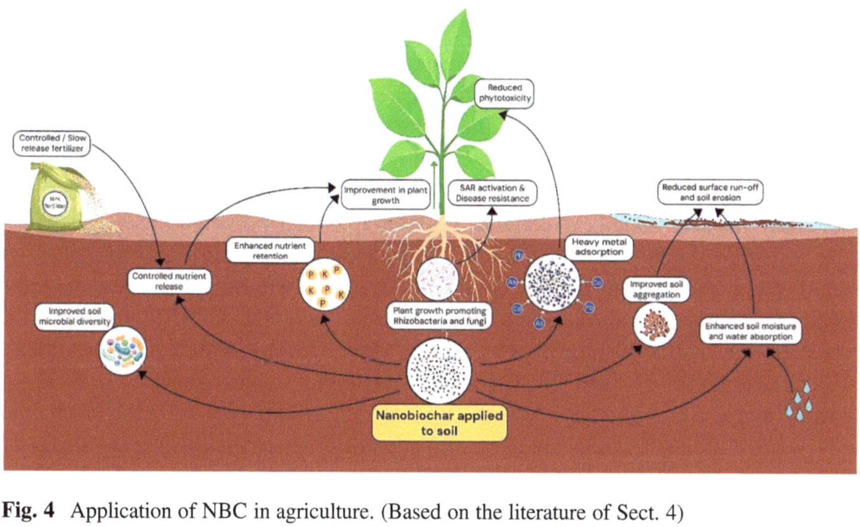

Fig. 4 Application of NBC in agriculture. (Based on the literature of Sect. 4)

water absorption capability. Enhanced micropore structure supports better stabilization of the soil aggregates, making it resistant to erosion and reducing the crack intensity factor in densely compacted soils to 3.9% from 0.4%, improving water-holding capacity and soil structure [63]. Zhou et al. (2020) reported improved soil health through increased soil carbon levels, beneficial microbial activity, and stabilization of soil aggregates necessary for soil fertility while promoting healthy plant growth by positively affecting the seed germination and root-shoot development at lower application rates of 0.184–0.268 tons per hectare, making it cheaper as compared to the conventional BC applied at a rate of 5–50 tons per hectare as per IBI-International BC Initiative [61].

Although NBC offers significant benefits, such as improving soil properties, increasing nutrient availability, and enhancing plant growth, there are potential risks associated with its application. These risks include the presence of heavy metals in nano-BC, which can lead to phytotoxicity, and the generation of free radicals and reactive oxygen species (ROS) during the pyrolysis process, which may harm plants. Additionally, changes in soil pH and contamination levels caused by NBC application can negatively impact plant growth and microbial populations [18]. To maximize its effectiveness while minimizing these risks, careful management and monitoring of NBC applications are essential.

4.2 *Remediation of Contaminants/Pollutants from Agricultural Soil*

Due to the enhanced physicochemical properties—high surface area, porosity, and cation exchange capacity—of NBC, it serves as one of the easiest materials to immobilize contaminants and pollutants, decrease bioavailability and phytotoxicity of the contaminants in the environment [27]. It facilitates contaminant removal through different mechanisms, viz., Surface complexation, ion exchange, electrostatic interactions, pi stacking and pi-pi stacking, and silicate binding.

Surface complexation is the primary mechanism involved in contaminant removal that enables effective binding with ionized pollutants owing to its high surface area and functional groups: carboxylic groups (–COOH) and hydroxide groups (–OH) [42]. Studies demonstrated that NBC can significantly reduce the bioavailability and toxicity of cadmium (Cd) in rice plants owing to its high surface area and pore volume, which improves Cd adsorption capacity and mitigates its negative impact on plant growth [61]. Furthermore, increasing the pyrolysis temperature enhances the surface area and adsorption potential of NBC, reducing Cd concentrations within plant tissues by 86.5–95.1% [62]. This reduction occurs due to the transformation of available Cd forms into more stable compounds. Such as $CdCO_3$ and $Cd(OH)_2$, which lower Cd uptake by plants [61, 62]. Ion exchange and electrostatic interactions are another mechanism where the surface chemistry of NBC enables it to attract and bind various ionic contaminants. For example, Zhou et al. (2017) demonstrated that functionalized NBC modified with metals or inorganic particles—manganese oxide (MnO_2)—can effectively adsorb copper ions with an adsorption capacity of up to 142 mg/g of Cu(II), highlighting its significant potential for heavy metal remediation [65]. Furthermore, NBC can effectively mitigate harmful organic toxic chemicals, reducing physiological damage in plants caused by soil contaminants. Its application has also been shown to enhance chlorophyll a and b contents by 69–79% and 79–91%, respectively, while lowering malondialdehyde (MDA) content, an indicator of oxidative stress, by 15% in cadmium-contaminated soils [66]. Additionally, NBC alleviates oxidative stress by downregulating the activity of antioxidative enzymes which are typically induced by heavy metal exposure [67]. Pi stacking and Pi-Pi stacking is another mechanism that enhances the NBC's adsorption capability for organic contaminants. These bonds occur between aromatic guest molecules and the aromatic moieties within NBC. This interaction is facilitated by the planar and rigid arrangement of aromatic compounds in NBC, making it highly effective for adsorbing aromatic agrochemicals like pesticides, herbicides, and fungicides, thereby supporting the removal of residual agrochemical contaminants from soil [68]. Silicate binding is yet another mechanism for contaminant removal where Silica (SiO_2), an anion, binds with various cations across a wide range of pH through ligand exchange. When modified by zwitterions, the anionic functionalities on silica surfaces can host both cations and anions more efficiently, enhancing its effectiveness in contaminant removal [42].

NBC not only aids in contaminant removal but also enhances soil health by promoting microbial diversity and biomass. In cadmium-contaminated soils, NBC has been shown to significantly increase the relative abundance of beneficial microbial communities, including *Proteobacteria, Actinobacteria,* and *Bacteroidetes.* This increase in microbial diversity and activity is due to reduced cadmium toxicity and the supportive habitat provided by the porous structure of NBC [64].

NBC offers a multifaceted approach to soil decontamination. Its high surface area, functionalized surface chemistry, and porous structure enable effective immobilization and removal of heavy metals and toxic chemicals. However, as NBC is still in its early stages of application, its long-term bioavailability and environmental impacts remain uncertain. Hence, further research is needed to understand its performance over long-term periods, as well as its effect on soil health and ecosystem dynamics.

4.3 *Slow/Controlled Fertilizer Releaser*

NBC is gaining attention as a promising material for the controlled release of fertilizers, helping to sustain nutrient levels in the soil over time. NBC and nanofertilizers derived from BC from various feedstocks such as corncob, wheat straw, pine needles, and Lantana camara release nutrients in controlled patterns, as reviewed by Chaubey et al., 2023 [27]. NBC tends to release most of the nutrient load in two phases: a rapidly available one in a relatively short period after application, and the other will be slower and uniformly available over time [59]. This can coincide with the fluctuating nutritional demands of the growing plants: that is, a maximum nutrient supply would be needed during the early stages of plant growth; a steady but moderate nutrient supply would be useful during later stages of growth. It starts off fast but slowly tapers down both on macro as well as micro-nutrients, such as NO_3^-, P_2O_5, K_2O, Mg_2^+, Ca_2^+, Fe_2^+, Fe_3^+, Zn_2^+, and Na_2O, and also stabilizes towards the end of the growth cycle. Particularly, the NO_3 and P_2O_5, are released rapidly in the initial stages but also reach a plateau sooner than other nutrients [59]. The nutrient release rate varies with the type of feedstock used for the production of NBC, as the pore structure and ion adsorption capacity differ with different types of feedstocks. NBC prepared from pine needles had better retention and sorption capacities for essential nutrients such as iron, magnesium, and sodium compared to those produced from wheat stalks and black gram. This difference in nutrient leaching capacity plays a crucial role in developing customized fertilizer amendments for specific plant needs and soil types [69].

In addition to controlled nutrient release, NBC also enhances nutrient uptake by plants and improves growth parameters such as shoot length, girth diameter, leaf area index, and crop yield [22]. For instance, the application of coconut shell-derived NBC increased maize crop yield by 4.4–5.7% compared to the control (unamended) treatment. Similarly, goat manure-derived NBC resulted in an increased plant (11%) and dry weight (25%), thereby enhancing nitrogen and copper uptake [48]. Additionally, a comparative study reported that NBC derived from wheat straw enhanced the germination percentages of cucumber, tomato, lettuce, and carrot by 11.0%, 10.0%, 17.5%, and 10.3%, respectively, compared to control (unamended) treatment [70].

NBC can be applied as foliar as well as soil application; both these application methods enhance root and shoot biomass, plant pigment content, nutrient status, and physiological indices. Foliar application of NBC boosts total soluble sugars, free amino acids, flavonoids, free proteins, and phenolics. While foliar application alone enhances plant growth, the most significant improvements are observed with a combination of foliar (0.3%) and soil (3%) applications [71].

Hence, NBC represents an innovative approach to controlled or slow-release fertilization, offering sustained nutrient supply, improved plant growth, and environmental benefits. Careful selection of feedstocks and optimization of application rates can effectively employ NBC for the stimulation of agricultural productivity as well as for sustainable practices (Fig. 4).

4.4 Stress Tolerance

NBC has demonstrated significant potential in boosting plant resilience to both biotic and abiotic stresses by enhancing soil health and promoting plant growth under adverse conditions. NBC improves plant resistance towards fungal foliar diseases by having an elicitor effect on the microbial communities associated with the plant [78]. Soil microbial diversity and activity increase with the application of NBC, thereby making nutrients more available for the plant to absorb and stimulating plant growth by modifying hormonal biosynthesis as well as enzymatic functions [78]. In addition, microbial activity has also led to the suppression of pathogens through actions like antibiosis, competition, and parasitism. Another role of NBC is the activation of systemic acquired resistance (SAR) in plants [26, 42] This increases the population of plant growth-promoting rhizobacteria and fungi, and it eventually triggers SAR mechanisms through the synthesis of phytohormones to provide plants with higher resistance against diseases [42]. For instance, NBC has been shown to significantly reduce lesions caused by *Phytophthora* on tobacco plants, highlighting its potent role in systemic acquired resistance (SAR). It has been reported that NBC stimulates plant immunity against pathogens by promoting growth and enhancing disease resistance [72]. Additionally, rice straw-derived NBC has demonstrated effectiveness in reducing disease incidence and severity caused by *Xanthomonas campestris pv. vesicatoria*, providing further evidence of its role in pathogen resistance [73].

Apart from its benefits for biotic stress tolerance, NBC has demonstrated considerable potential in terms of enhancing the resilience of plants towards abiotic stresses such as salinity and drought. Studies have shown that NBC effectively mitigates the adverse effects of salinity stress by promoting both shoot and root growth. For example, magnesium-based NBC has been shown to improve plant growth under salinity stress, while its application reduces sodium accumulation in plants like safflower, thereby increasing salinity tolerance. NBC achieves this by regulating soil pH and increasing cation exchange capacity [74, 75]. Furthermore, NBC improves drought tolerance by enhancing soil water use efficiency. It increases soil water retention, supporting plant growth even under water-limited conditions [76].

5 Challenges and Considerations

NBC exhibits a promising potential for tackling several agrarian and environmental challenges. However, its large-scale production and application bring along certain challenges and considerations. To begin with, the economic feasibility of NBC production and application may vary widely depending on the synthesis method. For example, reports estimate that output from ball milling may range from $8.2 to $103 per kilogram, carbonization at $6.6–22 per kilogram, centrifugation at $34–438 per kilogram, and sonication at $2.3–434 per kilogram [27]. An average farmer would consider a cost-benefit analysis before implementing additional steps required to convert bulk BC to NBC. For applications other than agriculture, NBC may still have an economic edge over alternatives like activated charcoal, graphene oxide, or carbon- based nanofibers. However, challenges such as low production yield, agglomeration, high energy consumption, and the heterogeneity of feedstocks may impede large-scale commercialization [42, 51, 80–82].

Another consideration is the limited availability of literature pertaining to the risks associated with the release of nanoparticles in the environment during field application or through the leaching of contaminants. Application-based, comprehensive lifecycle assessments are required to ensure the safe and sustainable use of nanomaterials. Organizations like The International Biochar Initiative (IBI), European Biochar Certificate (EBC), and World Biochar Certificate (WBC) have been instrumental in establishing guidelines and standards to ensure the sustainable production, quality, and application of BC. The IBI Guidelines focus on the production and characterization of BC, the EBC complements the IBI with stringent criteria for BC used in agriculture, construction, and environmental remediation. Meanwhile, the WBC focuses on BC's environmental performance, including its carbon footprint and lifecycle analysis. Such frameworks are essential to build consumer confidence, support regulatory compliance, and enable sustainable commercialization of such products globally [83, 84].

6 Conclusion

NBC is a promising material with applications across diverse fields [27, 85–87]. However, it is most popularly explored in the field of agriculture and soil remediation. In countries like India which generate a significant amount of agricultural biomass waste arising from wheat and rice husks, sugarcane bagasse, coconut shells, etc. can be efficiently engineered into NBC. This can prove to be very useful in tackling critical issues like declining soil fertility, water scarcity, and environmental pollution. For instance, it can help in curbing air pollution by reducing stubble burning in states like Punjab and Haryana by providing an eco-friendly soil amendment alternative. NBC can help improve nutrient use efficiency in areas such as the weathered soils of central and southern India, which suffer from a growing deficiency of secondary and micronutrients. Moreover, it can improve the water holding capacity in arid regions like Rajasthan and Gujarat, thereby supporting sustainable agriculture. Additionally, NBC can aid in the remediation of heavily polluted regions, such as those near industrial clusters in Maharashtra and West Bengal, by adsorbing heavy metals and other contaminants.

In summary, in this chapter, we have presented the multifaceted role of NBC in sustainable agriculture, emphasizing its potential to improve soil health, enhance crop productivity, and mitigate environmental challenges. While the potential of NBC is immense, several areas require further investigation to ensure its effective and sustainable use. A critical gap lies in understanding the long-term impact of NBC on soil ecosystems, particularly its interactions with soil microbiota and nutrient cycles. More research is also needed to optimize production methods, balancing cost reduction with the retention of its enhanced properties. Environmental and health risks associated with its nano-scale size need a thorough assessment to ensure safe agricultural application. Additionally, studies focusing on the mechanisms by which NBC mitigates soil stress conditions like salinity and drought, along with its potential to reduce greenhouse gas emissions, are crucial. Standardizing characterization techniques would enhance reproducibility and comparability across research. Investigating interactions between NBC and agrochemicals could reveal synergistic effects, and lifecycle assessments are necessary to evaluate its large-scale viability. Finally, policymakers and governing bodies must encourage research on NBC by offering incentives and subsidies in relevant areas. A well-planned approach can result in NBC emerging as a next-generation material that can remold future agricultural practices.

References

1. Department of Agriculture and Farmers' Welfare releases Final Estimates of major agricultural crops for 2023–24—Available online: https://pib.gov.in/Pressreleaseshare.aspx?PRID=2058534 (accessed on 22nd November 2024).
2. NPMCR. Available online: http://agricoop.nic.in/sites/default/files/NPMCR_1.pdf (accessed on 22nd November 2024).
3. Bhuvaneshwari, S., Hettiarachchi, H., & Meegoda, J. N. (2019). Crop residue burning in India: Policy challenges and potential solutions. *International Journal of Environmental Research and Public Health, 16*(5). https://doi.org/10.3390/ijerph16050832
4. Anuradha, S. K., Prashanth, C. S., & Bishnoi, R. (2021). Crop residue burning and its harmful impacts: A review. *Just Agriculture Multidisciplinary e-Newsletter, 1*(12).
5. Chaudhary, I. J., & Soni, S. (2020). Crop residue burning and its effects on the environment and microbial communities. In *Microbes in agriculture and environmental development* (pp. 87–106). CRC Press. https://doi.org/10.1201/9781003057819-5
6. Thandra, B., Sairam, M., Shankar, T., & Maitra, S. (n.d.). *Crop Residue Burning in India: Causes, Impacts, and, Solutions*. https://www.researchgate.net/publication/357428912
7. Talang, P. N., & R., Na Sorn, W., Polruang, S., & Sirivithayapakorn, S. (2024). Alternative crop residue management practices to mitigate the environmental and economic impacts of open burning of agricultural residues. *Scientific Reports, 14*(1), 14372. https://doi.org/10.1038/s41598-024-65389-3
8. Deshpande, M. V., Kumar, N., Pillai, D., Krishna, V., & v., & Jain, M. (2023). Greenhouse gas emissions from agricultural residue burning have increased by 75% since 2011 across India. *Science of the Total Environment, 904*, 166944. https://doi.org/10.1016/j.scitotenv.2023.166944
9. Lin, M., & Begho, T. (2022). Crop residue burning in South Asia: A review of the scale, effect, and solutions with a focus on reducing reactive nitrogen losses. *Journal of Environmental Management, 314*, 115104. https://doi.org/10.1016/j.jenvman.2022.115104
10. Yadav R, Bhat M.I., Sirazuddin, Singh R, Ahmad R, Pooniyan S (2020). Burning of crop residues: How does it affect soil health. Just Agriculture Multidisciplinary e-newsletter vol. 1 (1).
11. Khare, P., Deshmukh, Y., Yadav, V., Pandey, V., Singh, A., & Verma, K. (2021). BC production: A sustainable solution for crop residue burning and related environmental issues. *Environmental Progress and Sustainable Energy, 40*(2). https://doi.org/10.1002/ep.13529
12. Li, S., & Tasnady, D. (2023). BC for soil carbon sequestration: Current knowledge, mechanisms, and future perspectives. *C-Journal of Carbon Research, 9*(3) Multidisciplinary Digital Publishing Institute (MDPI). https://doi.org/10.3390/c9030067
13. Lehmann, J., Cowie, A., Masiello, C. A., Kammann, C., Woolf, D., Amonette, J. E., Cayuela, M. L., Camps-Arbestain, M., & Whitman, T. (2021). BC in climate change mitigation. *Nature Geoscience, 14*(12), 883–892). Nature Research. https://doi.org/10.1038/s41561-021-00852-8

14. Wang, J., Xiong, Z., & Kuzyakov, Y. (2016). BC stability in soil: Meta-analysis of decomposition and priming effects. *GCB Bioenergy, 8*(3), 512–523). Blackwell Publishing Ltd. https://doi.org/10.1111/gcbb.12266
15. Bekchanova, M., Campion, L., Bruns, S., Kuppens, T., Lehmann, J., Jozefczak, M., Cuypers, A., & Malina, R. (2024). BC improves the nutrient cycle in sandy-textured soils and increases crop yield: A systematic review. *Environmental Evidence, 13*(1) BioMed Central Ltd. https://doi.org/10.1186/s13750-024-00326-5
16. Hossain, M. Z., Bahar, M. M., Sarkar, B., Donne, S. W., Ok, Y. S., Palansooriya, K. N., Kirkham, M. B., Chowdhury, S., & Bolan, N. (2020). BC and its importance on nutrient dynamics in soil and plant. *BC, 2*(4), 379–420). Springer. https://doi.org/10.1007/s42773-020-00065-z
17. Azzi Adhikari, S., Timms, W., & Mahmud, M. A. P. (2022). Optimizing water holding capacity and hydrophobicity of BC for soil amendment—A review. *Science of the Total Environment, 851*, 158043. https://doi.org/10.1016/j.scitotenv.2022.158043
18. Lehmann, J., & Joseph, S. (Eds.). (2015). *BC for environmental management*. Routledge. https://doi.org/10.4324/9780203762264
19. Azzi, E. S., Karltun, E., & Sundberg, C. (2021). Small-scale BC production on Swedish farms: A model for estimating potential, variability, and environmental performance. *Journal of Cleaner Production, 280*, 124873. https://doi.org/10.1016/j.jclepro.2020.124873
20. Leng, L., Xiong, Q., Yang, L., Li, H., Zhou, Y., Zhang, W., Jiang, S., Li, H., & Huang, H. (2021). An overview on engineering the surface area and porosity of BC. *Science of the Total Environment, 763*, 144204. https://doi.org/10.1016/j.scitotenv.2020.144204
21. Bhandari, G., Gangola, S., Dhasmana, A., Rajput, V., Gupta, S., Malik, S., & Slama, P. (2023). Nano-BC: Recent progress, challenges, and opportunities for sustainable environmental remediation. *Frontiers in Microbiology, 14*). Frontiers Media SA. https://doi.org/10.3389/fmicb.2023.1214870
22. Rajput, V. D., Minkina, T., Ahmed, B., Singh, V. K., Mandzhieva, S., Sushkova, S., Bauer, T., Verma, K. K., Shan, S., van Hullebusch, E. D., & Wang, B. (2022). Nano-BC: A novel solution for sustainable agriculture and environmental remediation. *Environmental Research, 210*, 112891. https://doi.org/10.1016/j.envres.2022.112891
23. Huang, H., Reddy, N. G., Huang, X., Chen, P., Wang, P., Zhang, Y., Huang, Y., Lin, P., & Garg, A. (2021). Effects of pyrolysis temperature, feedstock type and compaction on water retention of BC amended soil. *Scientific Reports, 11*(1). https://doi.org/10.1038/s41598-021-86701-5
24. Nagaraju, K., Prasad, T. N. V. K. V., Chari, M. S., Ramu, Y. R., & Murthy, B. R. (2023). Synthesis and characterization of Rice husk NBC-based N and K fertilizers: Promoting environmental sustainability in agriculture. *International Journal of Plant & Soil Science, 35*, 254–261.
25. Hu, Y., Cao, Y., Ma, C., & Yan, W. (2024). Nano-BC as a potential amendment for metal(loid) remediation: Implications for soil quality improvement and stress alleviation. *Journal of Environmental Management, 351*, 119658. https://doi.org/10.1016/j.jenvman.2023.119658
26. Shafiq, F., Anwar, S., Firdaus-e-Bareen, Z. L., & Ashraf, M. (2023). Nano-BC: Properties and prospects for sustainable agriculture. *Land Degradation and Development, 34*(9), 2445–2463). John Wiley and Sons Ltd. https://doi.org/10.1002/ldr.4620
27. Chaubey, A. K., Pratap, T., Preetiva, B., Patel, M., Singsit, J. S., Pittman, C. U., & Mohan, D. (2023). Definitive review of NBC. *ACS Omega*. American Chemical Society, *9*, 12331. https://doi.org/10.1021/acsomega.3c07804
28. Bardi, M. J., Mutunga, J. M., Ndiritu, H., & Koch, K. (2023). Effect of pyrolysis temperature on the physiochemical properties of BC and its potential use in anaerobic digestion: A critical review. *Environmental Technology and Innovation, 32*, 103349. https://doi.org/10.1016/j.eti.2023.103349

29. Chacón, F. J., Sánchez-Monedero, M. A., Lezama, L., & Cayuela, M. L. (2020). Enhancing BC redox properties through feedstock selection, metal preloading and post- pyrolysis treatments. *Chemical Engineering Journal, 395*, 125100. https://doi.org/10.1016/j.cej.2020.125100
30. Murtaza, G., Usman, M., Iqbal, J., et al. (2024). Liming potential and characteristics of BC produced from woody and non-woody biomass at different pyrolysis temperatures. *Scientific Reports, 14*, 11469. https://doi.org/10.1038/s41598-024-61974-8
31. Tomczyk, A., Sokołowska, Z., & Boguta, P. (2020). BC physicochemical properties: Pyrolysis temperature and feedstock kind effects. In reviews in environmental science and biotechnology (Vol. 19, issue 1, pp. 191–215). Springer. https://doi.org/10.1007/s11157-020-09523-3
32. Kambo, H. S., & Dutta, A. (2015). A comparative review of BC and hydrochar in terms of production, physicochemical properties and applications. *Renewable and Sustainable Energy Reviews, 45*, 359–378., ISSN 1364-0321,. https://doi.org/10.1016/j.rser.2015.01.050
33. Ng, L. Y. F., Ariffin, H., Yasim-Anuar, T. A. T., Farid, M. A. A., & Hassan, M. A. (2022). High-energy ball milling for high productivity of NBC from oil palm biomass. *Nanomaterials (Basel, Switzerland), 12*(18). https://doi.org/10.3390/nano12183251
34. Yuan, Y., Zhang, N., & Hu, X. (2020). Effects of wet and dry ball milling on the physicochemical properties of sawdust derived-BC. *Instrumentation Science & Technology, 48*(3), 287–300. https://doi.org/10.1080/10739149.2019.1708751
35. Liu, G., Zheng, H., Jiang, Z., Zhao, J., Wang, Z., Pan, B., & Xing, B. (2018). Formation and physicochemical characteristics of Nano BC: Insight into chemical and colloidal stability. *Environmental Science and Technology, 52*(18), 10369–10379. https://doi.org/10.1021/acs.est.8b01481
36. Anupama, & Khare, P. (2021). A comprehensive evaluation of inherent properties and applications of nano-BC prepared from different methods and feedstocks. *Journal of Cleaner Production, 320*, 128759. https://doi.org/10.1016/j.jclepro.2021.128759
37. Xu, X., & Cölfen, H. (2021). Ultracentrifugation techniques for the ordering of nanoparticles. *Nanomaterials, 11*(2), 333. https://doi.org/10.3390/nano11020333
38. Song, B., Chen, M., Zhao, L., Qiu, H., & Cao, X. (2019). Physicochemical property and colloidal stability of micron- and nano-particle BC derived from a variety of feedstock sources. *Science of the Total Environment, 661*, 685–695. https://doi.org/10.1016/j.scitotenv.2019.01.193
39. Behnam, H., & Firouzi, A. F. (2023). Effects of synthesis method, feedstock type, and pyrolysis temperature on physicochemical properties of BC nanoparticles. *Biomass Conversion and Biorefinery, 13*(15), 13859–13869. https://doi.org/10.1007/s13399-021-02108-2
40. Guo, F., Bao, L., Wang, H., Larson, S. L., Ballard, J. H., Knotek-Smith, H. M., Zhang, Q., Su, Y., Wang, X., & Han, F. (2020). A simple method for the synthesis of BC nanodots using hydrothermal reactor. *MethodsX, 7*, 101022. https://doi.org/10.1016/j.mex.2020.101022
41. Xu, Y., Bai, T., Yan, Y., Zhao, Y., Yuan, L., Pan, P., & Jiang, Z. (2020). Enhanced removal of hexavalent chromium by different acid-modified BC derived from corn straw: Behavior and mechanism. *Water Science and Technology, 81*(10), 2270–2280. https://doi.org/10.2166/wst.2020.290
42. Sani, M. N. H., Amin, M., Siddique, A. B., Nasif, S. O., Ghaley, B. B., Ge, L., Wang, F., & Yong, J. W. H. (2023). Waste-derived NBC: A new avenue towards sustainable agriculture, environment, and circular bioeconomy. *Science of the Total Environment, 905*, 166881. https://doi.org/10.1016/j.scitotenv.2023.166881
43. Verheijen, F., Jeffery, S., Bastos, A. C., van der Velde, M., & Diafas, I. (n.d.). BC application to soils A critical scientific review of effects on soil properties. *Processes and Functions*. https://doi.org/10.2788/472
44. Jiang, M., He, L., Niazi, N. K., Wang, H., Gustave, W., Vithanage, M., Geng, K., Shang, H., Zhang, X., & Wang, Z. (2023). NBC for the remediation of contaminated soil and water: Challenges and opportunities. *BC, 5*(1) Springer. https://doi.org/10.1007/s42773-022-00201-x
45. Udomkun, P., Chandi, K., Boonupara, T., & Kaewlom, P. (2024). Innovative approaches: Exploring nano-BC technology's impact on soil properties, alachlor retention, and microbial

populations. *Environmental Technology and Innovation, 35*, 103659. https://doi.org/10.1016/j.eti.2024.103659

46. Xiao, J., Hu, R., & Chen, G. (2020). Micro-nano-engineered nitrogenous bone BC developed with a ball-milling technique for high-efficiency removal of aquatic Cd(II), Cu(II) and Pb(II). *Journal of Hazardous Materials, 387*, 121980. https://doi.org/10.1016/j.jhazmat.2019.121980
47. Ravindiran, G., Rajamanickam, S., Janardhan, G., Hayder, G., Alagumalai, A., Mahian, O., Lam, S. S., & Sonne, C. (2024). Production and modifications of BC to engineered materials and its application for environmental sustainability: A review. *BC, 6*(1) Springer. https://doi.org/10.1007/s42773-024-00350-1
48. Rashid, M. I., Shah, G. A., Iqbal, Z., Ramzan, M., Rehan, M., Ali, N., Shahzad, K., Summan, A., Ismail, I. M. I., & Ondrasek, G. (2023). NBC associated ammonia emission mitigation and toxicity to soil microbial biomass and corn nutrient uptake from farmyard manure. *Plants, 12*(9), 1740. https://doi.org/10.3390/plants12091740
49. Radwan, S., Hussien, E.-H. A., Ahmed, B., & Ali, N. (2024). Green solutions for soil contamination: Sustainable production and characterization of nano-BC from sugarcane bagasse and olive mill waste. https://doi.org/10.21203/rs.3.rs-4430795/v1
50. Brewer, C. E., Chuang, V. J., Masiello, C. A., Gonnermann, H., Gao, X., Dugan, B., Driver, L. E., Panzacchi, P., Zygourakis, K., & Davies, C. A. (2014). New approaches to measuring BC density and porosity. *Biomass and Bioenergy, 66*, 176–185. https://doi.org/10.1016/j.biombioe.2014.03.059
51. Ramanayaka, S., Vithanage, M., Alessi, D. S., Liu, W. J., Jayasundera, A. C. A., & Ok, Y. S. (2020). NBC: Production, properties, and multifunctional applications. *Environmental Science: Nano, 7*(11), 3279–3302). Royal Society of Chemistry. https://doi.org/10.1039/d0en00486c
52. Mahmoud, S. E. M. E., Ursueguia, D., Mahmoud, M. E., Abel-Fattah, T. M., & Díaz, E. (2023). Functional surface homogenization of NBC with cation exchanger for improved removal performance of methylene blue and lead pollutants. *Biomass Conversion and Biorefinery*. https://doi.org/10.1007/s13399-023-04098-9
53. Nataraja, K. C., Balaguravaiah, D., Rao, C. S., Krishna, T. G., Ramu, Y. R., & Kumari, P. L. (2021). Comparative analysis of chemical composition and spectral properties of BC produced from pigeonpea and cotton residues. *Journal of Research ANGRAU, 49*(2).
54. Naghdi, M., Taheran, M., Brar, S. K., Rouissi, T., Verma, M., Surampalli, R. Y., & Valero, J. R. (2017). A green method for production of NBC by ball milling- optimization and characterization. *Journal of Cleaner Production, 164*, 1394–1405.
55. Aziz, S., Uzair, B., Ali, M. I., et al. (2023). Synthesis and characterization of NBC from rice husk BC for the removal of safranin and malachite green from water. *Environmental Research, 238*, 116909.
56. Helal, M. I. D., Husein, M. E., & El, W. G. (2019). Characterization of agricultural residues-based Nano BC and its efficiency in adsorption/desorption of nutrients. *International Journal of Environment, 8*, 130–141.
57. Mahmoud, E., El Baroudy, A., Ali, N., & Sleem, M. (2020). Spectroscopic studies on the phosphorus adsorption in salt-affected soils with or without nano-BC additions. *Environmental Research, 184*, 109277.
58. Ng, L. Y. F., Ariffin, H., Yasim-Anuar, T. A. T., Farid, M. A. A., & Hassan, M. A. (2022). High-energy ball milling for high productivity of NBC from oil palm biomass. *Nanomaterials, 12*, 3251.
59. Lateef, A., Nazir, R., Jamil, N., Alam, S., Shah, R., Khan, M. N., Saleem, M., & Rehman, S.-. (2019). Synthesis and characterization of environmental friendly corncob BC based nano-composite—A potential slow release nano-fertilizer for sustainable agriculture. *Environmental Nanotechnology, Monitoring & Management, 11*, 100212. https://doi.org/10.1016/j.enmm.2019.100212
60. Chen, M., Alim, N., Zhang, Y., Xu, N., & Cao, X. (2018). Contrasting effects of BC nanoparticles on the retention and transport of phosphorus in acidic and alkaline soils. *Environmental Pollution, 239*, 562–570. https://doi.org/10.1016/j.envpol.2018.04.050

61. Zhou, b. B., chen, x. P., & henry, l. (2020). The effect of nano-BC on soil, water, and nutrient loss of a sloping land with different vegetation covers on loess plateau of China. *Applied Ecology and Environmental Research, 18*(2), 2845–2861. https://doi.org/10.15666/aeer/1802_28452861
62. Chen, X., Zhou, B., Wang, Q., Tao, W., & Lin, H. (2020). Nano-BC reduced soil erosion and nitrate loss in sloping fields on the loess plateau of China. *Catena, 187*, 104346. https://doi.org/10.1016/j.catena.2019.104346
63. Kumar, H., Cai, W., Lai, J., Chen, P., Ganesan, S. P., Bordoloi, S., Liu, X., Wen, Y., Garg, A., & Mei, G. (2020). Influence of in-house produced BCs on cracks and retained water during drying-wetting cycles: Comparison between conventional plant, animal, and nano-BCs. *Journal of Soils and Sediments, 20*(4), 1983–1996. https://doi.org/10.1007/s11368-020-02573-8
64. Liu, W., Li, Y., Feng, Y., Qiao, J., Zhao, H., Xie, J., Fang, Y., Shen, S., & Liang, S. (2020). The effectiveness of NBC for reducing phytotoxicity and improving soil remediation in cadmium-contaminated soil. *Scientific Reports, 10*(1), 858. https://doi.org/10.1038/s41598-020-57954-3
65. Zhou, L., Huang, Y., Qiu, W., Sun, Z., Liu, Z., & Song, Z. (2017). Adsorption properties of Nano-MnO2–BC composites for copper in aqueous solution. *Molecules, 22*(1). https://doi.org/10.3390/molecules22010173
66. Ramzan, M., Zia, A., Naz, G., Shahid, M., Shah, A. A., & Farid, G. (2023). Effect of NBC (nBC) on morpho-physio-biochemical responses of black cumin (Nigella sativa L.) in Cr-spiked soil. *Plant Physiology and Biochemistry, 196*, 859–867. https://doi.org/10.1016/j.plaphy.2023.02.037
67. Ren, T., Chen, N., Wan Mahari, W. A., Xu, C., Feng, H., Ji, X., Yin, Q., Chen, P., Zhu, S., Liu, H., Liu, G., Li, L., & Lam, S. S. (2021). BC for cadmium pollution mitigation and stress resistance in tobacco growth. *Environmental Research, 192*, 110273. https://doi.org/10.1016/j.envres.2020.110273
68. Wang, X. M., Kou, H., Wang, J., Teng, R., Du, X., & Lu, X. (2020). An octahedral magnetic metal organic frameworks for efficient extraction and enrichment of six pesticides with benzene ring prior to high performance liquid chromatography analysis. *Journal of Porous Materials, 27*(4), 1171–1177. https://doi.org/10.1007/s10934-020-00895-z
69. Das, S. K., & Ghosh, G. K. (2023). Development and evaluation of BC-based secondary and micronutrient enriched slow release nano-fertilizer for reduced nutrient losses. *Biomass Conversion and Biorefinery, 13*(13), 12193–12204. https://doi.org/10.1007/s13399-021-01880-5
70. Yang, y., zhou, b., hu, z., & lin, h. (2020). The effects of nano-BC on maize growth in northern Shaanxi province on the loess plateau. *Applied Ecology and Environmental Research, 18*(2), 2863–2877. https://doi.org/10.15666/aeer/1802_28632877
71. Khaliq, H., Anwar, S., Shafiq, F., Ashraf, M., Zhang, L., Haider, I., & Khan, S. (2023). Interactive effects of soil and foliar-applied NBC on growth, metabolites, and nutrient composition in Daucus carota. *Journal of Plant Growth Regulation, 42*(6), 3715–3729. https://doi.org/10.1007/s00344-022-10832-w
72. Kong, M., Liang, J., White, J. C., Elmer, W. H., Wang, Y., Xu, H., He, W., Shen, Y., & Gao, X. (2022). BC nanoparticle-induced plant immunity and its application with the elicitor methoxyindole in Nicotiana benthamiana. *Environmental Science: Nano, 9*(9), 3514–3524. https://doi.org/10.1039/D1EN00953B
73. Aftab, Z.-H., Aslam, W., Aftab, A., Shah, A. N., Akhter, A., Fakhar, U., Siddiqui, I., Ahmed, W., Majid, F., Wróbel, J., Ali, M. D., Aftab, M., Ahmed, M. A. A., Kalaji, H. M., Abbas, A., & Khalid, U. (2022). Incorporation of engineered nanoparticles of BC and fly ash against bacterial leaf spot of pepper. *Scientific Reports, 12*(1), 8561. https://doi.org/10.1038/s41598-022-10795-8
74. Ghassemi-Golezani, K., & Farhangi-Abriz, S. (2021). BC-based metal oxide nanocomposites of magnesium and manganese improved root development and productivity of safflower (Carthamus tinctorius L.) under salt stress. *Rhizosphere, 19*, 100416. https://doi.org/10.1016/j.rhisph.2021.100416

75. Zulfiqar, F., Wei, X., Shaukat, N., Chen, J., Raza, A., Younis, A., Nafees, M., Abideen, Z., Zaid, A., Latif, N., Naveed, M., & Siddique, K. H. M. (2021). Effects of BC and BC–compost mix on growth, performance and physiological responses of potted Alpinia zerumbet. *Sustainability, 13*(20), 11226. https://doi.org/10.3390/su132011226
76. Zhang, K., Wang, Y., Mao, J., & Chen, B. (2020). Effects of BC nanoparticles on seed germination and seedling growth. *Environmental Pollution, 256*, 113409. https://doi.org/10.1016/j.envpol.2019.113409
77. Mia, S., Dijkstra, F. A., & Singh, B. (2017). Aging induced changes in BC'S functionality and adsorption behavior for phosphate and ammonium. *Environmental Science & Technology, 51*, 8359–8367. https://doi.org/10.1021/acs.est.7b00647
78. Ramadan, M. M., Asran-Amal, & Abd-Elsalam, K. A. (2019). Micro/nano BC for sustainable plant health: Present status and future prospects. In *Carbon nanomaterials for agri-food and environmental applications* (pp. 323–357). Elsevier. https://doi.org/10.1016/B978-0-12-819786-8.00016-5
79. Yaashikaa, P. R., Kumar, P. S., Varjani, S., & Saravanan, A. (2020). A critical review on the BC production techniques, characterization, stability and applications for circular bioeconomy. *Biotechnology Reports, 28*). Elsevier B.V. https://doi.org/10.1016/j.btre.2020.e00570
80. Khurana, A., & Kumar, V. (2022). *State of biofertilizers and organic fertilizers in India.* Centre for Science and Environment. https://ruralindiaonline.org/en/library/resource/state-of-biofertilizers-and-organic-fertilizers-in-india/
81. Mukherjee, A., Maity, A., Pramanik, P., Shubha, K., Joshi, D. C., & Wani, S. H. (2019). Public perception about use of nanotechnology in agriculture. In *Advances in phytonanotechnology: From synthesis to application* (pp. 405–418). Elsevier. https://doi.org/10.1016/B978-0-12-815322-2.00019-5
82. Teli, S., Soni, S., Teli, P., & Agarwal, S. (2024). Recent insights into modified biochars: A half-decade study. *Journal of Materials Science, 59*, 18357. https://doi.org/10.1007/S10853-024-10272-9
83. Guidelines: World Biochar Certificate. (2023). https://www.european-biochar.org/media/doc/2/wbc_1_0b.pdf
84. International Biochar Initiative: Homepage. (n.d.). Retrieved November 25, 2024, from https://biochar-international.org/
85. Song, B., Cao, X., Gao, W., Aziz, S., Gao, S., Lam, C. H., & Lin, R. (2022). Preparation of nano-biochar from conventional biorefineries for high-value applications. *Renewable and Sustainable Energy Reviews, 157*, 112057. https://doi.org/10.1016/J.RSER.2021.112057
86. Pathak, H. K., Seth, C. S., Chauhan, P. K., Dubey, G., Singh, G., Jain, D., Upadhyay, S. K., Dwivedi, P., & Khoo, K. S. (2024). Recent advancement of nano-biochar for the remediation of heavy metals and emerging contaminants: Mechanism, adsorption kinetic model, plant growth and development. *Environmental Research, 255*, 119136. https://doi.org/10.1016/J.Envres.2024.119136
87. Ashfaq, M., Chauhan, D., & Talreja, N. (2024). Nanobiochar: An emerging material for the environment. *Energy, and Biomedical Applications., 1–16.* https://doi.org/10.1007/978-981-97-6544-7_1

Nanocoating in Postharvest Management of Crops, Floriculture and Fruits

Poorvaja Subramanian and Preethi Kathirvel

Contents

P. Subramanian · P. Kathirvel (✉)
Department of Microbial Biotechnology, Bharathiar University, Coimbatore, Tamil Nadu, India

R. S. Chaughule, A. V. Shelar (eds.), *Nanotechnology in Agriculture*,
https://doi.org/10.1007/978-3-031-97544-8_10

1 Introduction

Postharvest losses represent a significant challenge in the agricultural sector, particularly for fruits, vegetables, and flowers. These losses occur after harvest, extending through storage, transportation, and marketing until the produce finally reaches consumers. Unlike pre-harvest losses, which are often anticipated and mitigated through various agronomic practices, postharvest losses are more complex, involving both quantitative and qualitative dimensions [Fig. 1]. These losses are more expensive in terms of labor and economic value, in addition to being more agonizing. Vegetables, for example, have a high moisture content (usually 80–90%), which makes them particularly vulnerable to postharvest degradation. These crops are perishable, which emphasizes how crucial it is to implement efficient postharvest management techniques to protect their viability and marketability. India, with its diverse ago-climatic conditions, is well-positioned to produce various crops throughout the year [1]. The country's ability to grow numerous flowers, potted plants, foliage, and aromatic plants year-round is a testament to its agricultural potential These crops, however, are live commodities that continue their life processes even after harvesting.

Thus, posttreatments contribute to the depletion of the crop's food reserves and moisture content. Once separated from the parent plant, produce no longer receives the necessary water and nutrients, making it reliant on its reserves, which diminishes over time, leading to spoilage. Water loss due to transpiration is a primary factor affecting the quality of postharvest vegetables, leading to symptoms such as wilting, shriveling, soft texture, and a decline in nutritional value. Despite the robust agricultural output, India faces a significant challenge with postharvest losses. It is estimated that 30–40% of fresh produce is lost during the postharvest phase, representing a substantial economic burden and posing serious public health concerns [2]. This loss is particularly concerning given the vital role that fruits and vegetables play in providing essential nutrients, vitamins, minerals, and

Fig. 1 The key biological factors that cause spoilage

nutraceuticals necessary for preventing a range of diseases. The limited shelf life of produce exacerbates this issue, underscoring the urgent need for improved preservation techniques to enhance food security and nutritional adequacy.

Plant-pathogenic microorganisms, including bacteria and fungus, are the source of postharvest illnesses that worsen the situation. Produce quality and longevity during storage and transportation are greatly compromised by these diseases. To overcome these obstacles, creative solutions beyond standard postharvest management techniques are needed. In this regard, new methods for reducing postharvest losses are provided by nanotechnology, which has become a game-changing instrument in food science and agriculture [3].

Sustainable horticulture has unique prospects thanks to nanotechnology, which manipulates materials at the nanoscale. Because of their small size, nanomaterials have certain features that make it possible to design sophisticated postharvest management systems. Among these, functional food packaging, nanofilms, nanocoating, and nanosensors stand out as important developments. By increasing the shelf life of fruits, vegetables, and flowers, these nanotechnologies play a critical role in lowering postharvest losses and improving the overall quality of food. For example, fruits and vegetables can have a protective layer called nanocoating applied to their surface, which slows down the ripening process, prevents microbial development, and minimizes water loss. In a similar vein, retailers and customers can benefit from real-time food freshness monitoring through the integration of nanosensors into packaging. These developments contribute to preserving the nutritional content and sensory attributes of fresh produce in addition to extending its shelf life. Nonetheless, care must be taken when implementing nanotechnology in food science and agriculture. Although using nanomaterials has many advantages, there are also some drawbacks, especially when it comes to toxicity, health problems, and environmental effects. To guarantee that nanotechnology applications in agriculture are both efficient and secure, certain laws and safety precautions must be put in place.

The goal of this chapter is to give a thorough examination of the function of nanocoating in fruit, floriculture, and agricultural postharvest management. It looks at how postharvest losses can be addressed, shelf life can be increased, and creative packaging options can be used. By addressing the complexities of postharvest losses with cutting-edge solutions, the chapter will contribute to the broader goal of achieving sustainable agricultural practices that support food security and public health on a global scale [4].

2 Postharvesting of Crops

The majority of postproduction procedures, including cleaning, washing, choosing, grading, disinfecting, drying, packing, and storage, are included in postharvesting management [Fig. 2]. These actions are essential for removing unwanted components, enhancing the product's appearance, and guaranteeing that it satisfies quality

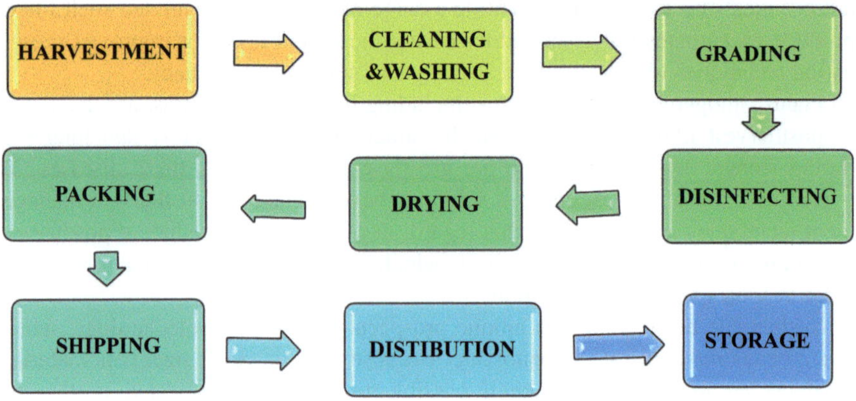

Fig. 2 The process involved in postharvesting

requirements for both processed and fresh commodities. Managing and regulating factors like temperature, relative humidity, packing choice, and the use of additional treatments like fungicides are all part of postharvest operations. Fruits and vegetables should have their worth increased gradually until they are consumed after they are gathered. Postharvest management aims to maximize additional value by minimizing losses from harvesting to distribution. In the end, this should help the whole community by boosting export revenue or prolonging the season for fresh product availability [5].

3 Impact of Biotic Factors on Crop Yields

The yield of horticultural crops is significantly reduced by various biotic and abiotic factors, with diseases and insects among the primary concerns. In India, these crops are particularly vulnerable to a wide range of globally prevalent diseases and pests. For example, crops in the Solanaceous family, such as potatoes and sweet potatoes, are especially susceptible to diseases like late blight, bacterial wilt, and early blight, as well as pest infestations from potato tuber moths, sweet potato weevils, and beetles. Additionally, horticultural crops can be affected by diseases caused by bacteria, nematodes, fungi, and viruses. Some specific instances include bacterial canker, speck and spot disease, bacterial wilt, fusarium wilt, early and late blight, powdery mildew, root-knot nematodes, and viral diseases such as the tomato yellow leaf curl virus. Furthermore, horticultural crops in India are also targeted by several insect pests that can cause significant damage and yield losses. Some pests are exclusive to a single crop, while others affect multiple crops [6].

4 The Limitations of Artificial Fungicides

Postharvest chemical fungicides, such as imazalil, thiabendazole, pyrimethanil, and fludioxonil, are used to citrus or boscalid and iprodione to control postharvest deterioration of fruit and vegetables. NJChloride-based chemicals can create chlorinated organic molecules, such as dichloramines, trichloromethanes, and chloramines, which are respiratory irritants when employed as sanitation agents. Due to the widespread use of chemical fungicides after harvest, pathogens have developed resistant strains, which has reduced the efficacy of fungicides. After some time, these methods are crucial to extending the technical life of fungicides, even though they are costly and time-consuming. Additionally, because chemical fungicides' active ingredients and co-formulants have been linked to health issues and environmental damage, consumers are concerned about their use.

5 Factors Affecting Postharvesting of Crops

Horticultural products are recognized as high-value crops, contributing both nutritious food and income-generating opportunities, such as through mushroom cultivation, floriculture, and the production of vegetable seeds. The perishability of fruits and vegetables varies significantly by crop, with some being highly perishable while others are less so. Postharvest losses in these crops affect not only their quantity and quality but also their visual appeal, which in turn impacts their marketability. A significant amount of resources, including seeds, planting supplies, land, fertilizers, water, energy, labor, and other inputs, are wasted as a result of these losses. Fruits and vegetables are generally acknowledged to experience larger postharvest losses than grains, oilseeds, and pulses. This is mostly because of their increased moisture content, bulkiness, weight, and rapid respiration rate of living tissues.

The factors leading to postharvest losses in horticultural produce can be categorized into primary and secondary factors. Primary factors include mechanical injuries, biological and chemical factors, biochemical reactions, microbial activity (such as fungi and bacteria), physiological processes, psychological influences, and environmental conditions. The postharvest degradation of fruits and vegetables is mostly caused by these factors. Secondary factors can exacerbate the effects of primary factors, leading to significant postharvest losses in horticultural products. Examples of these secondary factors include improper handling during harvesting and transportation, inadequate packaging, exposure to high temperatures during postharvest handling, and delays in getting the produce to market [7].

6 Nanotechnology in Postharvesting of Crops

The major objective of nanotechnology is the creation, characterization, manufacturing, and use of systems, devices, and structures through form and size manipulation at the nanoscale scale. These formations, called nanoparticles, exist on a scale known as a nanometer, which is one billionth of a meter (1×10^{-9} m). Nanoparticles are usually between one and one hundred nanometers in size. Because of the special qualities that nanoparticles possess, nanotechnology is a valuable instrument in many fields of study and business. In disciplines such as science, medicine, agriculture, and industry, there is a growing interest in understanding and manipulating matter at the nanoscale due to the distinct properties materials possess at this scale compared to their larger-scale counterparts. Revolutionary biodegradable packaging materials, such as edible and biodegradable films made from renewable resources, have been developed in response to the desire to increase food products' shelf lives and improve their quality while reducing packaging waste. These materials present a viable answer to the waste management problem because they are biodegradable. Nanotechnology has increasingly found applications within the food industry, providing significant benefits [8].

6.1 Antimicrobial Packing

The human food chain faces a variety of physical, chemical, and microbiological risks that pose serious safety concerns. One significant postharvest pathogen is *Botrytis cinerea*, which causes gray mold disease. This pathogen adversely affects many commercially important fruits and vegetables both during the growing season and in storage after harvest. Furthermore, it poses challenges for long-distance transportation of these products. Controlling this disease is especially crucial during storage, as it spreads rapidly among fruits and vegetables and can thrive at low temperatures (e.g., −0.5 °C). The application of packaging nanotechnology can enhance food safety by reducing material toxicity, controlling the flow of gases and moisture, and extending shelf life. Nanomaterials present an opportunity to improve the mechanical and thermal properties of packaging, thereby better-protecting foods from undesirable mechanical, thermal, chemical, or microbiological effects. Using coatings offers several benefits, including reduced water loss, slowed ripening, decreased risk of chilling and mechanical injury, diminished decay, and enhanced shine or gloss on coated products.

Antimicrobial packaging systems can include adding antimicrobial nanoparticle sachets to the package, dispersing bioactive agents within the packaging, coating the surface of the packaging with bioactive agents, and utilizing antimicrobial macromolecules with film-forming properties or edible matrices [9].

6.2 Nano-packing

In general, ingredients used in processing and packaging significantly improve the quality and shelf life of food items. Every year, a significant proportion of food spoils worldwide due to a lack of proper packaging expertise and technology. Nano-based technologies have already been proven to be the most successful of the cutting-edge food processing and packaging technologies. Nanomaterials used in nano wrapping and processing techniques can enhance and maintain the exceptional quality and shelf life of food. Gas scavengers, edible antimicrobial nanocomposite films, and metal nanoparticles including Ag, Cu, TiO_2, and MgO are examples of active food nano-packaging materials. Example: Since oxygen causes fruits and vegetables to deteriorate quickly, adding gaseous scavenger-NPs to processing and packing materials can stop oxidative processes and prolong the shelf life and quality of fruits and vegetables. It was discovered that nanocrystalline TiO_2 acts as an oxygen scavenger when food is packaged and processed. In addition, because of their antimicrobial properties, Ag-NPs, ZnO-NPs, TiO_2-NPs, MgO-NPs, CuO, chitosan nanoparticles, carbon nanotubes, quantum dots, etc., are commercially significant for enhancing the quality and shelf-life of fruits, vegetables, and imports [10].

6.3 Types of Nanomaterials Used in Postharvesting

Nanomaterials such as silver nanoparticles (AgNPs), nano-clay, nano-zinc oxide (nano-ZnO), nano-titanium dioxide (nano-TiO_2), and titanium nitride nanoparticles (nano-TiN) are being incorporated into food packaging as functional additives. Edible coatings and nano-lamination of fruits and vegetables can also be achieved using these materials, which offer various functional packaging applications [Fig. 3].

1. Silver nanoparticles (AgNPs): AgNPs were specifically designed for their antibacterial and sterilizing properties. They have a greater tendency to release

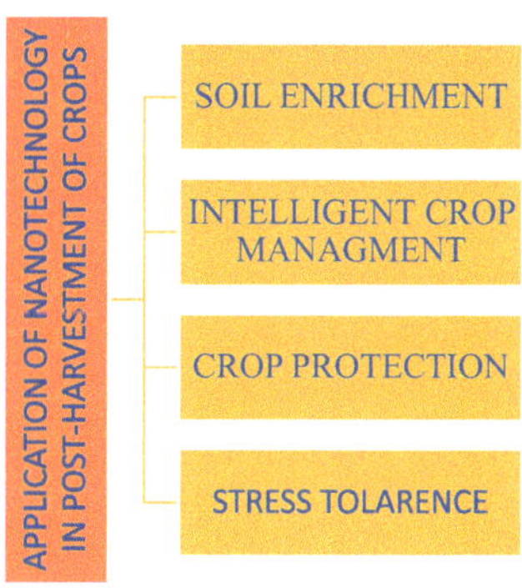

Fig. 3 The applications of nanotechnology in the postharvesting of crops [13]

silver ions compared to bulk silver, making them effective because of their larger surface area per mass.

2. Metal oxide nanoparticles: Metal oxide nanoparticles, such as titanium dioxide and zinc oxide, work as photocatalysts to destroy microbes and organic compounds. Reactive oxygen species (ROS) are produced when nano-ZnO and nano-TiO_2 interact, oxidizing the cytoplasm of bacteria and killing them. According to studies, ZnO is favored over silver since it is less costly and hazardous. The European Food Safety Authority (EFSA) authorized the use of Nano-TiN as a food contact material in 2012. Nano-TiN is created by heating TiO_2 particles in a nitrogen-containing gas at high temperatures [11].
3. Zeolite nanoparticles: As montmorillonite nanoclays function as a barrier to both moisture and oxygen, they can be applied as nano-coatings to prolong the shelf life of fruit and prevent spoiling. Coatings containing zeolite nanoparticles can absorb ethylene gas, promoting fruit ripening and extending shelf life. Additionally, the application of antioxidant-rich nanoparticles in nano-coatings can slow down oxidative processes, preserving fruit color and nutritional value [12].

Several types of nanocoating have been created to improve fruit quality and prolong fruit storage. An example of such innovation is chitosan-based nano-coatings, which are derived from the natural biopolymer chitin. These coatings help maintain fruit quality by offering protection against microbial growth. The use of eco-friendly nano-coatings, which eliminate the need for traditional chemical preservatives and packaging materials, also brings additional environmental benefits. Moreover, it is crucial to research consumer perception and acceptance of fruits coated with nanoparticles. These studies examine sensory aspects, safety concerns, and labeling to reassure consumers.

6.4 Applications of Nanotechnology in Postharvesting of Crops

6.4.1 Soil Enrichment

Nanotechnology contributes to greater agricultural sustainability by providing creative ways to enrich the soil by improving its physical and hydrological characteristics. Nanomaterials (NMs) have a major impact on the bulk density, porosity, structure, and texture of soil. For example, adding nanoparticles (NPs) improves water drainage and aeration by forming nanoscale holes and channels between soil aggregates. By decreasing the space between soil particles and enhancing the stiffness of the soil structure, platinum nanoparticles have been demonstrated to raise the saturated hydraulic conductivity. Similarly, soil aggregation is improved by carbon nanotubes, which can increase the stability of sandy loam soils by up to 35% due to their high aspect ratios and elastic qualities. Additionally, NMs are essential for water management because they can increase the soil's capacity to store water by improving its surface area and hydrophilicity or by creating water-retentive hydrogels. These characteristics aid plants in resisting drought stress and water loss. Additionally, applying NMs increases soil stability by lowering settlement and liquefaction. Utilizing nanotechnology, soil may be improved to support sustainable farming methods, maximize water use, and boost crop yields [14].

Nanotechnology also plays a pivotal role in enhancing nutrient retention and availability in the soil, ensuring that plants absorb essential minerals more effectively. Nanomaterials like nano-zeolites act as slow-release carriers for fertilizers, reducing nutrient leaching and environmental pollution. Moreover, nanoparticles can facilitate the remediation of contaminated soils by adsorbing heavy metals and toxins, thus restoring soil health and fertility. Nanoclays, with their high cation exchange capacities, further aid in binding nutrients and preventing their loss through runoff. The incorporation of biochar nanoparticles can improve soil organic matter, promoting microbial activity and enhancing soil fertility over time. Additionally, nanoscale sensors embedded in the soil can provide real-time monitoring of soil health, including moisture content, pH, and nutrient levels, enabling precise interventions for optimal crop growth. Together, these advancements make nanotechnology a cornerstone for revolutionizing soil management in postharvest agriculture.

6.4.2 Intelligent Crop Management and Crop Protection

A revolutionary approach to crop management is provided by nanocoating technology, which creates protective coatings that increase plant yield and resilience. Applying these incredibly thin coatings to seeds, leaves, or fruits creates a barrier that protects them from environmental stresses including dehydration, diseases, and pests. By permitting the regulated release of fertilizers and agrochemicals, nanocoating increases the efficiency of nutrient intake while lowering waste and its negative effects on the environment. They can also improve the shelf life of harvested crops, control moisture retention, and provide UV protection. Farmers may manage crops sustainably by using nanocoating in farming methods while reducing resource inputs and increasing output.

6.4.3 Stress Tolerance

Nanocomposites significantly improve postharvest crops' ability to tolerate stress because of their special structural and functional characteristics. Nanocomposites can reduce oxidative stress brought on by the buildup of reactive oxygen species (ROS) during handling and storage by modifying plant metabolic activities. These substances engage in apoplastic and symplastic interactions with biological processes, and their efficacy is greatly influenced by their size and zeta potential. By regulating ROS levels, they may protect photosynthetic pigments, preserve cellular integrity, and lessen lipid peroxidation. Furthermore, their impact on plant hormones linked to stress, such as cytokinins and abscisic acid (ABA), enhances stress resistance [15]. By these means, nanocomposites can increase crop nutritional value, reduce postharvest losses, and prolong shelf life, providing a long-term solution to the problems.

Nanocomposites also play a key role in enhancing enzymatic activities that counteract stress-induced damage, such as superoxide dismutase (SOD) and catalase (CAT), which neutralize ROS and maintain cellular homeostasis. Their nano-dimensions allow for efficient penetration into plant tissues, enabling localized protection and repair. Additionally, nanocomposites can interact with cell wall components to reinforce structural integrity, helping crops withstand mechanical stress during transportation and storage. The incorporation of nanoparticles like silica and titanium dioxide can create a protective barrier on crop surfaces, reducing water loss and microbial infections. Moreover, nanocomposites can regulate ion channels, minimizing ion toxicity and osmotic stress in harvested crops. By integrating these multifunctional properties, nanocomposites offer a comprehensive approach to improving postharvest stress tolerance, thereby ensuring better quality and extended marketability of agricultural produce.

7 Nano-structured Conductive Polymers

DeBerry first proposed the idea of employing conductive polymer coatings to stop metal corrosion in 1985. According to DeBerry's research, even after being exposed to sulfuric acid for an extended period, stainless steel coated with polyaniline maintained its passivation. Since this discovery, conducting polymer coatings—which offer unique qualities including high conductivity, simplicity in synthesis, affordability, environmental sustainability, non-toxicity, and particular redox chemistries—have drawn a lot of attention for their potential in corrosion prevention. Although the exact protective mechanisms of these coatings are not fully understood, it is widely accepted that they protect metals by forming physical barriers, oxide layers, and passive oxide layers that hinder corrosion. These coatings work well to keep ions, water, and oxygen from penetrating the metal surface. Coatings with certain properties, like long-term corrosion resistance, high adhesion, durability, and resistance to mechanical abrasion, are required in many different industries. Studies have indicated that the integration of nanoparticles into conducting polymer coatings can improve a range of characteristics, including mechanical strength, adhesion, thermal stability, hydrophobicity, magnetostatic shielding, and electrochemical anticorrosion. Nanoparticles also help to fill pores in the coating, thereby increasing its barrier effect. Polymer coatings embedded with nanoparticles such as TiO2, Zn, ZnO, Si, SiO_2, Al_2O_3, and CeO_2, have been found to offer superior protection compared to traditional polymer coatings [16].

In addition to monitoring crop health and soil conditions, nanocoating-based sensors can also enhance postharvest management by detecting changes in the storage environment. These sensors can monitor humidity, temperature, and gas levels, such as ethylene and carbon dioxide, which are critical indicators of ripening, spoilage, or degradation in stored crops. By providing real-time data, nanocoating sensors help ensure optimal storage conditions, reducing postharvest losses and extending the shelf life of agricultural products. Furthermore, these sensors can be incorporated into packaging materials, creating intelligent packaging that alerts consumers and producers to spoilage or contamination risks. This integration not only improves food safety but also reduces waste by enabling timely interventions. As nanotechnology continues to advance, these nanosensors will become more sensitive, cost-effective, and accessible, revolutionizing both precision agriculture and postharvest monitoring systems.

8 Nanocoating in Postharvesting

8.1 Nanocoating

A coating is a cohesive layer formed when a coating material is applied to a substrate one or more times. Nanocoating is extremely thin films at the nanoscale, typically ranging in thickness from 1 to 100 nanometers. Unlike traditional paint, nanocoating does not alter the surface's topography, fill in imperfections, or create a smooth surface; it also does not provide resistance to wear and abrasion. Instead, nanocoating imparts specific chemical or physical functions to a surface. Examples include gas-barrier coatings, oleophobic or hydrophilic/hydrophobic properties, enhanced corrosion resistance, and improved conductive or insulating qualities.

Nanocoating can be constructed using either a single molecule or multiple molecular layers and can be applied to a variety of substrates, including metals, glass, ceramics, and polymers. Some nanocoating is made from polymers either in situ or prior to application. "Smart coatings" refer to those with multifunctional properties, such as self-healing capabilities, controlled release of active substances, and thermal insulation. Very small amounts of additives are incorporated into the coating material to modify various characteristics, including structure, gloss, surface tension, flow behavior, and resistance to weather and UV light. Depending on the intended application, certain nanoparticles may be used in functional coatings, such as antimicrobial coatings. Coatings that contain nanomaterials generally offer superior processing qualities and high-performance characteristics compared to traditional coatings [17].

8.2 Vegetables and Fruits

Fruits are vital components of a healthy diet, offering rich sources of vitamins, minerals, antioxidants, and dietary fiber. However, their high moisture content significantly limits their shelf life, leading to rapid degradation. In consequence, numerous types of edible coatings and nanocomposite films have been created and are currently routinely used to increase the shelf life of fresh produce. Modified environment techniques, e.g., have proven to be effective in preserving the quality of these perishable goods. One key indicator of fruit and vegetable maturity is the total soluble solids content (°Brix), and edible coatings have proven effective in reducing TSS and slowing the ripening process. Antimicrobial nanoparticles, which are used to increase the shelf life and quality of fruits and vegetables during storage, are carried by these coatings. The type of food, its surface features, the rheological qualities of the coating solution, and the coating's main objective all influence the choice of coating technique. These factors collectively determine the effectiveness of the coating in preserving fresh produce. The adhesion of the coating to the food surface is particularly important, as is wettability, which assesses the interfacial contact between the food surface and the coating. Common methods for applying edible coatings to fresh fruits and vegetables include dipping, spraying, or hand-coating [18]. Though techniques like fluidized bed processing and foaming exist, they are primarily confined to research environments. The effectiveness of these coatings is influenced by the type of coating used, the quantity applied, as well as its viscosity and surface tension.

8.3 Flowers

Standardizing appropriate postharvest methods is essential to preserving cut flowers' quality and lifespan, particularly to increase their export potential in India. Developing suitable solutions to prolong the vase life of flowers, a crucial quality metric must be the main focus of postharvest management research and innovation. The quality of cut flowers after harvest is greatly influenced by variables including temperature, humidity, and water relations. In tackling these issues, nanotechnology specifically, the use of silver nanoparticles has shown a great solution [19]. As an ethylene-binding inhibitor, silver nitrate ($AgNO_3$), a silver salt frequently found in commercial flower preservation treatments, successfully decreases ethylene-induced senescence in flowers. As seen in rose cut flowers, pulsing cut flowers with AgNO3 solutions has been shown to improve vase life and solution absorption. Further, silver nanoparticles' antibacterial qualities offer defense within plant tissues, which is particularly advantageous for ethylene-sensitive flowers like carnations. The quality and marketability of cut flowers can be greatly increased by including nanocoating, such as silver nanoparticles, in postharvest management techniques.

8.4 Nanofoods

The agri-food sectors have been significantly promoting research in nanotechnology. When food has been raised, generated processed, or packaged using nanoparticle tools, or methods from nanotechnology, the food is considered nanofood. It excludes food that has undergone atomic modification or food made using nanotechnology. Many of the major problems that the world's food supply is now experiencing might be resolved by nanotechnology. Food polymers and polymeric assemblages can be precisely modified to improve food safety and quality through the application of nanotechnology. Moreover, nano teas and canola active oil, a kind of canola cooking oil, are two new consumer products created with nanotechnology. An addition known as "nano drops" is included in canola oil and is intended to deliver phyto-chemicals, vitamins, and minerals [20].

Nanotechnology has also enabled the development of intelligent and functional nano-foods that address nutritional deficiencies and provide health benefits beyond basic nutrition. For example, nano-encapsulation techniques allow for the delivery of bioactive compounds such as probiotics, antioxidants, and omega-3 fatty acids in a controlled and targeted manner, enhancing their bioavailability and stability [21]. Because of their antioxidant properties that enhance food quality, the usage of nanoliposomes that encapsulate phenolic chemicals in EC has recently been investigated with several food applications. The physical characteristics of cellulose-based biopolymer coatings are altered by the addition of nanoliposomes that contain rutin or quercetin. The outcomes of an in vitro release analysis employing a fatty food simulant demonstrate the value of using nanoliposomes as an encapsulation technology that has antioxidant activity to extend food shelf life and preserve polyphenols while regulating their release from biodegradable matrices. Nano-coatings on fruits and vegetables can extend their shelf life by preventing microbial growth and oxidation. In addition, nanotechnology has facilitated the creation of nano-sensors embedded in food packaging, which can monitor freshness, detect contamination, and provide real-time feedback to consumers. Advanced nano-carriers are also being employed to mask unpleasant flavors and odors in fortified foods without altering their sensory qualities. Furthermore, the integration of nanoparticles into food production processes enables the precise manipulation of food textures, colors, and flavors to meet consumer preferences. As a result, nano-technology is reshaping the food industry by offering innovative solutions for enhancing food safety, sustainability, and nutritional value.

8.5 Nanoparticles: Automated Released

Nanoparticles and nanocarriers facilitate the regulated release of growth regulators, nutrients, and other agricultural byproducts to maximize the growth and development of plants. The idea of controlled release has been completely transformed by nanotechnology, which makes it possible to administer various substances—including medications, chemicals, and nutrients—precisely and effectively. Nanocarriers, which can encapsulate and release chemicals at predefined rates and under certain conditions, are at the core of this breakthrough. These nanocarriers, which are frequently made of nanoparticles or nanocapsules, provide several benefits. The use of nanotechnology in agriculture allows for controlled release, where nutrients are enclosed in nanoscale containers. By releasing nutrients to plants gradually in response to environmental variables like pH or soil moisture, these carriers make sure that plants get the proper amount of nutrients at the correct time. In addition to increasing agricultural productivity, this accuracy lowers fertilizer runoff and pollution in the environment [22].

8.6 Nanocoating as Sensor

The application of biosensors has important consequences for increasing crop yield, especially when combined with developments in molecular biology, nanomaterials, and microfluidics. These biosensors can be used to track soil microbial activity and anticipate possible illnesses.

When examining soil with biosensors, the fundamental principle is to measure the differences in oxygen consumption between positive and negative microorganisms. Additionally, biosensors can be utilized to detect contaminants and prevent their spread, leveraging the unique properties of nanomaterials. For example, biosensors can be employed to detect nitrate concentrations in plants and identify infected plants by detecting specific markers such as methyl salicylate and azelaic acid. Furthermore, biosensors have been reported to monitor the infection of citrus fruit by *Penicillium digitatum*. A gold nanoparticle-based electrochemical evaluation employing differential pulse voltammetry and a nanoparticle-based electrochemical biosensor that targets pathogen DNA via recombinase polymerase amplification was used to quickly and sensitively detect plant pathogens. By enabling precise farming techniques like satellite monitoring, geographic systems, and remote sensing technologies for identifying pests or stress indicators like drought, smart delivery systems, and nanosensors are essential to enhancing natural agricultural processes [23]. One of the main features of tools supported by nanotechnology is anticipated to be the integration of independent sensors with real-time GPS tracking. Moreover, nanosensors can be strategically placed throughout fields to monitor crop growth and soil parameters. Nanoparticle-based sensors were 10,000 times more sensitive than traditional PCR and gel electrophoresis techniques in detecting infected plants before any disease signs were apparent.

8.7 Properties of Edible Nanocoating

Molecular structure, rather than chemical composition or molecular size, is the primary factor that determines the efficiency of an edible coating. An effective edible coating needs to fulfil several requirements, according to Arvanitoyannis and Gorris (1999).

- It should be water-resistant and impermeable to water vapor to protect the product and form a proper barrier.
- The coating must allow for adequate oxygen levels (1–3%) around the product to prevent anaerobic respiration while avoiding excessive carbon dioxide buildup.
- It should enhance the product's structural integrity, appearance, and mechanical handling properties.
- The coating should be capable of carrying active ingredients like vitamins and antioxidants and help preserve volatile flavor compounds.
- The material used must be non-sticky, easily emulsifiable, and efficient in drying without melting at temperatures below 40 °C.
- It must not release any unwanted odors or degrade the fresh product's freshness.
- The coating should be of low viscosity, ranging from translucent to opaque, and capable of withstanding minor pressure.
- Finally, the coating must be cost-effective for practical use [24].

8.8 Methods of Nanocoating Synthesis

The methods for nanocoating synthesis are summarised below [25].

1. **The Sol–Gel Method**: Sol–gel methods create a gel-like structure by hydrolyzing and condensing precursors, usually metal alkoxides. To produce a thin, consistent nanolayer, the gel is coated and allowed to cure. It is frequently used in coatings that are hydrophobic, antibacterial, or UV-resistant.
2. **Chemical Vapor Deposition (CVD):** This process creates a thin nanocoating by reacting or breaking down gaseous precursors at high temperatures on the surface of a substrate. It is employed to create coatings with specific chemical and physical qualities that are incredibly consistent and adherent.
3. **Physical vapor deposition (PVD):** Materials are evaporated in a vacuum and then deposited as a thin layer on a substrate using a process known as PVD. PVD includes processes like evaporation and sputtering. It works well for producing long-lasting, resilient nanocoating.
4. **Layer-by-Layer (LbL) Assembly:** Electrostatic interactions are used to deposit alternating layers of materials with opposing charges on a substrate. Multilayered, functional nanocoating with exact thickness control may be produced using this technique.
5. **Dip-Coating and Spin-Coating:** To equally apply the coating material, a substrate is either submerged in a solution containing nanoparticles (dip-coating) or spun rapidly (spin-coating). These are easy and affordable ways to create homogeneous, thin nanolayers.
6. **Electrochemical Deposition:** Nanoparticles are transferred from a solution onto a substrate by use of electrochemical processes. This technique works well for producing composite and metal nanocoating.

7. **Spray coating:** The process of spraying a solution containing nanoparticles onto a substrate; when the solvent evaporates, a thin layer is created. It works well with vast surface areas and is scalable.
8. **Atomic Layer Deposition (ALD):** A regulated and accurate process that uses alternating exposure to chemical precursors to deposit atomic layers of materials on a substrate. Ultra-thin and conformal coatings are perfect for ALD.
9. **Plasma-Enhanced Coating:** Before or during the application of nanocoating, surfaces are activated or modified using plasma. Coating functionalization and adhesion are improved by this technique. Depending on the use, each technique has unique benefits, such as UV protection, antibacterial qualities, or regulated nutrient release for crop management.

8.9 Characterization of Nanocoating on Floriculture, Crops, Vegetables, and Fruits

The characterization of nanocoating involves various analytical techniques to evaluate their properties and effectiveness. **Zeta potential and particle size measurements** are performed at ambient temperature (25 ± 2 °C) using dynamic light scattering (DLS) with a zeta sizer laser diffractometer operating at 633 nm. DLS measures the scattering intensity caused by the Brownian motion of nanoparticles in suspension, providing insights into droplet size and stability. For example, the nanoparticle analyzer (SZ-100) can determine the size of nanoemulsion droplets, a critical parameter for ensuring uniform application and effective coating on crops and products [26]. Additionally, **SEM analysis** allows for detailed observation of the dried coatings on fruits and vegetables. Samples are prepared by placing them on postharvest crops using symmetrical adhesive tape, followed by platinum sputtering. These are then examined under an SEM operating at an acceleration voltage of 5 kV, providing high-resolution images of the coating structure and distribution [27].

Sensory evaluation techniques complement physical characterization to assess the quality and appeal of nanocoating. **Sensory visual shine analysis** is conducted using a reflectometer to measure the gloss or shine of coated fruits, which can enhance consumer appeal [28]. Additionally, **sensory flavor evaluation** is performed to assess the impact of nanocoating on fruit flavor. Frozen fruit juice samples are thawed and evaluated by trained panelists using William's design, under controlled conditions at 14 ± 1 °C. To calibrate their palates, panelists first taste a "warm-up" control sample before ranking coded samples based on flavor and rating off-flavors on a scale from 1 (no off-flavors) to 10 (extreme off-flavor). The testing is conducted in isolated booths under red lighting to eliminate visual bias, with data recorded using Compu Sense Cloud software. These sensory analyses ensure that nanocoating not only preserves the physical attributes of produce but also maintains or enhances their organoleptic qualities.

8.10 Physiochemical Characteristics

Nanocoating exhibits unique physicochemical properties that enhance their functional applications. They possess nanometer-scale dimensions, with particle sizes typically ranging from 1 to 100 nanometers, providing a high surface-to-volume ratio that improves their performance such as better adhesion, durability, and functional effectiveness. These coatings are optically transparent, ensuring that the natural appearance of coated materials, such as flowers and fruits, remains unaltered—an essential feature for applications in floriculture and fruit preservation. Additionally, weight reduction during storage can be monitored to evaluate the coating's efficiency. This is calculated as a percentage of the initial weight, allowing for the assessment of coated versus uncoated samples over time, as described in relevant studies.

This method provides valuable insights into the protective role of nanocoating in prolonging shelf life and maintaining product quality.

Nanocoating also exhibits exceptional mechanical, thermal, and chemical stability, which enhances its resistance to environmental stresses such as UV radiation, oxidation, and microbial growth. The high tensile strength and flexibility of these coatings allow them to form uniform, thin layers over complex surfaces without cracking or peeling. Furthermore, their hydrophobic or hydrophilic properties can be tailored depending on the application, aiding in moisture regulation, which is particularly beneficial for preserving perishable items. Nanocoating can also act as barriers to gas exchange, slowing down the ripening process by limiting oxygen and ethylene permeability in fruits and vegetables. Additionally, the incorporation of antimicrobial agents, such as silver or zinc oxide nanoparticles, into the nanocoating matrix enhances its ability to prevent microbial contamination, ensuring product safety. These multifunctional properties make nanocoating indispensable for a wide range of applications, from food preservation to advanced agricultural tools.

9 Toxicity and Safety and Consumer Acceptability

Though nanotechnology has many benefits for the food business, there are safety concerns that need to be addressed. Scholars have conducted a thorough investigation of the prospective health hazards and hazards related to nanoparticles transferring from packaging materials into food. As soon as the nanoparticles are introduced into the agricultural environment, several changes start to occur, Environmental factors can cause nanocomposites to degrade by releasing incorporated nanoparticles from the polymeric texture into the surrounding environment. Certain food packaging materials, such as low-density polyethylene, lose strength when exposed to environmental factors, UV radiation, or ozone in humid environments [23]. Even substances classified as Generally Regarded as Safe (GRAS) require further investigation when in their nano forms, as their physicochemical properties can differ significantly from those in their larger counterparts. Due to their minute size, nanomaterials are more prone to bioaccumulation in bodily tissues and organs. Assessing the safety of nanoparticles in food packaging involves evaluating exposure risks. If migration tests indicate no transfer of nanomaterials into food, the risk of human exposure is eliminated. However, if nanoparticles are ingested through food or packaging, they pose a significant exposure risk. Toxicity is typically assessed through in vivo and in vitro studies, with mice commonly used in the former and human and rodent cell lines in the latter [29]. Understanding the absorption, distribution, metabolism, and excretion of nanomaterials is essential for calculating their kinetics and potential accumulation in the body. Developing nanocoating techniques to control the surface characteristics of substrates for use in the environment, biosystems, food systems, etc., has practical limits in addition to nanotoxicity. Fruit coating takes a lot of time, and the proper coating cannot be achieved. Therefore, there is a great need for quick and innovative coating techniques. The size and quantity of NP-loaded cargoes for CRISPR DNA and protein payloads have been well-established following plant genetic editing. Furthermore, whether NPs can transport CRISPR chemicals to plant mitochondria for genome alteration is unclear. Whether these NPs can transport CRISPR reagents to plant plastids and mitochondria for genome alteration is yet uncertain [30].

Research into the application of nanotechnology in food packaging has largely centered on antimicrobial packaging, which has sparked consumer concerns regarding its safety and potential long-term effects on health due to the release of silver ions. Nonetheless, nanotechnology in food packaging also includes innovations like oxygen-absorbing films, oxygen-blocking plastic bottles, decay-inhibiting films, antibacterial food containers, stronger packaging films, UV-protection packaging, and barcodes for food security. Studies suggest that consumers may be more accepting of packaging improvements rather than direct changes to food, as the use of nanotechnology in packaging is generally perceived as less problematic by the public [31].

10 Future Prospects and Research Requirements

Nanotechnology has the potential to revolutionize the horticulture sector by providing innovative solutions to crop security challenges. It plays a crucial role in the management of nutrients and postharvest quality. However, several research gaps need to be addressed to maximize the benefits of nanotechnology in gardening.

One critical area of research involves developing secure and long-lasting nanoparticles for agricultural use. This includes creating and combining nanoparticles that are non-toxic to plants biocompatible and biodegradable for humans and animals [32]. Additionally, further investigation is necessary to understand the behavior and fate of nanomaterials in the environment to ensure their safety in horticulture.

There is a persistent risk of nanomaterials entering the food chain through air, water, and soil during their production and application. This contamination could result in DNA damage, disruption of cell membranes, and cell death. Currently, very few studies have examined the effects of nanofoods on human and animal health. To enhance consumer acceptance, proper labeling and regulations for nanofoods are essential. When controlled and regulated appropriately, the application of nanotechnologies has the potential to significantly improve food processing and product quality, thereby promoting human health and well-being.

Despite extensive research, the commercialization of biopolymers or their composites enhanced with nanomaterials has been slow. Scale-up studies should be included in research projects to facilitate the transition of products from the laboratory to safe industrial settings. However, before these materials can be widely used, further studies on nanobiocomposites are necessary to investigate their effects on human health and environmental safety, including aspects such as transport, cytotoxicity, and ecotoxicity. Therefore, products containing nanomaterials must be thoroughly examined for potential adverse effects on humans, animals, and plants [33].

11 Conclusion

Human life is significantly influenced by the food and agricultural sectors, which is why regulatory bodies have recently allowed the careful application of nanotechnology in these areas. The limitations of traditional farming practices have prevented available farmlands from being fully utilized. Additionally, there is always a risk of widespread foodborne illnesses, as agricultural raw materials undergo a lengthy processing chain that can lead to contamination by pathogenic bacteria and spoilage agents. Effective food preservation can only be achieved through the detection and sterilization of microbes. Current challenges include the delayed detection and measurement of toxins, spoilage, or pathogenic organisms, and the estimation of shelf life [34].

Nanotechnology emerges as one of the most promising solutions to address the limitations of conventional agricultural methods. It has the potential to enhance crop yield and quality at every stage of agriculture, including seed priming, germination, fertigation, and postharvest management. However, challenges still persist during the postharvest period, and existing technologies have not fully resolved these issues. The application of nanomaterials (NMs) in postharvest management of horticultural crops has proven to be an intriguing alternative. Research published by Maldonado in 2003 indicated that these nanomaterials can be utilized in various ways, such as being applied directly to agricultural products or incorporated into new materials used for coatings, packaging, and other applications [35]. The results have been promising, demonstrating that modifications in postharvest physiology and microbial management through nanotechnology can extend the shelf life of horticultural products. This chapter focuses on how the innovative method of nanocoating can address current challenges associated with postharvest loss. These coatings aim to enhance the preservation of crops against spoilage and microbial contamination, ultimately reducing postharvest losses.

References

1. Nath, A., Meena, L. R., Kumar, V., & Panwar, A. S. (2018). Postharvest management of horticultural crops for doubling farmerâ€™ s income. *Journal of Pharmacognosy and Phytochemistry, 7*(1S), 2682–2690.
2. Gajanana, T. M., Murthy, D. S., & Sudha, M. (2011). Post harvest losses in fruits and vegetables in South India–A review of concepts and quantification of losses. *Indian Food Packer, 65*(6), 178–187.
3. Prabhu, K., Jayakumari, S., Devi, D. G., & Rahe, R. (2022). A review of bio-Nanocoatings used to extend fruits shelf life. *Journal Of Food Chemistry And Nanotechnology, 7*(4), 78–84.
4. Ruffo Roberto, S., Youssef, K., Hashim, A. F., & Ippolito, A. (2019). Nanomaterials as alternative control means against postharvest diseases in fruit crops. *Nanomaterials, 9*(12), 1752.
5. Janghu, S., Kumar, V., & Yadav, A. K. (2024). Post-harvest management of fruits and vegetables.
6. Ashinie, S. K., & Tefera, T. T. (2019). Horticultural crops research and development in Ethiopia: Review on current status. *Journal of Biology, Agriculture and Healthcare, 9*(13), 1–14.
7. Singh, D., Sharma, R. R., & Kesharwani, A. K. (2021). Postharvest losses of horticultural produce. In *Postharvest handling and diseases of horticultural produce* (pp. 1–24). CRC Press.
8. Yadollahi, A., Arzani, K., & Khoshghalb, H. (2009, August). The role of nanotechnology in horticultural crops postharvest management. In *Southeast Asia symposium on quality and safety of fresh and fresh-cut produce 875* (pp. 49–56).
9. Singh, G., & Rattanpal, H. (2014). Use of nanotechnology in horticulture: A review. *International Journal of Agricultural Sciences and Veterinary Medicine, 2*, 34–42.
10. Rana, R. A., Siddiqui, M. N., Skalicky, M., Brestic, M., Hossain, A., & Kayesh, I. T. (2021). Prospects of nanotechnology in improving the productivity and quality of horticultural crops. *Horticulturae, 7*(10), 332.
11. Bumbudsanpharoke, N., & Ko, S. (2015). Nano-food packaging: An overview of market, migration research, and safety regulations. *Journal of Food Science, 80*(5), R910–R923.
12. Kumar, S. (2024). Nano-technology enhanced edible coating application on climacteric and non-climacteric fruits: A review. *International Journal of Advanced Biochemistry Research*, 58–68.
13. Babu, P. J., Saranya, S., Longchar, B., & Rajasekhar, A. (2022). Nanobiotechnology-mediated sustainable agriculture and post-harvest management. *Current Research in Biotechnology, 4*, 326–336.
14. Lopez-Polo, J., Monasterio, A., Cantero-López, P., & Osorio, F. A. (2021). Combining edible coatings technology and nanoencapsulation for food application: A brief review with an emphasis on nanoliposomes. *Food Research International, 145*, 110402.
15. Manzoor, M. A., Xu, Y., Xu, J., Wang, Y., Sun, W., Liu, X., & Zhang, C. (2024). Nanotechnology-based approaches for promoting horticulture crop growth, antioxidant response and abiotic stresses tolerance. *Plant Stress, 11*, 100337.
16. Xu, H., & Zhang, Y. (2019). A review on conducting polymers and nanopolymer composite coatings for steel corrosion protection. *Coatings, 9*(12), 807.
17. Vasile, C. (2018). Polymeric nanocomposites and nanocoatings for food packaging: A review. *Materials, 11*(10), 1834.

18. Tumuluru, J. S. (Ed.). (2023). Food Processing and Packaging Technologies: Recent Advances.
19. Bhattacharjee, S. K. (1999). Post harvest management of cut flowers, cut foliages and post production management of potted plants. *Journal of Ornamental Horticulture, 2*(1), 32–39.
20. Sekhon, B. S. (2014). Nanotechnology in agri-food production: An overview. *Nanotechnology, Science and Applications, 7*, 31–53.
21. Nile, S. H., Baskar, V., Selvaraj, D., Nile, A., Xiao, J., & Kai, G. (2020). Nanotechnologies in food science: Applications, recent trends, and future perspectives. *Nano-Micro Letters, 12*, 1–34.
22. Chaupoo, A. S., Dhakad, A., Beese, S., Kumar, A., Kumar, P., Pandey, S. K., & Kurdekar, A. K. (2023). Transforming horticulture: The influence of nanotechnology on crop enhancement. *International Journal of Environment and Climate Change, 13*(11), 3381–3400.
23. Neme, K., Nafady, A., Uddin, S., & Tola, Y. B. (2021). Application of nanotechnology in agriculture, postharvest loss reduction and food processing: Food security implication and challenges. *Heliyon, 7*(12), e08539.
24. Mahela, U., Rana, D. K., Joshi, U., & Tariyal, Y. S. (2020). Nano edible coatings and their applications in food preservation. *Journal of Postharvest Technology, 8*(4), 52–63.
25. Srinivasan, L. V., & Rana, S. S. (2024). A critical review of various synthesis methods of nanoparticles and their applications in biomedical, regenerative medicine, food packaging, and environment. *Discover Applied Sciences, 6*(7), 371.
26. Amer, M. A. (2023). Using effect of nano emulsion based on plant extracts as edible coating on prolonging the shelf life of fresh zucchini. *Al-Azhar Journal of Agricultural Research, 48*(1), 12–30.
27. Xing, Y., Yang, H., Guo, X., Bi, X., Liu, X., Xu, Q., & Zheng, Y. I. (2020). Effect of chitosan/Nano-TiO2 composite coatings on the postharvest quality and physicochemical characteristics of mango fruits. *Scientia Horticulturae, 263*, 109135.
28. McAvoy, G., & Williamson, J. (2020). Proceedings of the 133rd annual meeting, Florida State Horticultural Society, 1 May 2021.
29. Momin, J. K., Jayakumar, C., & Prajapati, J. B. (2013). Potential of nanotechnology in functional foods. *Emirates Journal of Food and Agriculture, 25*(1), 10–19.
30. Ansari, M. A. (2023). Nanotechnology in food and plant science: Challenges and future prospects. *Plants, 12*(13), 2565.
31. Sarfraz, J., Gulin-Sarfraz, T., Nilsen-Nygaard, J., & Pettersen, M. K. (2020). Nanocomposites for food packaging applications: An overview. *Nanomaterials, 11*(1), 10.
32. Bhat, N. R., Bhatt, A., & Suleiman, M. K. (2024). *Physiology of growth and development in horticultural plants*. CRC Press.
33. Wypij, M., Trzcińska-Wencel, J., Golińska, P., Avila-Quezada, G. D., Ingle, A. P., & Rai, M. (2023). The strategic applications of natural polymer nanocomposites in food packaging and agriculture: Chances, challenges, and consumers' perception. *Frontiers in Chemistry, 10*, 1106230.
34. Sahoo, M., Vishwakarma, S., Panigrahi, C., & Kumar, J. (2021). Nanotechnology: Current applications and future scope in food. *Food Frontiers, 2*(1), 3–22.
35. Shukla, P., Chaurasia, P., Younis, K., Qadri, O. S., Faridi, S. A., & Srivastava, G. (2019). Nanotechnology in sustainable agriculture: Studies from seed priming to post-harvest management. *Nanotechnology For Environmental Engineering, 4*, 1–15.

Harnessing Nanotechnology for Enhanced CRISPR-Cas Plant Genome Editing

Harshali Shende and Rajendra H. Patil

Contents

H. Shende
Department of Biotechnology (Formerly Institute of BioInformatics and Biotechnology), Savitribai Phule Pune University, Pune, MH, India

R. H. Patil (✉)
Department of Biotechnology, Savitribai Phule Pune University, Pune, MH, India
e-mail: rpatil@unipune.ac.in

R. S. Chaughule, A. V. Shelar (eds.), *Nanotechnology in Agriculture*,
https://doi.org/10.1007/978-3-031-97544-8_11

1 Introduction

By 2050, the world's population is projected to grow to around 9.6 billion, leading to a 60% increase in the demand for staple crops [1]. Food safety has become a global concern due to rising food demand coupled with declining crop yields, driven by climate change, soil degradation, and the spread of crop diseases [2]. Agriculture is vital for feeding the global population, but current methods must be modernized to meet increasing food demands. Although traditional plant breeding methods have enhanced crop traits, they are increasingly limited by reduced genetic variation, which poses challenges for meeting future production needs [3].

In the past decade, there have been substantial advancements in genome editing techniques. Genome editing (GE) technology has allowed the development of crops with higher yields, better nutritional value, and enhanced resistance to herbicides, pests, and diseases. Genome editing (GE) tools are generally classified into three generations: first-generation tools include meganucleases (MegaN) and zinc finger nucleases (ZFNs), second-generation tools are represented by transcription activator-like effector nucleases (TALENs), and the third-generation is characterized by the CRISPR/Cas9 system.

[4–7]. Early genome editing approaches, such as meganucleases, zinc finger nucleases (ZFNs), and TALENs, each offered unique benefits but also had notable drawbacks. Meganucleases and ZFNs required extensive customization and were prone to off-target effects, while TALENs, though more flexible, still faced challenges in design and efficiency. The CRISPR/Cas9 system, the latest advancement in genome editing technology, addresses many of these issues with its ease of use, precision, and adaptability, representing a significant leap forward in genetic engineering [8].

The CRISPR/Cas9 genome editing system, derived from the type II CRISPR adaptive immunity system of the bacterium *Streptococcus pyogenes*, is known for its simplicity and high efficiency. It outperforms other programmable nucleases like ZFNs and TALENs in terms of ease of use, precision, and effectiveness [9]. The CRISPR/Cas9 system is poised to transform plant research and crop breeding by enabling precise DNA modifications. This technology allows molecular breeders to eliminate undesirable DNA sequences and make targeted improvements or adjustments to genomic and transcriptional regulatory elements. CRISPR-Cas9 has emerged as a critical tool in plant genome editing, offering unprecedented precision and efficiency in modifying specific genomic sequences. This technology allows for the targeted manipulation of genes within plant genomes, enabling the development of crops with enhanced traits such as improved yield, disease resistance, and abiotic stress tolerance. The ability of CRISPR-Cas9 to induce site-specific double-strand breaks, followed by precise gene editing through homologous recombination or non-homologous end joining, has revolutionized plant biotechnology. Its role in accelerating the breeding of genetically superior crops is of paramount importance, as it provides a robust platform for addressing global agricultural challenges and advancing sustainable food production.

Plant genetic transformation consists of two key phases: delivering genetic material into plant cells and regenerating the transformed plants [10]. There are several methods for introducing genetic cargo into plant cells, including biological techniques like Agrobacterium-mediated transformation (AMT), viral-mediated methods, chemical approaches such as polymer-mediated delivery (e.g., polyethylene glycol [PEG]), physical methods like particle bombardment (gene gun-mediated or biolistic transformation), and electroporation. Among these, gene gun-mediated transformation and AMT are among the most effective and widely used techniques for plant genetic transformation [11–16]. Traditional methods for delivering biomolecules into plants face significant limitations, including low gene transmission efficiency, a restricted range of applicable species, limited types of cargo that can be delivered, and potential tissue damage. Nanotechnology offers significant advantages in enhancing the efficiency and precision of CRISPR-Cas9 gene editing in plants. One of the primary benefits is the improved delivery of CRISPR-Cas9 components into plant cells.

Nanotechnology is the science and engineering of manipulating materials at the nanoscale, typically between 1 and 100 nanometers. Nanoparticles are ultra-small particles, ranging from 1 to 100 nanometers, with unique physical and chemical properties due to their high surface area and quantum effects. Advancements in nanotechnology have opened up new possibilities to address these challenges. Nanoparticles (NPs) offer a promising solution by enabling species-independent passive delivery of DNA, RNA, and proteins, thus overcoming many of the drawbacks associated with conventional delivery techniques [10]. Incorporating nanotechnology into the development of genetically modified (GM) organisms (GMOs) offers a powerful approach. This involves using nanoparticles (NPs) as nanocarriers to create binding complexes with biomodifier molecules, such as the CRISPR/Cas system, and delivering them into plant cells effectively [17, 18]. Nanoparticles can be engineered to cross plant cell walls more effectively, delivering the CRISPR-Cas9 complex directly to the target cells with minimal disruption. This not only increases the success rate of genetic modifications but also allows for the precise targeting of specific tissues or cell types, making the gene editing process more accurate and less prone to off-target effects.

This chapter delves into the cutting-edge integration of nanotechnology to enhance the CRISPR-Cas system for plant genome editing. We will explore a range of nanomaterials, including metal nanoparticles, carbon-based nanomaterials, and polymeric nanoparticles, as carriers for CRISPR-Cas components. The chapter will examine the mechanisms by which these nanoparticles facilitate the delivery of gene-editing tools, focusing on how nanotechnology can improve the uptake, stability, and target specificity of CRISPR-Cas components. By minimizing off-target effects and boosting editing efficiency, nanotechnology holds the potential to significantly advance the precision and effectiveness of plant genome editing. This chapter aims to highlight the powerful synergy between nanotechnology and CRISPR-Cas, underscoring its transformative potential for the future of agricultural biotechnology.

2 Basics of CRISPR-Cas System

2.1 Understanding CRISPR-Cas9: What It Is and Why It Matters

Crop failure is primarily caused by biotic and abiotic factors, which significantly impact the economic value of crops [19]. Environmental and climate changes, such as frequent floods, droughts, temperature fluctuations, increased soil salinity, the use of harmful chemical additives and fertilizers, and pathogen-triggered diseases, all contribute to the deterioration of plant health, directly affecting crop quality and yield [20]. Improving crop refinement without genomic engineering would be an extremely challenging task. Conventional breeding programs would take significantly longer to introduce new quality-related or disease-tolerance traits. As a result, genomic engineering stands out as a crucial technology for the future development of nutritionally enhanced crops.

CRISPR-Cas9 is a genome editing technology that enables the insertion, deletion, or modification of genetic information at specific loci within the genome. Its reliance on Watson–Crick base pairing makes CRISPR-Cas9 a highly versatile and powerful tool with a wide range of applications [8]. The development of the CRISPR/Cas9 technique has significantly advanced crop breeding and gene function studies, thanks to its simplicity, precision, ease of use, and stable genome editing with high target efficiency [21]. Due to its simplicity and high precision in targeting specific genes, CRISPR has emerged as the preferred tool for researchers. It also offers extensive potential for introducing and controlling desirable traits in crops, making it a revolutionary resource for crop enhancement. Furthermore, CRISPR has been widely employed to investigate gene functions and integrate beneficial traits into crop plants. Traits such as resistance to biotic and abiotic stresses, as well as improvements in yield and quality, have been enhanced through genome editing [22] [23, 24]. This approach also facilitates hybrid breeding by directly introducing desired traits or removing undesirable ones in elite varieties, making it a straightforward and effective method [25].

2.2 *Discovery of CRISPR: A Breakthrough in Genetic Engineering*

In 1987, Ishino et al. from Osaka University (Japan) made the initial discovery of CRISPR in the DNA sequences of Escherichia coli bacteria [26]. In 2007, Barrangou and Horvath demonstrated the CRISPR system's mechanism in *Streptococcus thermophilus* [27]. The Cas9 protein was initially characterized in relation to its association with CRISPR repeats in a paper by Bolotin et al. [28] A crucial element of the CRISPR–Cas9 system is crRNA, a short RNA molecule derived from the CRISPR locus that guides immune proteins to foreign DNA. Identified by John van der Oost's team, crRNA is essential for directing the prokaryotic immune response [29]. Virginijus Siksnys's research showed that a 20-base pair "guide" crRNA sequence, complementary to the target DNA, is both necessary and sufficient for CRISPR–Cas9 nuclease activity, regardless of the length of the spacer sequence [30]. The final essential component for a functional CRISPR–Cas9 system, identified by Emmanuelle Charpentier's team in 2011, is tracrRNA (trans-activating CRISPR RNA). This short RNA molecule is crucial for processing crRNA [31]. Further research, which won a Nobel Prize, revealed tracrRNA's crucial role in target DNA cleavage and proposed that combining crRNA and tracrRNA into a single chimeric molecule, sgRNA (single guide RNA), greatly simplified CRISPR–Cas9 applications.

2.3 *The Spectrum of CRISPR Systems: Classification and Characteristics*

CRISPR-Cas systems are classified into three major types—Type I, Type II, and Type III—based on their phylogeny, sequence, and locus organization. Type I systems, with six subtypes, are characterized by the Cas3 protein, which has helicase and DNase domains. Type II systems, including subtypes II-A and II-B, are notable for the Cas9 protein, which is involved in adaptation, crRNA processing, and DNA cleavage, often with the assistance of tracrRNA. Type III systems, featuring the Cas10 protein, target both DNA and RNA. Type II systems are the most studied due to their ability to induce double-strand breaks in DNA [32].

2.4 *Mechanism of CRISPR Cas9*

Figure 1 illustrates the mechanism of action of CRISPR-Cas9, demonstrating how the guide RNA (gRNA) directs the Cas9 protein to a specific target DNA sequence, resulting in a double-strand break. The figure also depicts the cell's repair pathways,

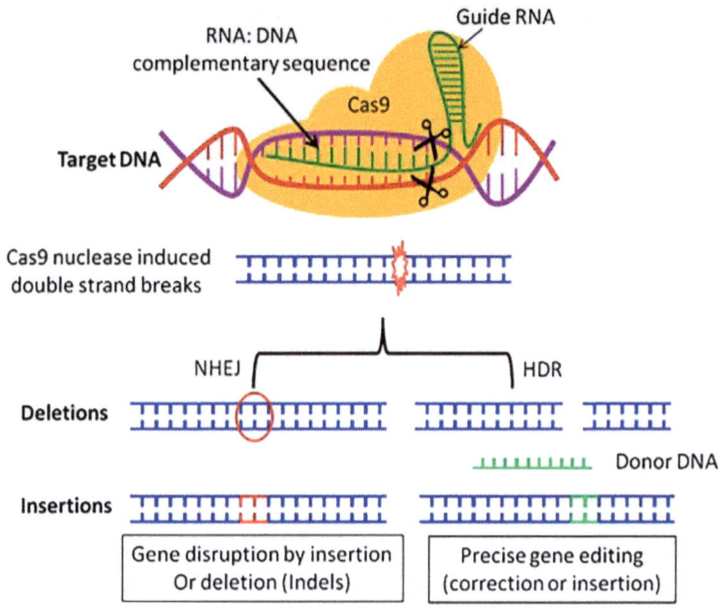

Fig. 1 Mechanism of action of CRISPR CAS9. (*Source*: The potential of CRISPR/Cas9 gene editing as a treatment strategy for inherited diseases [33] Licensed under CC BY)

including non-homologous end joining (NHEJ) for gene disruption and homology-directed repair (HDR) for precise gene editing.

The CRISPR/Cas9 system consists of two key components: the Cas9 endonuclease and a single-guide RNA (sgRNA). The first 20 nucleotides of the sgRNA's 5′ end are designed to specifically recognize and bind to a short three-base-pair sequence known as the protospacer-adjacent motif (PAM), which is located downstream of the target site. The remaining 80 nucleotides of the sgRNA, a conserved fragment, form a secondary structure that binds to the Cas9 protein [34]. This 20-nucleotide sequence directs the Cas9 endonuclease to precisely and accurately cleave the desired DNA sequence [35]. The resulting genomic break triggers the cell's natural repair mechanisms, either through homologous directed repair (HDR) or non-homologous end joining (NHEJ). In homologous directed repair (HDR), the repair of a double-strand break (DSB) requires a homologous template that matches the sequence at the break site. This template guides the accurate repair of the DNA, allowing for precise edits or the introduction of new genetic material [36]. The homologous directed repair (HDR) pathway utilizes donor DNA sequences, either from sister chromatids or foreign DNA, to achieve precise insertions and substitutions at the sites of DNA double-strand breaks (DSBs), enabling further genetic alterations [37]. The NHEJ pathway, which is more common in higher plants, often leads to insertions or deletions (indels) that cause frameshift mutations, resulting in the loss of gene function at the repair site [38]. The non-homologous end joining (NHEJ) repair pathway is a widely used method to create small deletions or insertions that disrupt the function of a specific gene. This approach can be carried out at any point during the cell cycle and does not require a homologous repair template [39].

2.5 CRISPR Technology in Agriculture

Since 2013, significant advancements have been achieved in plant genome editing due to the development and application of CRISPR/Cas systems [40]. CRISPR/Cas-based gene editing offers a range of applications, such as enhancing crop yield, improving resistance to diseases and pathogens, increasing tolerance to herbicides, and strengthening resilience to environmental stress. In 2020, Kashaf Zafar et al. employed CRISPR-Cas9-mediated genome editing to confer resistance to bacterial blight in Super Basmati rice. The study specifically targeted four effector-binding elements (EBEs) within the promoter region of the *OsSWEET14* gene, which is associated with susceptibility to *Xanthomonas oryzae pv. oryzae* (Xoo) infections. The results indicated that CRISPR-Cas9 can be utilized to develop durable resistance against bacterial blight in locally cultivated rice varieties by disrupting the recognition sites used by Xoo strains prevalent in the region [41]. In 2020, Sukumaran Sunitha et al. used CRISPR/Cas9, delivered via Agrobacterium, to edit the *TAS4* (Trans-Acting Small-interfering locus 4) and *MYBA7* (Transcription factor) genes in grapes. This genetic modification resulted in enhanced tolerance to both biotic and abiotic stresses. The edited genes demonstrated significant resistance against the bacterium *Xylella fastidiosa* and the Grapevine Red Blotch Virus (GRBV), the causative agent of Red Blotch Disease [42]. In 2019, Fei Yunyan et al. utilized CRISPR/Cas9 technology to knock out the *Wx* gene in two elite japonica rice lines, with the goal of developing superior sticky rice varieties. As a result of the *Wx* gene knockout, the amylose content in the rice endosperm was reduced, leading to the successful creation of two elite sticky rice varieties [43]. In 2019, Rui Yan et al. targeted the *LpPDS* gene in *Lilium* species, a genus of over 100 flowering plant species that grow from bulbs, using CRISPR/Cas9 technology. The gene editing was aimed at enhancing the beauty and color variety of the flowers, resulting in the development of attractive pale yellow to albino-green color shades [44]. In 2015, Xiang Ji et al. conducted a study using the geminivirus, beet severe curly top virus (BSCTV). Transient assays in *Nicotiana benthamiana* demonstrated that sgRNA-Cas9 constructs effectively inhibited viral accumulation and introduced mutations at the target sequences. Furthermore, transgenic *Arabidopsis* and *N. benthamiana* plants overexpressing sgRNA-Cas9 showed a high level of resistance to virus infection [45]. In 2017, Vladimir Nekrasov et al. reported the successful development of "Tomelo," a non-transgenic tomato variety resistant to the powdery mildew fungal pathogen, using CRISPR/Cas9 technology [46]. In 2016, Jinrui Shi et al. utilized CRISPR-Cas9-based advanced breeding technology to create novel variants of the *ARGOS8* gene in maize. The results highlight the effectiveness of the CRISPR-Cas9 system in producing new allelic variations for breeding drought-tolerant crops [47]. In 2015, Yasuhiro Ito et al. applied the CRISPR/Cas9 system to efficiently induce mutations in the tomato genome. Their study specifically targeted the *RIN* gene, which encodes a MADS-box transcription factor involved in regulating fruit ripening [48]. In 2019, Qingbo Ke et al. created transgenic sweet potato plants (referred to as RLE plants) with down-regulated expression of *IbLCY-ε*. These RLE plants exhibited increased levels of total carotenoids and β-carotene, and demonstrated enhanced tolerance to oxidative stress induced by methyl viologen (MV), as well as to salt and drought conditions [49]. In 2021, Rukmini Mishra et al. developed T-DNA-free homozygous mutant lines using a single transcript CRISPR/Cas9 system. These lines, featuring the desired modification in the *CaERF28* locus, showed markedly increased resistance to the anthracnose pathogen *Colletotrichum truncatum* and exhibited enhanced expression of defense-related genes [50].

2.6 CRISPR in the Global Market

In 2011, Caribou Biosciences secured $11 million to focus on applying CRISPR technology, with a particular emphasis on agriculture [51]. Funding for CRISPR research from the US National Institutes of Health (NIH) saw a significant increase between 2013 and 2014 [51]. The start-up Arvegenix is applying CRISPR technology to improve the quality of oil and meal in pennycress. The objective is to transform pennycress into a dual-purpose crop: a cover crop for use between main growing seasons and a valuable product similar to canola for the oil and feed industries. Success in this endeavor would enable farmers to increase their income while continuing to benefit from sustainable cover cropping practices [52]. The startup Inari, founded in 2016, has raised $55 million to develop seeds and plants optimized using CRISPR technology for growth in adverse conditions [53]. Currently, 25% of biochemical research companies utilizing CRISPR technology are focusing on its applications in plants, aiming to advance agricultural sustainability and crop improvements [54]. In January 2017, Monsanto announced a deal with Harvard and the Broad Institute of MIT to use CRISPR-Cpf1, a tool similar to CRISPR-Cas9, but potentially offering greater efficiency, fewer off-target effects, and a smaller size. In November 2017, Arcadia Biosciences and Syngenta secured licenses for CRISPR-Cas research in agriculture.

2.7 CRISPR-Cas9 Delivery in Plants: Methods and Challenges

For genome editing with the CRISPR/Cas9 system, delivering the sgRNA and Cas9 proteins into the target cells is crucial. This is typically achieved using expression vectors or microinjected RNA/mRNA (for Cas9) to express the sgRNA and Cas9 protein within plant cells. CRISPR/Cas9 technology has been successfully applied in plant cells using various methods, including electroporation, Agrobacterium-mediated transformation, biolistic (shotgun) approaches, and polyethylene glycol (PEG)-mediated delivery [55–59]. Agrobacterium-mediated transformation employs the bacterium *Agrobacterium tumefaciens* to deliver CRISPR-Cas9 components into plant cells, proving effective for many dicotyledonous plants but often exhibiting reduced efficiency in monocots and certain recalcitrant species [60]. This method can also result in variable integration of CRISPR constructs, leading to inconsistent gene editing outcomes. Biolistic delivery, or particle bombardment, uses a gene gun to shoot gold or tungsten particles coated with CRISPR-Cas9 DNA into plant cells. Although applicable to a broad range of plant species, including monocots, it frequently results in multiple copies of the CRISPR construct being integrated, complicating the selection of stable transformants and potentially causing physical damage to cells [61]. PEG-mediated methods involve using polyethylene glycol to facilitate the uptake of CRISPR-Cas9 components into plant protoplasts, allowing transient expression in various plant species. However, this technique is often hampered by low transformation efficiency and challenges in regenerating whole plants from protoplasts, limiting its utility for stable, long-term gene editing [62].

3 Fundamentals of Nanotechnology

3.1 Definition and Scope

Nobel laureate Richard Feynman is recognized for introducing the idea of nanotechnology during a lecture he delivered in 1959. The US Environmental Protection Agency defines nanotechnology as the science of understanding and controlling matter at dimensions of approximately 1–100 nm, where unique physical properties enable novel applications. However, this definition is somewhat rigid concerning size dimensions. Nanoparticles are fundamental to nanotechnology. These extremely small particles, measuring less than 100 nm, can be composed of carbon, metals, metal oxides, or organic materials [63]. Nanoparticles exhibit unique physical, chemical, and biological properties compared to their larger counterparts. These include enhanced chemical reactivity, increased stability, a greater surface area-to-volume ratio, and improved mechanical strength. Due to these distinct properties, nanoparticles have a wide range of applications. Additionally, nanoparticles vary in size, shape, and material, which further influences their functionality [64]. Nanoparticles exhibit a diverse range of sizes and can take on various shapes, including spherical, cylindrical, tubular, conical, hollow core, spiral, and flat. Their surfaces may be either smooth or irregular, and they can have crystalline or amorphous structures, existing as single or multiple crystals, and may be dispersed or aggregated [65]. Major institutions and organizations are heavily investing in nanotechnology research and development. This advanced field is actively explored for new applications aimed at enhancing efficiency, performance, and cost-effectiveness, while minimizing environmental impact, highlighting its promising future.

3.2 Synthesis

Nanomaterials can be synthesized using either a bottom-up or top-down approach. The bottom-up approach builds nanoparticles from atomic or molecular precursors, employing methods such as sol-gel processing, spinning, chemical vapor deposition (CVD), pyrolysis, and biosynthesis. Conversely, the top-down approach involves reducing bulk materials to nanoscale particles through techniques like mechanical milling, nanolithography, laser ablation, sputtering, and thermal decomposition [64]. The synthesis of nanoparticles can be achieved through three primary methods: physical, chemical, and biological. Physical methods, such as laser ablation or vapor deposition, offer precise control over particle size and shape but often require high-energy inputs and sophisticated equipment. Chemical methods, including sol–gel processes and chemical vapor deposition, allow for scalable production and can produce a wide range of nanoparticles with tailored properties. However, they often involve hazardous chemicals and complex reaction conditions. Biological methods, leveraging microorganisms or plant extracts, provide environmentally friendly and cost-effective alternatives, and can produce nanoparticles with unique properties [63]. Nonetheless, they generally have slower production rates and may suffer from inconsistencies in particle size and quality. Each method has its own set of advantages and limitations, making the choice of synthesis technique dependent on the desired application and scalability requirements.

3.3 Characterization Techniques

Particle size is a critical parameter for characterizing nanoparticles and influencing their distribution and scale. Electron microscopy, such as SEM or TEM, is commonly used for particle size analysis [66]. The surface area of nanoparticles is essential for characterization, as its ratio to volume significantly influences their properties and performance. BET analysis is commonly used to measure surface area, [67] while NMR spectroscopy and differential mobility analyzers (DMA) or modified SMPS are employed for particles in liquid and gaseous phases, respectively. X-ray photoelectron spectroscopy (XPS) is commonly used to measure the composition of nanoparticles [68]. A zeta potentiometer is commonly used to measure surface charge and dispersion stability of nanoparticles in solution [69] Crystallography, using powder X-ray, electron, or neutron diffraction, is employed to determine the structural organization of nanoparticles [70].

3.4 The Significance of Nanotechnology in Modern Agriculture

Nanotechnology holds the potential for a promising and sustainable future. Nanotechnology has demonstrated significant potential in advancing sustainable agriculture. Recent studies suggest that nanotechnology can enhance the agricultural sector by boosting the efficiency of agricultural inputs and providing solutions to environmental and agricultural challenges, ultimately improving food productivity and security. As a result, there has been growing interest in exploring the applications of nanotechnology in agriculture in recent years. Nanotechnology has numerous applications in agriculture, including the development of nanofertilizers, nanopesticides, and nanobiosensors. Nanofertilizers are nanomaterials that serve as nutrients (either micro- or macronutrients) or function as carriers or additives, such as when they are combined with minerals, to deliver nutrients effectively. Nanotechnology is employed to minimize the loss of mobile nutrients, create slow-release fertilizers, and enhance the availability of nutrients that are otherwise difficult to access [71]. The application of nanotechnology in plant nutrition enables the creation of slow or controlled-release fertilizers, which enhance fertilizer use efficiency and minimize nutrient loss to the environment, thereby making them more environmentally friendly [72]. Pesticides are commonly used in agriculture, but only a small fraction reaches the target pests, leading to environmental contamination and the development of pesticide resistance [73]. While biopesticides offer a safer alternative, their effectiveness is limited. Nanopesticides provide a promising solution by enabling slow degradation and controlled release of active ingredients, enhancing pest control, and reducing environmental impact [74]. Biosensors are hybrid receptor-transducer systems that detect specific biological analytes by sensing physical and chemical properties in a medium [75]. Nanobiosensors represent an advanced form of biosensors, capable of detecting target analytes at ultra-low concentrations due to their compact size, high sensitivity, and stability. This technology enables early detection and improved management of agricultural resources, potentially enhancing crop yields. The use of nano-sized materials as bioreceptors on transducers offers a significant advantage over traditional sensors. Nanosensors offer high sensitivity and selectivity for detecting analytes at ultra-low concentrations, along with rapid response times and enhanced stability due to their high surface-to-volume ratio and efficient electron-transfer kinetics. They provide precise monitoring capabilities for various agricultural applications.

Figure 2 highlights various applications of nanotechnology in modern agriculture, including its role in improving yield and quality, enhancing plant growth, and aiding genetic engineering. It also showcases uses in postharvest preservation, stress mitigation, and the development of agrochemicals for sustainable farming.

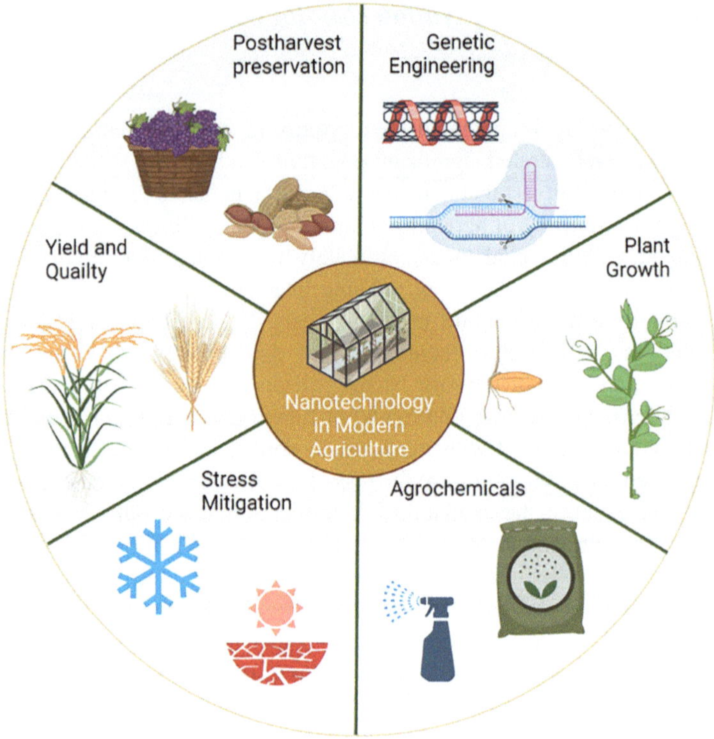

Fig. 2 Various applications of nanotechnology in agriculture. (*Source*: Phytonanotechnology applications in modern agriculture [76] Licensed under CC BY)

3.5 The Role of Nanotechnology in Revolutionizing Plant Genome Editing

Despite over three decades of progress, plant transformation and regeneration continue to pose challenges for many crop species. Traditional biomolecule delivery methods face limitations such as restricted cargo types, limited species applicability, low efficiency, and potential tissue damage. To overcome these challenges, new strategies are required for effective gene delivery in crop plants. Tissue culture and regeneration processes remain major obstacles in plant transformation. Nanoparticles are increasingly used to improve gene editing by enabling efficient and targeted delivery of plasmids, RNA, and ribonucleoproteins (RNPs). Nanotechnology-based methods offer a cost-effective, straightforward, and robust approach for gene or molecule transfer into plants, achieving high efficiency with minimal toxicity [61]. Nanoparticles (NPs) enhance plant transformation by shielding genetic material from cellular enzymatic degradation, such as nucleases [77, 78]. NPs used for gene delivery are categorized based on their base material, including carbon-based, silicon-based, metallic, and polymer-based nanoparticles.

Asitha T. Silva et al. (2010) used fluorescent conjugated polymer nanoparticles (CPNs) to deliver siRNAs and achieve gene knockdown in tobacco plant protoplasts [79]. Feng-Peng Chang et al. (2013) utilized functionalized mesoporous silica nanoparticles (MSNs) to create an MSN-mediated system for transient gene expression in plants. In this system, MSNs acted as carriers for delivering foreign DNA into intact *Arabidopsis thaliana* roots without requiring mechanical force [80]. Arezoo Golestanipour et al. (2018) investigated the use of arginine-functionalized single-walled carbon nanotubes for gene delivery in tobacco root cells. The study demonstrated that the cationic and hydrophobic properties of the engineered peptide, combined with the nanotubes' nano-cylindrical shape, facilitated passage through plant cell barriers and effective condensation of plasmid DNA encoding the green fluorescent protein. Successful gene delivery was confirmed through fluorescence microscopy and western blot analysis [81].

Nanotechnology has revolutionized genetic transformation in plants by offering highly efficient and targeted delivery mechanisms for genetic material. This approach enhances precision, minimizes toxicity, and shields nucleic acids from enzymatic degradation. By addressing limitations of conventional methods, such as inefficient gene delivery and potential tissue damage, nanotechnology provides a powerful tool for advancing crop improvement and plant biotechnology, paving the way for more robust and sustainable agricultural solutions.

4 Integration of Nanotechnology with CRISPR-Cas

4.1 Understanding the CRISPR Barriers

CRISPR/Cas9 technology offers several significant advantages. The CRISPR/Cas9 system only requires a single sgRNA for RNA-DNA recognition, making it more specific, efficient, and cost-effective compared to the protein-DNA binding mechanism [82]. The system is easily customizable by simply altering the sgRNA sequence, enabling the simultaneous editing of multiple independent sites [83]. Additionally, CRISPR/Cas9 can modify chromosomal targets with high precision and low cytotoxicity. The widespread use of CRISPR technologies in plants faces several challenges, including difficulties in CRISPR cargo delivery, limitations of plant tissue and cell culture, and lack of universal methods across species. Additionally, our limited understanding of plant genetic and metabolic networks impedes trait development, while regulatory hurdles and societal acceptance may further complicate market introduction. Despite the transformative impact of genome-editing tools and the significant advancements they have brought to plant genctic engineering, CRISPR applications in plant biotechnology still face several challenges. A major challenge for CRISPR genetic engineering in plants is the delivery of CRISPR reagents through the plant cell wall, a rigid and thick extracellular matrix mainly composed of polysaccharides, which is not present in most other biological systems where CRISPR has been successful. Using protoplasts—plant cells with their cell walls removed via enzymatic digestion—can bypass this barrier and facilitate CRISPR reagent delivery. However, regenerating whole plants from protoplasts remains technically difficult and has not been achieved in most plant species [84]. To successfully deliver biomacromolecules into a walled plant cell, both the cargo and carrier must overcome two major barriers: the plant cell wall and the plasma membrane. The plant cell wall is an intricate network of biopolymers forming a multilayered matrix hydrogel, which limits access to the plasma membrane [85].

4.2 The Three Key Forms of CRISPR Cas

Gene editing with CRISPR/Cas9 typically involves three main formats: plasmids that encode both the Cas9 protein and sgRNA, a mixture of Cas9 mRNA and sgRNA, and ribonucleoprotein (RNP) complexes consisting of Cas9 protein and sgRNA [86]. Using a plasmid is a more stable and cost-effective method compared to Cas9 mRNA or protein, as it allows continuous production of Cas9 nuclease for ongoing gene editing. Additionally, plasmids can easily pass through cellular and nuclear membranes, enabling Cas9 expression in eukaryotic cells. Plasmids encoding CRISPR–Cas reagents can be delivered into plant cells using various methods for stable or transient transformation. Transient methods are often preferred as they avoid chromosomal integration and rely on temporary expression by plant machinery. This approach has led to fewer off-target mutations in bread wheat and reduces regulatory challenges due to the lack of heritable DNA integration. However, the applicability of transient plasmid delivery is limited, as many plant species are not easily susceptible to these DNA-delivery techniques [87]. Cas9 mRNA offers advantages like efficient nuclear entry, rapid gene editing onset, and bypassing the need for DNA transcription. Since Cas9 is translated directly in the cytoplasm, it ensures quick protein expression and minimizes off-target effects and genomic integration [88]. The simplest method involves directly delivering Cas protein with sgRNA, allowing for temporary expression and quick gene editing without the need for mRNA transcription and translation [89, 90]. This ribonucleoprotein approach also provides safe delivery, protecting Cas9 and RNA from degradation, though it faces challenges like low cellular uptake efficiency [91].

4.3 Nanotechnology: A Catalyst for CRISPR-Cas Precision and Efficacy

In the past decade, combining plant biotechnology with nanomaterials has proven highly effective. An ideal advanced delivery system should meet several criteria, including biocompatibility, the ability to load high doses of cargo, and controlled release functionality. Additionally, the carrier materials must be chemically and

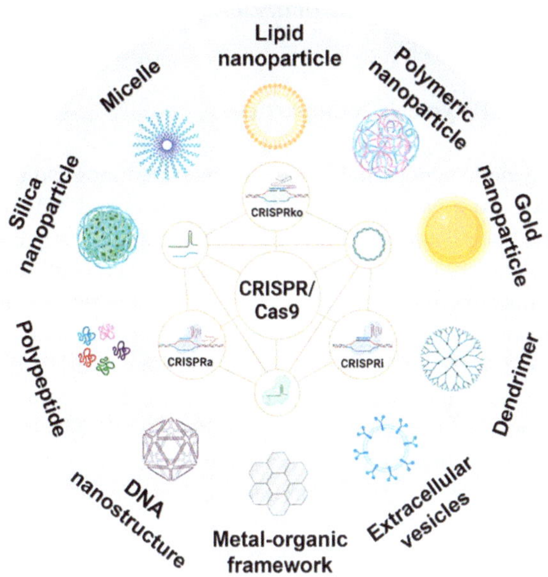

Fig. 3 Different nanocarriers for the delivery of CRISPR CAS9. (*Source*: A nanotechnology-based CRISPR/Cas9 delivery system for genome editing in cancer treatment [92] Lincesed under CC BY)

biologically inert and should not interact with plant cells. Nanomaterials facilitate targeted delivery and release of cargoes while protecting them from degradation. Moreover, nanoparticles can typically accomplish these tasks across different plant species. Nano-delivery systems enhance gene delivery by masking the negative charge of DNA, thereby protecting it from degradation by intracellular nucleases. This is accomplished through electrostatic interactions between the negatively charged DNA and polycations, by encapsulating DNA in biodegradable polymers, or through adsorption. Nanomaterials have the potential to overcome key obstacles in CRISPR genome editing in plants by enhancing cargo delivery, enabling species-independent applications, improving germline transformation, and increasing gene editing efficiency. Nanotechnology-based delivery systems are gaining growing interest in gene editing. Recent interest in nanotechnology-based vectors for CRISPR/Cas9 delivery has spurred extensive research, highlighting their promising potential. Nanocarriers can be developed and optimized through precise nanoparticle design and chemical modifications. Due to the strong negative charge of CRISPR/Cas9 components, the most common method for encapsulating these large molecules is through electrostatic attraction with positively charged nanoparticles. Various positively charged materials, including cationic lipid-based nanoparticles, polymers, and polypeptides, have been developed for packaging CRISPR/Cas9.

Figure 3 showcases a range of nanocarriers utilized for the efficient delivery of CRISPR-Cas9 systems. These include lipid and polymeric nanoparticles, gold nanoparticles, and silica nanoparticles, along with advanced carriers like micelles, dendrimers, extracellular vesicles, DNA nanostructures, and metal-organic frameworks. Each carrier offers unique properties to enhance the stability, targeting, and functionality of CRISPR-Cas9 delivery.

4.4 *Potential Nanocarriers for Efficient CRISPR-Cas Delivery*

4.4.1 Polymeric Nanoparticles

Polymeric nanoparticles have diverse structures and characteristics because they are made from natural or synthetic monomers and polymers, allowing precise control over their features. They are biocompatible, simple, and effective drug delivery vehicles, [93] capable of encapsulating therapeutics within their core, chemically conjugating them to the polymer, or binding them to the nanoparticle surface. This versatility enables the delivery of hydrophilic and hydrophobic macromolecules, including proteins, [94–97] making them ideal carriers for CRISPR/Cas cargo. Polymeric nanoparticles are excellent candidates for delivering CRISPR/Cas cargo because they are biodegradable, water-soluble, biocompatible, stable during storage, and their surfaces can be easily modified for targeted delivery. These properties enable them to deliver proteins and genetic material to specific cells or tissues. Polymer-based nanoparticles are widely used for nucleic acid delivery due to their strong electrostatic interactions with negatively charged genetic materials and high loading capacity [98]. Cationic polymers like chitosan and polyethyleneimine are particularly effective for CRISPR/Cas9 delivery [99]. Ribonucleoprotein (RNP) delivery offers high target specificity compared to vector-based systems [100]. However, delivering genetic materials to mature plants remains challenging due to the cell wall, necessitating further research on nanocarriers that can penetrate thick membranes and cell walls without external force.

Fluorescent conjugated polymer nanoparticles (CPNs) were used by Asitha T. Silva et al. in 2010 for transferring siRNAs and silencing a target gene in plant protoplasts. CPNs exhibited minimal to no effect on viability over 72 hours and were not toxic to protoplasts. Within 2 h of delivery, CPNs may penetrate protoplasts, as shown by microscopy and flow cytometry [79].

4.4.2 Dendrimers

Dendrimers are synthetic, hyperbranched nano-polymers with multiple branches extending outward from a central core, creating an ideal three-dimensional geometry [101]. These dendritic polymers consist of three main components: a central core, an interior section made up of repeating branches that define each generation, and an outer shell characterized by terminal functional groups. Studies have shown that dendrimers are nanoscale molecules with diameters ranging from a few nanometers to several tens of nanometers.

Poly(amidoamine) dendrimers were successfully used by Karthikeyan Pasupathy et al. in 2008 to transport plasmid DNA encoding GFP to turfgrass cells. To transfer genes directly and non-invasively, this innovative method made use of the physiochemical characteristics and nanosize of the poly(amidoamine) dendrimer.

Confocal fluorescence microscopy revealed that the GFP gene was expressed in plant cells [102]. Hyperbranched PAMAM (hPAMAM)-G2 with a diethylenetriamine core was examined by Amin amani et al. for its capacity to interact with DNA, resistance to ultrasonic degradation, and intake by alfalfa cells. Additionally, it was looked at how ultrasound affected hPAMAM-G2's ability to transport and express the gusA gene in alfaalfa cells [103].

4.4.3 Lipid-Based Nanoparticles

Lipid-based delivery systems have gained significant attention due to their ability to permeate physiological barriers, ease of synthesis, low cost, and scalability [104]. Based on their preparation and properties, lipid nanoparticles are classified into various types, such as solid lipid nanoparticles, nanostructured lipid carriers, liposomes, and niosomes [105]. Lipid-based nanoparticles are spherical structures with at least one lipid bilayer surrounding an internal aqueous compartment [93]. They are categorized into liposomes and lipid nanoparticles (LNPs). Liposomes are small lipid vesicles, typically 50–1000 nm in size, [106] composed of a hydrophilic head and a hydrophobic tail. Their ability to protect drugs from degradation, deliver them to specific sites, and reduce toxicity makes liposomes valuable for drug delivery applications [107]. Liposomes are sacs or spheres composed of at least one phospholipid bilayer, designed to deliver target molecules into cells. Typically, they are colloidal particles in the nanometer range [108]. Their amphiphilic nature allows them to easily pass through cell membranes [109]. Liposomes, with flexible biophysical properties, can carry hydrophilic, hydrophobic, and lipophilic drugs. LNPs, differing from liposomes in their micellar structure [110] can be tailored during synthesis to modify size, surface charge, and composition, making them highly effective for drug delivery, especially in nucleic acid delivery [107]. LNPs are particularly suited for delivering CRISPR/Cas cargo due to their efficient and customizable structure [111].

Liposomes are promising nanodrug delivery systems due to their resemblance to the plasma membrane [112]. Lipid-based nanoparticles are among the most commonly used for nucleic acid delivery, as their lipid layer improves the CRISPR cargo's ability to cross cell membranes and protects it from RNases and enzymatic degradation [113]. Among various lipid-based delivery systems, solid lipid nanoparticles (SLNs) are composed of solid lipids. SLNs are spherical particles ranging in size from 50 to 1000 nm [114].

To mediate the delivery of donor DNA into citrus protoplasts, Lamiaa M. Mahmoud et al. 2022 identified the most effective cationic lipid nanoparticles and provided a technique employing LipofectamineTM LTX Reagent and PLUS Reagent. Citrus protoplasts were transfected with a Cas9 construct harboring a gRNA targeting the CsNPR3 gene using Lipofectamine, a cationic lipid transfection method, with or without polyethylene glycol (PEG, MW 6000) [115].

4.4.4 Gold Nanoparticles

Gold nanoparticles (AuNPs) are widely used as multifunctional gene carriers due to their easy synthesis, biocompatibility, well-defined surface chemistry, and ability to facilitate molecular imaging via FRET. Their low dispersity allows the formation of monolayers for aggregating multiple moieties. Additionally, their low cytotoxicity and limited bio-distribution contribute to their broad application [116–119].

According to Mortazavi and Zohrabi (2018), the rice genome has been integrated effectively after the delivery of the plasmids pUBC (carries cryIA(c) gene) and pTra132 (carries hph gene) through gold nanoparticles into embryogenic rice lines [120].

4.4.5 Mesoporous Silica Nanoparticles

Silica exists in various forms, such as silica particles, hollow particles, nanotubes, and mesoporous silica nanoparticles (MSNs). Surface-modified silica nanoparticles are effective gene delivery carriers due to their inert nature, large surface area, and cationic charge, which facilitates conjugation with target molecules [121]. Mesoporous silica nanoparticles (MSNs) are porous particles with pore sizes ranging from 2 to 50 nm, designed to hold drug molecules in delivery systems [122]. They offer advantages such as rich mesoporosity, adjustable particle size (10–1000 nm), large pore volume (2–20 nm), flexible morphology, high surface area, easy functionalization, and good biocompatibility. Their high porosity allows for substantial drug loading [123]. The ability to customize pore sizes, volumes, and surface areas, along with the ease of encapsulating drugs, proteins, and biogenetic materials, makes MSNs a highly versatile delivery vehicle. These advantages allow MSNs to transport both small and large molecules effectively. Due to their low cost, scalability, potential for surface functionalization, and high biocompatibility, developing new MSNs for cargo delivery is increasingly prioritized in nano-delivery applications. MSNs provide superior internalization efficiency compared to traditional methods like microinjection, gene guns, and electroporation, thanks to their high surface area and small size [80].

Feng-Peng Chang and colleagues in 2013 developed a plant transient gene expression system using functionalized mesoporous silica nanoparticles (MSNs). MSNs were used as carriers in this approach to transfer foreign DNA into intact Arabidopsis thaliana roots without applying mechanical force [80]. Brian G. Trewyn and colleagues in 2013 investigated direct protein delivery to plant cells utilizing mesoporous silica nanoparticles (MSNs) as carriers to transfer Cre recombinase protein into maize (Zea mays) cells in their experiment. Cre protein was loaded into the pores of gold-plated MSNs, and these particles were delivered to plant cells with loxP sites flanking a selection gene and a reporter gene using the biolistic technique [124].

4.5 Mechanism of Nano-CRISPR Uptake in Plants

According to earlier research, there are two main routes for NP transport: up-to-down and down-to-up. When NPs are sprayed aboveground, they enter the plant via the stomata and proceed through the phloem as well as the xylem. This is known as the up-to-down transport pathway. The process by which NPs are taken up by plant roots and moved to stems and leaves is known as "down-to-up transportation." NPs can be transferred through symbiotic pathways. One is that NPs enter the cytoplasm by piercing the cell membrane. The second is that NPs are taken up by cells and passed on to nearby cells via plasmodesmata. The apoplast pathway is thought to be an additional major route for the transportation of NPs. Specifically, NPs are typically absorbed by the endoderm of roots via this pathway and are subsequently transported to the stems, leaves, and fruits of plants via the xylem or phloem. Additionally, it has been noted that endocytosis, carrier proteins, and connected channels can all aid in the transportation of NPs. The size, surface charge, and species of the plants may have a greater impact on the absorption and transportation of nanoparticles in plants. Variations in a plant's physiology and metabolism can influence how NPs are absorbed and transported. The size of nanoparticles is a crucial factor that influences how well plants absorb NPs. Due to constraints around biological barriers, plants may be able to absorb smaller NPs more readily. Plants typically have a pore size of only 10 nm for their cell walls, 40 nm for the width of their plasmodesmata, 1 nm for their aquaporin pores, and 1 μm for their endocytosis NPs. The root surface absorbs positively charged NPs more readily than negatively charged ones. As a result, there were substantially more positively charged NPs in roots than negatively charged NPs. However, internalization rates and transmission efficiencies of positively charged NPs in plants were far inferior to those of negatively charged NPs. The explanation could be that positively charged NPs are more likely to accumulate in roots because the mucous layer of marginal cells that covers the root cap of plants is made up of a layer of negatively charged root exudates. Complex interactions among various factors affect how NPs are absorbed by plant roots. In actuality, there is currently a lack of information on the transportation of NPs in plants; more research is required to fully comprehend their absorption and distribution [125].

5 Concerns and Regulatory Frameworks in CRISPR-Cas Nanotechnology

5.1 Concerns and Regulatory Considerations Related to CRISPR

The CRISPR-Cas system has become the preferred method for crop improvement due to its robustness, speed, and precision in genome manipulation. However, the commercial use of newly developed genome-edited crops has raised moral, regulatory, and biosafety concerns, as they impact the living conditions of current and future moral entities, including the non-human environment. A few studies have shown that CRISPR-Cas9-based genome editing in plants can lead to off-target effects. In 2018, Qiang Zhang and colleagues reported unexpectedly high levels of off-target mutagenesis induced by CRISPR/Cas9 in T1 Arabidopsis mutants, despite the sgRNA being predicted to have a high specificity score [126].

Regulating genome-edited crops is complex. While regulations are crucial for public safety, genetically engineered crops may be over-regulated, as they pose no greater risks than conventionally bred crops [127]. In the United States, safety testing for such crops takes over a decade and costs millions, hindering innovation. Moreover, regulatory practices differ significantly around the world [128]. GMOs are created by inserting exogenous genes into the host genome, while genome editing methods like CRISPR-Cas, using the NHEJ repair mechanism, induce small indels within the endogenous target gene. These modifications resemble natural variations and those from chemical or physical mutagenesis but are achieved with greater specificity and efficiency through genome editing [125, 129, 130]. From a utilitarian perspective, manipulating CRISPR-Cas without inserting a transgene is considered less risky compared to earlier GM techniques, making it morally more acceptable [131].

Regulatory approaches to genome-edited plants vary internationally. In the EU and New Zealand, a process-based approach is used, where gene-edited organisms are regulated as GMOs according to the techniques used for their creation. In contrast, the United States and most other countries adopt a product-based approach, where regulation depends on the final characteristics of the product. If the edited plant does not contain foreign DNA, it is not classified as a GMO and is exempt from GMO regulations [132]. In the United States, biotechnology products are regulated under a Coordinated Framework involving the EPA, USDA, and FDA, each addressing the environmental, agricultural, and human health impacts of bioengineered crops [133]. The USDA's APHIS specifically regulates genome-edited crops through the "Am I Regulated?" portal, where companies can inquire if their products fall under USDA regulation [134]. Recently, the USDA announced that it will not regulate CRISPR-Cas-edited plants like Camelina sativa and a drought-tolerant soybean, allowing them to be cultivated and sold without the same regulatory hurdles as GMOs. This lenient approach enables biotechnology companies to introduce new plant varieties faster and at lower costs. In the past 3 years, at least five CRISPR-Cas9-modified organisms have avoided USDA regulation [135]. Overall, these trends suggest that the CRISPR-Cas technique can be used to develop plants that are not classified as GMOs in the United States. The international regulatory criteria for CRISPR-Cas genome-edited plants are not well-defined, particularly regarding whether these plants should be governed as GMOs. Ongoing global discussions are exploring how genome editing differs from other biotechnologies, both in principle and in practice [136].

The ethical debate on CRISPR-Cas in plant breeding extends beyond biotechnology acceptance, addressing concerns related to modern agriculture's ecological and societal impact. CRISPR-Cas9 raises questions about justice and fairness across generations. In pluralist societies, there should be a societal discussion on setting guiding standards for the development and use of new technologies [131]. While the CRISPR-Cas system enables effective manipulation of genomic DNA, it raises numerous moral, ethical, political, and scientific concerns that could potentially result in a moratorium, halting this area of research. Therefore, it is crucial to identify the challenges, clarify concepts, and encourage dialogue among all relevant stakeholders, including universities, societies, and science and technology systems, to develop an optimal ethical and legal framework [137].

5.2 Concerns and Regulatory Considerations Related to Nanotechnology

Nanoparticle-based delivery systems offer several advantages over traditional methods, including extended shelf life, enhanced stability, ease of use, and improved solubility and permeability [138]. As nanotechnologies for plant systems advance, it is crucial to simultaneously investigate their safety implications and develop effective regulatory policies. In the United States, nanotechnologies used in genetic engineering are regulated under the Toxic Substances Control Act, which requires producers to prove safety [139]. However there is currently no government oversight of nanomaterials in laboratories, greenhouses, or fields. Both the US FDA and EPA do not recognize nanomaterials as new chemicals requiring additional oversight, despite their potential risks at the nanoscale. This contrasts with their requirement for manufacturers to prove that food ingredients are safe [140]. Similarly, in the European Union, nanotechnology is primarily regulated under REACH, which may not fully address nanoscale-specific issues. Globally, clear regulations for the production, use, labeling, and disposal of nanomaterials are lacking. As nanosystems become more prevalent in agriculture, there is a pressing need for clear legislation to ensure public health and safety.

The "nanoness" of nanoparticles, including their surface structure and reactivity, drives processes such as dissolution, redox reactions, and the generation of reactive oxygen species (ROS). These properties can lead to biological and toxicological effects that differ from those produced by bulk particles with the same chemical composition. Some researchers argue that the novel size-dependent properties of nanoparticles, rather than just particle size, should be the primary criterion for their regulation due to environmental, health, and safety concerns. Early research on nanotechnologies, such as carbon nanotubes, highlighted concerns about their non-biodegradability and compared them to carcinogenic asbestos fibers [141]. Other studies linked nanoparticle toxicity to heavy-metal impurities introduced during synthesis [142]. Although subsequent research has alleviated some of these initial concerns, toxicity and safety issues continue to be addressed within the nanomaterial community. In many cases, nanoparticles struggle to cross intracellular physiological barriers even after internalization, and they often become agglomerated within vacuoles. This aggregation can lead to cytotoxicity, presenting challenges in developing an effective delivery system [143]. It is important to recognize that both the long-term and short-term effects of engineered nanoparticles used for advancing sustainable agriculture and biomedical applications on the environment remain largely undocumented [144]. Most toxicity studies focus on short-term exposures to model systems, often neglecting the ecological impact of the accumulation of these man-made engineered nanoparticles [145] [146]. For example, while wastewater treatment can recycle polluted water, similar principles do not apply to contaminated soils, where reclamation is much slower. The accumulation of man-made nanoparticles is often unregulated and overlooked, leading to runoff into water bodies or fields. Consequently, regulating and restoring nanoparticle-infested environments is complex and requires significant maintenance. Data on the fate of internalized engineered nanoparticles (ENPs) accumulated by exposed plants is lacking. Given these concerns, strict regulations for disposal and monitoring are essential to safeguard the environment from potential harm. Despite these emerging issues, it is undeniable that ENPs hold significant promise as delivery systems in plant biotechnology's future [147].

Nanotechnology in agriculture is still in its early stages compared to other fields due to concerns about consumer health risks and the lack of unified regulations for risk assessment. The introduction of nanoparticles into human biology through agriculture requires careful risk evaluation. Without standardized guidelines, high-risk quotients may deter industry and consumers. To harmonize risk assessment data and improve regulatory frameworks, clear guidelines are needed. Currently, conflicting toxicity results from different studies add to the uncertainty, highlighting the need for specific guidelines to reduce uncertainty and facilitate effective regulation, especially in agriculture [148].

For CRISPR applications, it is essential to understand the lifecycle of nanomaterials post-delivery to ensure that edited plants, their residues, progeny, and consumers remain free of these materials. This understanding could enable the development of safer nanotechnologies and reduce regulatory barriers [18].

6 Conclusion and Future Prospects

The integration of nanotechnology with CRISPR-Cas systems represents a promising frontier in the field of plant genome editing. This chapter has elucidated the revolutionary impact of CRISPR technology on genetic modifications, emphasizing its precision and efficiency. However, as highlighted, the challenge of delivering CRISPR components effectively into plant cells remains a significant hurdle. By exploring various nanocarrier options, including metal nanoparticles, carbon-based nanomaterials, and polymeric nanoparticles, we have demonstrated how these innovative delivery systems can enhance the uptake, stability, and specificity of CRISPR-Cas components. The potential to minimize off-target effects and improve overall editing efficiency underscores the critical role of nanotechnology in advancing the application of CRISPR in agriculture.

Looking forward, the future prospects of CRISPR-Cas and nanotechnology in plant genome editing are vast and transformative. As research continues to evolve, we anticipate significant advancements in the development of more sophisticated nanocarriers that can facilitate the precise and efficient delivery of CRISPR components. However, the regulatory frameworks surrounding gene editing technologies remain uncertain and fragmented, which may pose challenges for the widespread adoption of these innovations in crop improvement and agricultural sustainability. Addressing ethical concerns and promoting responsible use will be crucial as we navigate this complex landscape. Despite these challenges, the synergistic relationship between nanotechnology and CRISPR-Cas holds immense potential to revolutionize agriculture, ultimately leading to enhanced food security and resilience in the face of global challenges.

References

1. Bajželj, B., Richards, K. S., Allwood, J. M., Smith, P., Dennis, J. S., Curmi, E., & Gilligan, C. A. (2014). Importance of food-demand management for climate mitigation. *Nature Climate Change, 4*(10), 924–929. https://doi.org/10.1038/nclimate2353
2. Shaheen, A., & Abed, Y. (2018). Knowledge, attitude, and practice among farmworkers applying pesticides in cultivated area of the Jericho district: A cross-sectional study. *The Lancet, 391*, S3. https://doi.org/10.1016/s0140-6736(18)30328-3
3. Chen, K., & Gao, C. (2014). Targeted genome modification technologies and their applications in crop improvements. *Plant Cell Reports, 33*, 575–583. https://doi.org/10.1007/s00299-013-1539-6
4. Butler, N. M., Baltes, N. J., Voytas, D. F., & Douches, D. S. (2016). Geminivirus-mediated genome editing in potato (*Solanum tuberosum l.*) using sequence-specific nucleases. *Frontiers in Plant Science, 7*, 1045. https://doi.org/10.3389/fpls.2016.01045
5. Christian, M., Cermak, T., Doyle, E. L., Schmidt, C., Zhang, F., Hummel, A., Bogdanove, A. J., & Voytas, D. F. (2010). Targeting DNA double-strand breaks with TAL effector nucleases. *Genetics, 186*(2), 757–761. https://doi.org/10.1534/genetics.110.120717
6. Lin, Q., Zong, Y., Xue, C., Wang, S., Jin, S., Zhu, Z., Wang, Y., Anzalone, A. V., Raguram, A., Doman, J. L., Liu, D. R., & Gao, C. (2020). Prime genome editing in rice and wheat. *Nature Biotechnology, 38*(5), 582–585. https://doi.org/10.1038/s41587-020-0455-x
7. Manghwar, H., Lindsey, K., Zhang, X., & Jin, S. (2019). CRISPR/Cas system: Recent advances and future prospects for genome editing. *Trends in Plant Science, 24*(12), 1102–1125. https://doi.org/10.1016/j.tplants.2019.09.006
8. Doudna, J. A., & Charpentier, E. (2014). The new frontier of genome engineering with CRISPR-Cas9. *Science, 346*(6213), 1258096. https://doi.org/10.1126/science.1258096
9. Ma, X., Zhu, Q., Chen, Y., & Liu, Y. G. (2016). CRISPR/Cas9 platforms for genome editing in plants: Developments and applications. *Molecular Plant, 9*(7), 961–974. https://doi.org/10.1016/j.molp.2016.04.009
10. Cunningham, F. J., Goh, N. S., Demirer, G. S., Matos, J. L., & Landry, M. P. (2018). Nanoparticle-mediated delivery towards advancing plant genetic engineering. *Trends in Biotechnology, 36*(9), 882–897. https://doi.org/10.1016/j.tibtech.2018.03.009

11. Ahmed, R. I., Ding, A., Xie, M., & Kong, Y. (2018). Progress in optimization of *Agrobacterium*-mediated transformation in sorghum (*Sorghum* bicolor). *International Journal of Molecular Sciences, 19*(10), 2983. https://doi.org/10.3390/ijms19102983
12. Imai, R., Hamada, H., Liu, Y., Linghu, Q., Kumagai, Y., Nagira, Y., Miki, R., & Taoka, N. (2020). In planta particle bombardment (IPB): A new method for plant transformation and genome editing. *Plant Biotechnology, 37*(2), 171–176. https://doi.org/10.5511/PLANTBIOTECHNOLOGY.20.0206A
13. Mohammed, S., Samad, A. A., & Rahmat, Z. (2019). *Agrobacterium*-mediated transformation of rice: Constraints and possible solutions. *Rice Science, 26*(3), 133–146. https://doi.org/10.1016/j.rsci.2019.04.001
14. Ozyigit, I. I. (2020). Gene transfer to plants by electroporation: Methods and applications. *Molecular Biology Reports, 47*(4), 3195–3210. https://doi.org/10.1007/s11033-020-05343-4
15. Tian, Y., Chen, K., Li, X., Zheng, Y., & Chen, F. (2020). Design of high-oleic tobacco (*Nicotiana tabacum L.*) seed oil by CRISPR-Cas9-mediated knockout of NtFAD2-2. *BMC Plant Biology, 20*, 1–12. https://doi.org/10.1186/s12870-020-02441-0
16. Shin, J. H., Han, J. H., Park, H. H., Fu, T., & Kim, K. S. (2019). Optimization of polyethylene glycol-mediated transformation of the pepper anthracnose pathogen *colletotrichum scovillei* to develop an applied genomics approach. *The Plant Pathology Journal, 35*(6), 575. https://doi.org/10.5423/PPJ.OA.06.2019.0171
17. Abd-Elsalam, K. A. (2020). Multifunctional hybrid nanomaterials for sustainable agri-food and ecosystems: A note from the editor. In *Multifunctional hybrid nanomaterials for sustainable agri-food and ecosystems* (pp. 1–19). Elsevier. https://doi.org/10.1016/B978-0-12-821354-4.00001-7
18. Demirer, G. S., Silva, T. N., Jackson, C. T., Thomas, J. B., W. Ehrhardt, D., Rhee, S. Y., Mortimer, J. C., & Landry, M. P. (2021). Nanotechnology to advance CRISPR–Cas genetic engineering of plants. *Nature Nanotechnology, 16*(3), 243–250. https://doi.org/10.1038/s41565-021-00854-y
19. Naik, B. J., Shimoga, G., Kim, S. C., Manjulatha, M., Subramanyam Reddy, C., Palem, R. R., Kumar, M., Kim, S. Y., & Lee, S. H. (2022). CRISPR/Cas9 and nanotechnology pertinence in agricultural crop refinement. *Frontiers in Plant Science, 13*, 843575. https://doi.org/10.3389/fpls.2022.843575
20. Yang, B. (2020). Grand challenges in genome editing in plants. *Frontiers in Genome Editing, 2*, 2. https://doi.org/10.3389/fgeed.2020.00002
21. Pan, C., Ye, L., Qin, L., Liu, X., He, Y., Wang, J., Chen, L., & Lu, G. (2016). CRISPR/Cas9-mediated efficient and heritable targeted mutagenesis in tomato plants in the first and later generations. *Scientific Reports, 6*(1), 24765. https://doi.org/10.1038/srep24765
22. Liu, L., Gallagher, J., Arevalo, E. D., Chen, R., Skopelitis, T., Wu, Q., Bartlett, M., & Jackson, D. (2021). Enhancing grain-yield-related traits by CRISPR–Cas9 promoter editing of maize CLE genes. *Nature Plants, 7*(3), 287–294. https://doi.org/10.1038/s41477-021-00858-5
23. Farooq, M., Bashir, M., Khan, M., Iqbal, B., & Ali, Q. (2021). Role of CRISPR to improve abiotic stress tolerance in crop plants. *Biological and Clinical Sciences Research Journal, 2021*. https://doi.org/10.54112/bcsrj.v2021i1.69
24. Usman, B., Nawaz, G., Zhao, N., Liao, S., Qin, B., Liu, F., Liu, Y., & Li, R. (2021). Programmed editing of rice (*Oryza sativa l.*) osspl16 gene using crispr/cas9 improves grain yield by modulating the expression of pyruvate enzymes and cell cycle proteins. *International Journal of Molecular Sciences, 22*(1), 249. https://doi.org/10.3390/ijms22010249
25. Liu, J., Wang, S., Wang, H., Luo, B., Cai, Y., Li, X., Zhang, Y., & Wang, X. (2021). Rapid generation of tomato male-sterile lines with a marker use for hybrid seed production by CRISPR/Cas9 system. *Molecular Breeding, 41*, 1–12. https://doi.org/10.1007/s11032-021-01215-2
26. Ishino, Y., Shinagawa, H., Makino, K., Amemura, M., & Nakatura, A. (1987). Nucleotide sequence of the iap gene, responsible for alkaline phosphatase isoenzyme conversion in *Escherichia coli*, and identification of the gene product. *Journal of Bacteriology, 169*(12), 5429–5433. https://doi.org/10.1128/jb.169.12.5429-5433.1987

27. Barrangou, R., Fremaux, C., Deveau, H., Richards, M., Boyaval, P., Moineau, S., Romero, D. A., & Horvath, P. (2007). CRISPR provides acquired resistance against viruses in prokaryotes. *Science, 315*(5819), 1709–1712. https://doi.org/10.1126/science.1138140
28. Bolotin, A., Quinquis, B., Sorokin, A., & Dusko Ehrlich, S. (2005). Clustered regularly interspaced short palindrome repeats (CRISPRs) have spacers of extrachromosomal origin. *Microbiology, 151*(8), 2551–2561. https://doi.org/10.1099/mic.0.28048-0
29. Brouns, S. J. J., Jore, M. M., Lundgren, M., Westra, E. R., Slijkhuis, R. J. H., Snijders, A. P. L., Dickman, M. J., Makarova, K. S., Koonin, E. V., & Van Der Oost, J. (2008). Small CRISPR RNAs guide antiviral defense in prokaryotes. *Science, 321*(5891), 960–964. https://doi.org/10.1126/science.1159689
30. Gasiunas, G., Barrangou, R., Horvath, P., & Siksnys, V. (2012). Cas9-crRNA ribonucleoprotein complex mediates specific DNA cleavage for adaptive immunity in bacteria. *Proceedings of the National Academy of Sciences, 109*(39), E2579–E2586. https://doi.org/10.1073/pnas.1208507109
31. Deltcheva, E., Chylinski, K., Sharma, C. M., Gonzales, K., Chao, Y., Pirzada, Z. A., Eckert, M. R., Vogel, J., & Charpentier, E. (2011). CRISPR RNA maturation by trans-encoded small RNA and host factor RNase III. *Nature, 471*(7340), 602–607. https://doi.org/10.1038/nature09886
32. Rouillon, C., Zhou, M., Zhang, J., Politis, A., Beilsten-Edmands, V., Cannone, G., Graham, S., Robinson, C., Spagnolo, L., White, M. (2013). Structure of the CRISPR interference complex CSM reveals key similarities with cascade. *Molecular Cell, 52*(1), 124–134. https://doi.org/10.1016/j.molcel.2013.08.020
33. Abdelnour, S., Xie, L., Hassanin, A., Zuo, E., Lu, Y. (2021). The potential of CRISPR/Cas9 gene editing as a treatment strategy for inherited diseases. *Frontiers in Cell and Developmental Biology, 9*, 699597. https://doi.org/10.3389/fcell.2021.699597
34. Bortesi, L., & Fischer, R. (2015). The CRISPR/Cas9 system for plant genome editing and beyond. *Biotechnology Advances, 33*(1), 41–52. https://doi.org/10.1016/j.biotechadv.2014.12.006
35. Harrison, M. M., Jenkins, B. V., O'Connor-Giles, K. M., & Wildonger, J. (2014). A CRISPR view of development. *Genes & Development, 28*(17), 1859–1872. https://doi.org/10.1101/gad.248252.114
36. Salsman, J., & Dellaire, G. (2017). Precision genome editing in the CRISPR era. *Biochemistry and Cell Biology, 95*(2), 187–201. https://doi.org/10.1139/bcb-2016-0137
37. Tang, X. D., Gao, F., Liu, M. J., Fan, Q. L., Chen, D. K., & Ma, W. T. (2019). Methods for enhancing clustered regularly interspaced short palindromic repeats/Cas9-mediated homology-directed repair efficiency. *Frontiers in Genetics, 10*, 551. https://doi.org/10.3389/fgene.2019.00551
38. Belhaj, K., Chaparro-Garcia, A., Kamoun, S., Patron, N. J., & Nekrasov, V. (2015). Editing plant genomes with CRISPR/Cas9. *Current Opinion in Biotechnology, 32*, 76–84. https://doi.org/10.1016/j.copbio.2014.11.007
39. Rao, M. J., & Wang, L. (2021). CRISPR/Cas9 technology for improving agronomic traits and future prospective in agriculture. *Planta, 254*, 1–16. https://doi.org/10.1007/s00425-021-03716-y
40. Ricroch, A., Clairand, P., & Harwood, W. (2017). Use of CRISPR systems in plant genome editing: Toward new opportunities in agriculture. *Emerging Topics in Life Sciences, 1*(2), 169–182. https://doi.org/10.1042/ETLS20170085
41. Zafar, K., Khan, M. Z., Amin, I., Mukhtar, Z., Yasmin, S., Arif, M., Ejaz, K., & Mansoor, S. (2020). Precise CRISPR-Cas9 mediated genome editing in super basmati rice for resistance against bacterial blight by targeting the major susceptibility gene. *Frontiers in Plant Science, 11*, 575. https://doi.org/10.3389/fpls.2020.00575
42. Sunitha, S., & Rock, C. D. (2020). CRISPR/Cas9-mediated targeted mutagenesis of TAS4 and MYBA7 loci in grapevine rootstock 101-14. *Transgenic Research, 29*(3), 355–367. https://doi.org/10.1007/s11248-020-00196-w

43. Yunyan, F., Jie, Y., Fangquan, W., Fangjun, F., Wenqi, L., Jun, W., Yang, X., Jinyan, Z., & Weigong, Z. (2019). Production of two elite glutinous rice varieties by editing Wx gene. *Rice Science, 26*(2), 118–124. https://doi.org/10.1016/j.rsci.2018.04.007
44. Yan, R., Wang, Z., Ren, Y., Li, H., Liu, N., & Sun, H. (2019). Establishment of efficient genetic transformation systems and application of crispr/cas9 genome editing technology in *lilium pumilum dc. Fisch.* and *lilium longiflorum* white heaven. *International journal of molecular sciences, 20*(12), 2920. https://doi.org/10.3390/ijms20122920
45. Ji, X., Zhang, H., Zhang, Y., Wang, Y., & Gao, C. (2015). Establishing a CRISPR–Cas-like immune system conferring DNA virus resistance in plants. *Nature Plants, 1*(10), 1–4. https://doi.org/10.1038/NPLANTS.2015.144
46. Nekrasov, V., Wang, C., Win, J., Lanz, C., Weigel, D., & Kamoun, S. (2017). Rapid generation of a transgene-free powdery mildew resistant tomato by genome deletion. *Scientific Reports, 7*(1), 482. https://doi.org/10.1038/s41598-017-00578-x
47. Shi, J., Gao, H., Wang, H., Lafitte, H. R., Archibald, R. L., Yang, M., Hakimi, S. M., Mo, H., & Habben, J. E. (2017). ARGOS8 variants generated by CRISPR-Cas9 improve maize grain yield under field drought stress conditions. *Plant Biotechnology Journal, 15*(2), 207–216. https://doi.org/10.1111/pbi.12603
48. Ito, Y., Nishizawa-Yokoi, A., Endo, M., Mikami, M., & Toki, S. (2015). CRISPR/Cas9-mediated mutagenesis of the RIN locus that regulates tomato fruit ripening. *Biochemical and Biophysical Research Communications, 467*(1), 76–82. https://doi.org/10.1016/j.bbrc.2015.09.117
49. Ke, Q., Kang, L., Kim, H. S., Xie, T., Liu, C., Ji, C. Y., Kim, S. H., Park, W. S., Ahn, M. J., Wang, S., Li, H., Deng, X., & Kwak, S. S. (2019). Down-regulation of lycopene ε-cyclase expression in transgenic sweetpotato plants increases the carotenoid content and tolerance to abiotic stress. *Plant Science, 281*, 52–60. https://doi.org/10.1016/j.plantsci.2019.01.002
50. Mishra, R., Mohanty, J. N., Mahanty, B., & Joshi, R. K. (2021). A single transcript CRISPR/Cas9 mediated mutagenesis of CaERF28 confers anthracnose resistance in chilli pepper (*Capsicum annuum L.*). *Planta, 254*(1), 5. https://doi.org/10.1007/s00425-021-03660-x
51. Ledford, H. (2015). CRISPR, the disruptor. *Nature, 522*(7554), 20. https://doi.org/10.1038/522020a
52. Gao, C. (2018). The future of CRISPR technologies in agriculture. *Nature Reviews Molecular Cell Biology, 19*(5), 275–276. https://doi.org/10.1038/nrm.2018.2
53. Eş, I., Gavahian, M., Marti-Quijal, F. J., Lorenzo, J. M., Mousavi Khaneghah, A., Tsatsanis, C., Kampranis, S. C., & Barba, F. J. (2019). The application of the CRISPR-Cas9 genome editing machinery in food and agricultural science: Current status, future perspectives, and associated challenges. *Biotechnology Advances, 37*(3), 410–421. https://doi.org/10.1016/j.biotechadv.2019.02.006
54. Brinegar, K., Yetisen, A. K., Choi, S., Vallillo, E., Ruiz-Esparza, G. U., Prabhakar, A. M., Khademhosseini, A., & Yun, S. H. (2017). The commercialization of genome-editing technologies. *Critical Reviews in Biotechnology, 37*(7), 924–932. https://doi.org/10.1080/07388551.2016.1271768
55. Jiang, W., Zhou, H., Bi, H., Fromm, M., Yang, B., & Weeks, D. P. (2013). Demonstration of CRISPR/Cas9/sgRNA-mediated targeted gene modification in Arabidopsis, tobacco, sorghum and rice. *Nucleic Acids Research, 41*(20), e188–e188. https://doi.org/10.1093/nar/gkt780
56. Li, J. F., Norville, J. E., Aach, J., McCormack, M., Zhang, D., Bush, J., Church, G. M., & Sheen, J. (2013). Multiplex and homologous recombination-mediated genome editing in *Arabidopsis* and *Nicotiana benthamiana* using guide RNA and Cas9. *Nature Biotechnology, 31*(8), 688–691. https://doi.org/10.1038/nbt.2654
57. Mao, Y., Zhang, H., Xu, N., Zhang, B., Gou, F., & Zhu, J. K. (2013). Application of the CRISPR-Cas system for efficient genome engineering in plants. *Molecular Plant, 6*(6), 2008–2011. https://doi.org/10.1093/mp/sst121

58. Nekrasov, V., Staskawicz, B., Weigel, D., Jones, J. D. G., & Kamoun, S. (2013). Targeted mutagenesis in the model plant *Nicotiana benthamiana* using Cas9 RNA-guided endonuclease. *Nature Biotechnology, 31*(8), 691–693. https://doi.org/10.1038/nbt.2655
59. Shan, Q., Wang, Y., Li, J., Zhang, Y., Chen, K., Liang, Z., Zhang, K., Liu, J., Xi, J. J., Qiu, J. L., & Gao, C. (2013). Targeted genome modification of crop plants using a CRISPR-Cas system. *Nature Biotechnology, 31*(8), 686–688. https://doi.org/10.1038/nbt.2650
60. Sood, P., Bhattacharya, A., & Sood, A. (2011). Problems and possibilities of monocot transformation. *Biología plantarum, 55*(1), 1–15. https://doi.org/10.1007/s10535-011-0001-2
61. Chandrasekaran, R., Rajiv, P., & Abd-Elsalam, K. A. (2019). Carbon nanotubes: Plant gene delivery and genome editing. In *Carbon nanomaterials for agri-food and environmental applications* (pp. 279–296). https://doi.org/10.1016/B978-0-12-819786-8.00014-1
62. Ahmar, S., Mahmood, T., Fiaz, S., Mora-Poblete, F., Shafique, M. S., Chattha, M. S., & Jung, K. H. (2021). Advantage of nanotechnology-based genome editing system and its application in crop improvement. *Frontiers in Plant Science, 12*, 663849. https://doi.org/10.3389/fpls.2021.663849
63. Hasan, S. (2014). A review on nanoparticles: Their synthesis and types. *Research Journal of Recent Sciences, 2277*, 2502.
64. Khan, Y., Sadia, H., Ali Shah, S. Z., Khan, M. N., Shah, A. A., Ullah, N., Ullah, M. F., Bibi, H., Bafakeeh, O. T., Ben Khedher, N., Eldin, S. M., Fadhl, B. M., & Khan, M. I. (2022). Classification, synthetic, and characterization approaches to nanoparticles, and their applications in various fields of nanotechnology: A review. *Catalysts, 12*(11), 1386. https://doi.org/10.3390/catal12111386
65. Machado, S., Pacheco, J. G., Nouws, H. P. A., Albergaria, J. T., & Delerue-Matos, C. (2015). Characterization of green zero-valent iron nanoparticles produced with tree leaf extracts. *Science of the Total Environment, 533*, 76–81. https://doi.org/10.1016/j.scitotenv.2015.06.091
66. Marsalek, R. (2014). Particle size and zeta potential of ZnO. *APCBEE Procedia, 9*, 13–17. https://doi.org/10.1016/j.apcbee.2014.01.003
67. Raja, P. M. V., & Barron, A. R. (2019). 2.3: BET surface area analysis of nanoparticles. physical methods in chemistry and nanoscience.
68. Greczynski, G., & Hultman, L. (2021). The same chemical state of carbon gives rise to two peaks in X-ray photoelectron spectroscopy. *Scientific Reports, 11*(1), 11195. https://doi.org/10.1038/s41598-021-90780-9
69. Marsalek, R., Kotyrba, M., Volna, E., & Jarusek, R. (2021). Neural network modelling for prediction of zeta potential. *Mathematics, 9*(23), 3089. https://doi.org/10.3390/math9233089
70. Meng, C., Das, P., Shi, X., Fu, Q., Müllen, K., & Wu, Z. S. (2021). In situ and operando characterizations of 2D materials in electrochemical energy storage devices. *Small Science, 1*(4), 2000076. https://doi.org/10.1002/smsc.202000076
71. Kah, M., Kookana, R. S., Gogos, A., & Bucheli, T. D. (2018). A critical evaluation of nanopesticides and nanofertilizers against their conventional analogues. *Nature Nanotechnology, 13*(8), 677–684. https://doi.org/10.1038/s41565-018-0131-1
72. Liu, R., & Lal, R. (2014). Synthetic apatite nanoparticles as a phosphorus fertilizer for soybean (*Glycine max*). *Scientific Reports, 4*(1), 5686. https://doi.org/10.1038/srep05686
73. Carriger, J. F., Rand, G. M., Gardinali, P. R., Perry, W. B., Tompkins, M. S., & Fernandez, A. M. (2006). Pesticides of potential ecological concern in sediment from South Florida canals: An ecological risk prioritization for aquatic arthropods. *Soil & Sediment Contamination, 15*(1), 21–45. https://doi.org/10.1080/15320380500363095
74. Chhipa, H. (2017). Nanofertilizers and nanopesticides for agriculture. *Environmental Chemistry Letters, 15*, 15–22. https://doi.org/10.1007/s10311-016-0600-4
75. Sun, Y. P., Li, X.-q., Cao, J., Zhang, W.-x., & Wang, H. P. (2006). Characterization of zero-valent iron nanoparticles. *Advances in Colloid and Interface Science, 120*(1–3), 47–56. https://doi.org/10.1016/j.cis.2006.03.001

76. Jiang, M., Song, Y., Kanwar, K., Ahammed, G., Shao, S., Zhou, J. (2021). Phytonanotechnology applications in modern agriculture. *Journal of Nanobiotechnology, 19*, 1–20. https://doi.org/10.1186/s12951-021-01176-w
77. Finiuk, N., Buziashvili, A., Burlaka, O., Zaichenko, A., Mitina, N., Miagkota, O., Lobachevska, O., Stoika, R., Blume, Y., & Yemets, A. (2017). Investigation of novel oligoelectrolyte polymer carriers for their capacity of DNA delivery into plant cells. *Plant Cell, Tissue and Organ Culture (PCTOC), 131*, 27–39. https://doi.org/10.1007/s11240-017-1259-7
78. Joldersma, D., & Liu, Z. (2018). Plant genetics enters the nano age? *Journal of Integrative Plant Biology, 60*(6), 446–447. https://doi.org/10.1111/jipb.12646
79. Silva, A. T., Nguyen, A., Ye, C., Verchot, J., & Moon, J. H. (2010). Conjugated polymer nanoparticles for effective siRNA delivery to tobacco BY-2 protoplasts. *BMC Plant Biology, 10*, 1–14. https://doi.org/10.1186/1471-2229-10-291
80. Chang, F. P., Kuang, L. Y., Huang, C. A., Jane, W. N., Hung, Y., Hsing, Y. I. C., & Mou, C. Y. (2013). A simple plant gene delivery system using mesoporous silica nanoparticles as carriers. *Journal of Materials Chemistry B, 1*(39), 5279–5287. https://doi.org/10.1039/c3tb20529k
81. Golestanipour, A., Nikkhah, M., Aalami, A., & Hosseinkhani, S. (2018). Gene delivery to tobacco root cells with single-walled carbon nanotubes and cell-penetrating fusogenic peptides. *Molecular Biotechnology, 60*, 863–878. https://doi.org/10.1007/s12033-018-0120-5
82. Liu, B., Saber, A., & Haisma, H. J. (2019). CRISPR/Cas9: A powerful tool for identification of new targets for cancer treatment. *Drug Discovery Today, 24*(4), 955–970. https://doi.org/10.1016/j.drudis.2019.02.011
83. Yi, L., & Li, J. (2016). CRISPR-Cas9 therapeutics in cancer: Promising strategies and present challenges. *Biochimica et Biophysica Acta (BBA)-Reviews on Cancer, 1866*(2), 197–207. https://doi.org/10.1016/j.bbcan.2016.09.002
84. Eeckhaut, T., Lakshmanan, P. S., Deryckere, D., Van Bockstaele, E., & Van Huylenbroeck, J. (2013). Progress in plant protoplast research. *Planta, 238*, 991–1003. https://doi.org/10.1007/s00425-013-1936-7
85. Cosgrove, D. J. (1997). Assembly and enlargement of the primary cell wall in plants. *Annual Review of Cell and Developmental Biology, 13*(1), 171–201. https://doi.org/10.1146/annurev.cellbio.13.1.171
86. Xu, X., Wan, T., Xin, H., Li, D., Pan, H., Wu, J., & Ping, Y. (2019). Delivery of CRISPR/Cas9 for therapeutic genome editing. *The Journal of Gene Medicine, 21*(7), e3107. https://doi.org/10.1002/jgm.3107
87. Liang, Z., Chen, K., Li, T., Zhang, Y., Wang, Y., Zhao, Q., Liu, J., Zhang, H., Liu, C., Ran, Y., & Gao, C. (2017). Efficient DNA-free genome editing of bread wheat using CRISPR/Cas9 ribonucleoprotein complexes. *Nature Communications, 8*(1), 14261. https://doi.org/10.1038/ncomms14261
88. Hesaraki, M., & Doroudian, M. (2012). *Candidate Vaccine., 1*, 27–35.
89. Zhang, S., Shen, J., Li, D., & Cheng, Y. (2020). Strategies in the delivery of Cas9 ribonucleoprotein for CRISPR/Cas9 genome editing. *Theranostics, 11*(2), 614. https://doi.org/10.7150/thno.47007
90. Yan, J., Kang, D. D., Turnbull, G., & Dong, Y. (2022). Delivery of CRISPR-Cas9 system for screening and editing RNA binding proteins in cancer. *Advanced Drug Delivery Reviews, 180*, 114042. https://doi.org/10.1016/j.addr.2021.114042
91. Chen, F., Alphonse, M., & Liu, Q. (2020). Strategies for nonviral nanoparticle-based delivery of CRISPR/Cas9 therapeutics. *Wiley Interdisciplinary Reviews: Nanomedicine and Nanobiotechnology, 12*(3), e1609. https://doi.org/10.1002/wnan.1609
92. Zhou, S., Li, Y., Wu, Q., Gong, C. (2024). Nanotechnology-based CRISPR/Cas9 delivery system for genome editing in cancer treatment. *MedComm – Biomaterials and Applications, 3*(1), e70. https://doi.org/10.1002/mba2.70

93. Mitchell, M. J., Billingsley, M. M., Haley, R. M., Wechsler, M. E., Peppas, N. A., & Langer, R. (2021). Engineering precision nanoparticles for drug delivery. *Nature Reviews Drug Discovery, 20*(2), 101–124. https://doi.org/10.1038/s41573-020-0090-8
94. Zhang, L., Beatty, A., Lu, L., Abdalrahman, A., Makris, T. M., Wang, G., & Wang, Q. (2020). Microfluidic-assisted polymer-protein assembly to fabricate homogeneous functional nanoparticles. *Materials Science and Engineering: C, 111*, 110768. https://doi.org/10.1016/j.msec.2020.110768
95. Knight, F. C., Gilchuk, P., Kumar, A., Becker, K. W., Sevimli, S., Jacobson, M. E., Suryadevara, N., Wang-Bishop, L., Boyd, K. L., Crowe, J. E., Joyce, S., & Wilson, J. T. (2019). Mucosal immunization with a pH-responsive nanoparticle vaccine induces protective CD8+ lung-resident memory T cells. *ACS Nano, 13*(10), 10939–10960. https://doi.org/10.1021/acsnano.9b00326
96. Strand, M. S., Pan, H., Krasnick, B., Zhang, X., Goedegebuure, P. S., Fleming, T., DeNardo, D. G., Gillanders, W. E., Hawkins, W. G., Wickline, S. A., & Fields, R. C. (2017). Precision delivery of RAS-inhibiting siRNA to pancreatic cancer via peptide-based nanoparticles. *Journal of Clinical Oncology, 35*, 287. https://doi.org/10.1200/jco.2017.35.4_suppl.287
97. Jose, S., Cinu, T. A., Sebastian, R., Shoja, M. H., Aleykutty, N. A., Durazzo, A., Lucarini, M., Santini, A., & Souto, E. B. (2019). Transferrin-conjugated Docetaxel-PLGA nanoparticles for tumor targeting: Influence on MCF-7 cell cycle. *Polymers, 11*(11), 1905. https://doi.org/10.3390/polym11111905
98. Zuris, J. A., Thompson, D. B., Shu, Y., Guilinger, J. P., Bessen, J. L., Hu, J. H., Maeder, M. L., Joung, J. K., Chen, Z. Y., & Liu, D. R. (2015). Cationic lipid-mediated delivery of proteins enables efficient protein-based genome editing in vitro and in vivo. *Nature Biotechnology, 33*(1), 73. https://doi.org/10.1038/nbt.3081
99. Ragelle, H., Riva, R., Vandermeulen, G., Naeye, B., Pourcelle, V., Le Duff, C. S., D'Haese, C., Nysten, B., Braeckmans, K., De Smedt, S. C., Jérôme, C., & Préat, V. (2014). Chitosan nanoparticles for siRNA delivery: Optimizing formulation to increase stability and efficiency. *Journal of Controlled Release, 176*, 54–63. https://doi.org/10.1016/j.jconrel.2013.12.026
100. Liu, C., Zhang, L., Liu, H., & Cheng, K. (2017). Delivery strategies of the CRISPR-Cas9 gene-editing system for therapeutic applications. *Journal of Controlled Release, 266*, 17–26. https://doi.org/10.1016/j.jconrel.2017.09.012
101. Noopur, P., & Kamal, M. (2017). Dendrimers: A novel carrier for drug delivery system. *PharmaTutor, 5*(10), 48–54.
102. Pasupathy, K., Lin, S., Hu, Q., Luo, H., & Ke, P. C. (2008). Direct plant gene delivery with a poly(amidoamine) dendrimer. *Biotechnology Journal: Healthcare Nutrition Technology, 3*(8), 1078–1082. https://doi.org/10.1002/biot.200800021
103. Amani, A., Zare, N., Asadi, A., & Asghari-Zakaria, R. (2018). Ultrasound-enhanced gene delivery to alfalfa cells by hPAMAM dendrimer nanoparticles. *Turkish Journal of Biology, 42*(1), 63–75. https://doi.org/10.3906/biy-1706-6
104. Ranjbar, S., Emamjomeh, A., Sharifi, F., Zarepour, A., Aghaabbasi, K., Dehshahri, A., Mohammadinejad, R. (2023). Lipid-Based Delivery Systems for Flavonoids and Flavonolignans: Liposomes, Nanoemulsions, and Solid Lipid Nanoparticles. *Pharmaceutics, 15*(7), 1944. https://doi.org/10.3390/pharmaceutics15071944
105. Khosa, A., Reddi, S., & Saha, R. N. (2018). Nanostructured lipid carriers for site-specific drug delivery. *Biomedicine & Pharmacotherapy, 103*, 598–613. https://doi.org/10.1016/j.biopha.2018.04.055
106. Yanar, F., Mosayyebi, A., Nastruzzi, C., Carugo, D., & Zhang, X. (2020). Continuous-flow production of liposomes with a millireactor under varying fluidic conditions. *Pharmaceutics, 12*(11), 1001. https://doi.org/10.3390/pharmaceutics12111001
107. Sercombe, L., Veerati, T., Moheimani, F., Wu, S. Y., Sood, A. K., & Hua, S. (2015). Advances and challenges of liposome assisted drug delivery. *Frontiers in Pharmacology, 6*, 286. https://doi.org/10.3389/fphar.2015.00286

108. Akbarzadeh, A., Rezaei-Sadabady, R., Davaran, S., Joo, S. W., Zarghami, N., Hanifehpour, Y., Samiei, M., Kouhi, M., & Nejati-Koshki, K. (2013). Liposome: Classification, preparation, and applications. *Nanoscale Research Letters, 8*, 1–9. https://doi.org/10.1186/1556-276X-8-102
109. Wiesman, Z., Ben, D. N., Sharvit, E., Grinberg, S., Linder, C., Heldman, E., & Zaccai, M. (2007). Novel cationic vesicle platform derived from vernonia oil for efficient delivery of DNA through plant cuticle membranes. *Journal of Biotechnology, 130*(1), 85–94. https://doi.org/10.1016/j.jbiotec.2007.01.040
110. Leung, A. K. K., Tam, Y. Y. C., Chen, S., Hafez, I. M., & Cullis, P. R. (2015). Microfluidic mixing: A general method for encapsulating macromolecules in lipid nanoparticle systems. *The Journal of Physical Chemistry B, 119*(28), 8698–8706. https://doi.org/10.1021/acs.jpcb.5b02891
111. Zhen, S., & Li, X. (2020). Liposomal delivery of CRISPR/Cas9. *Cancer Gene Therapy, 27*(7), 515–527. https://doi.org/10.1038/s41417-019-0141-7
112. Filipczak, N., Pan, J., Yalamarty, S. S. K., & Torchilin, V. P. (2020). Recent advancements in liposome technology. *Advanced Drug Delivery Reviews, 156*, 4–22. https://doi.org/10.1016/j.addr.2020.06.022
113. Hejabi, F., Abbaszadeh, M. S., Taji, S., O'Neill, A., Farjadian, F., & Doroudian, M. (2022). Nanocarriers: A novel strategy for the delivery of CRISPR/Cas systems. *Frontiers in Chemistry, 10*, 957572. https://doi.org/10.3389/fchem.2022.957572
114. Ghasemiyeh, P., & Mohammadi-Samani, S. (2018). Solid lipid nanoparticles and nanostructured lipid carriers as novel drug delivery systems: Applications, advantages and disadvantages. *Research in pharmaceutical sciences, 13*(4), 288–303. https://doi.org/10.4103/1735-5362.235156
115. Mahmoud, L. M., Kaur, P., Stanton, D., Grosser, J. W., & Dutt, M. (2022). A cationic lipid mediated CRISPR/Cas9 technique for the production of stable genome edited citrus plants. *Plant Methods, 18*(1), 33. https://doi.org/10.1186/s13007-022-00870-6
116. Kim, Y., Zhu, J., Yeom, B., Di Prima, M., Su, X., Kim, J. G., Yoo, S. J., Uher, C., & Kotov, N. A. (2013). Stretchable nanoparticle conductors with self-organized conductive pathways. *Nature, 500*(7460), 59–63. https://doi.org/10.1038/nature12401
117. Rana, S., Bajaj, A., Mout, R., & Rotello, V. M. (2012). Monolayer coated gold nanoparticles for delivery applications. *Advanced Drug Delivery Reviews, 64*(2), 200–216. https://doi.org/10.1016/j.addr.2011.08.006
118. She, W., Li, N., Luo, K., Guo, C., Wang, G., Geng, Y., & Gu, Z. (2013). Dendronized heparin-doxorubicin conjugate based nanoparticle as pH-responsive drug delivery system for cancer therapy. *Biomaterials, 34*(9), 2252–2264. https://doi.org/10.1016/j.biomaterials.2012.12.017
119. Zhu Y, Qian H, Drake BA, Jin R (2010) Atomically precise Au25(SR)18 nanoparticles as catalysts for the selective hydrogenation of α,β-unsaturated ketones and aldehydes. Angewandte Chemie International Edition, 7(49), 1295–1298, doi: https://doi.org/10.1002/anie.200906249.
120. Mortazavi, S., & Zohrabi, Z. (2018). Biolistic co-transformation of rice using gold nanoparticles. *Iran Agricultural Research, 37*(1), 75–82. https://doi.org/10.22099/iar.2018.4755
121. Kneuer, C., Sameti, M., Bakowsky, U., Schiestel, T., Schirra, H., Schmidt, H., & Lehr, C. M. (2000). A nonviral DNA delivery system based on surface modified silica-nanoparticles can efficiently transfect cells in vitro. *Bioconjugate Chemistry, 11*(6), 926–932. https://doi.org/10.1021/bc0000637
122. Wang, Y., Zhao, Q., Han, N., Bai, L., Li, J., Liu, J., Che, E., Hu, L., Zhang, Q., Jiang, T., & Wang, S. (2015). Mesoporous silica nanoparticles in drug delivery and biomedical applications. *Nanomedicine: Nanotechnology, Biology and Medicine, 11*(2), 313–327. https://doi.org/10.1016/j.nano.2014.09.014

123. Iturrioz-Rodríguez, N., Correa-Duarte, M. A., & Fanarraga, M. L. (2019). Controlled drug delivery systems for cancer based on mesoporous silica nanoparticles. *International Journal of Nanomedicine, 3389-3401*, 3389. https://doi.org/10.2147/IJN.S198848
124. Martin-Ortigosa, S., Peterson, D. J., Valenstein, J. S., Lin, V. S. Y., Trewyn, B. G., Alexander Lyznik, L., & Wang, K. (2014). Mesoporous silica nanoparticle-mediated intracellular cre protein delivery for maize genome editing via loxP site excision. *Plant Physiology, 164*(2), 537–547. https://doi.org/10.1104/pp.113.233650
125. Wang, X., Xie, H., Wang, P., Yin, H. (2023). Nanoparticles in plants: uptake, transport and physiological activity in leaf and root. *Materials, 16*(8), 3097. https://doi.org/10.3390/ma16083097
126. Zhang, Q., Xing, H. L., Wang, Z. P., Zhang, H. Y., Yang, F., Wang, X. C., & Chen, Q. J. (2018). Potential high-frequency off-target mutagenesis induced by CRISPR/Cas9 in *Arabidopsis* and its prevention. *Plant Molecular Biology, 96*, 445–456. https://doi.org/10.1007/s11103-018-0709-x
127. National Academies of Sciences, Engineering, and Medicine, Division on Earth and Life Studies, Board on Agriculture and Natural Resources, & Committee on Genetically Engineered Crops: Past Experience and Future Prospects. (2017). *Genetically engineered crops: Experiences and prospects*. National Academies Press.
128. Prado, J. R., Segers, G., Voelker, T., Carson, D., Dobert, R., Phillips, J., Cook, K., Cornejo, C., Monken, J., Grapes, L., Reynolds, T., & Martino-Catt, S. (2014). Genetically engineered crops: From idea to product. *Annual Review of Plant Biology, 65*(1), 769–790.
129. Huang, S., Weigel, D., Beachy, R. N., & Li, J. (2016). A proposed regulatory framework for genome-edited crops. *Nature Genetics, 48*(2), 109–111. https://doi.org/10.1038/ng.3484
130. Li, G., Liu, Y. G., & Chen, Y. (2019). Genome-editing technologies: The gap between application and policy. *Science China. Life Sciences, 62*(11), 1534–1538. https://doi.org/10.1007/s11427-019-1566-1
131. Bartkowski, B., Theesfeld, I., Pirscher, F., & Timaeus, J. (2018). Snipping around for food: Economic, ethical and policy implications of CRISPR/Cas genome editing. *Geoforum, 96*, 172–180. https://doi.org/10.1016/j.geoforum.2018.07.017
132. Zhang, D., Hussain, A., Manghwar, H., Xie, K., Xie, S., Zhao, S., Larkin, R. M., Qing, P., Jin, S., & Ding, F. (2020). Genome editing with the CRISPR-Cas system: An art, ethics and global regulatory perspective. *Plant Biotechnology Journal, 18*(8), 1651–1669. https://doi.org/10.1111/pbi.13383
133. Wolt, J. D. (2017). Safety, security, and policy considerations for plant genome editing. *Progress in Molecular Biology and Translational Science, 149*, 215–241. https://doi.org/10.1016/bs.pmbts.2017.03.005
134. Waltz, E. (2018). With a free pass, CRISPR-edited plants reach market in record time. *Nature Biotechnology, 36*(1), 6–8. https://doi.org/10.1038/nbt0118-6b
135. Waltz, E. (2016). Gene-edited CRISPR mushroom escapes US regulation. *Nature, 532*(7599), 293.
136. Fears, R., & Ter Meulen, V. (2017). How should the applications of genome editing be assessed and regulated? *eLife, 6*, e26295. https://doi.org/10.7554/eLife.26295
137. De, L. I., Casado, M., Marfany, G., Baroni, M. L., & Escarrabill, M. (2017). Gene editing in humans: Towards a global and inclusive debate for responsible research. *The Yale Journal of Biology and Medicine, 90*(4), 673.
138. Kumari, A., Yadav, S. K., & Yadav, S. C. (2010). Biodegradable polymeric nanoparticles based drug delivery systems. *Colloids and Surfaces B: Biointerfaces, 75*(1), 1–18. https://doi.org/10.1016/j.colsurfb.2009.09.001
139. Heller, D. A., Jena, P. V., Pasquali, M., Kostarelos, K., Delogu, L. G., Meidl, R. E., Rotkin, S. V., Scheinberg, D. A., Schwartz, R. E., Terrones, M., Wang, Y. H., Bianco, A., Boghossian, A. A., Cambré, S., Cognet, L., Corrie, S. R., Demokritou, P., Giordani, S., Hertel, T., Ignatova, T., Islam, M. F., Iverson, N. M., Jagota, A., Janas, D., Kono, J., Kruss, S., Landry, M. P., Li, Y., Martel, R., Maruyama, S., Naumov, A. V., Prato, M., Quinn, S. J., Roxbury, D., Strano,

M. S., Tour, J. M., Weisman, R. B., Wenseleers, W., & Yudasaka, M. (2020). Banning carbon nanotubes would be scientifically unjustified and damaging to innovation. *Nature Nanotechnology, 15*(3), 164–166. https://doi.org/10.1038/s41565-020-0656-y

140. Alfadul, S., & Elneshwy, A. (2010). Use of nanotechnology in food processing, packaging and safety—Review. *African Journal of Food, Agriculture, Nutrition and Development, 10*(6). https://doi.org/10.4314/ajfand.v10i6.58068
141. Kostarelos, K. (2008). The long and short of carbon nanotube toxicity. *Nature Biotechnology, 26*(7), 774–776. https://doi.org/10.1038/nbt0708-774
142. Pikula, K., Chaika, V., Zakharenko, A., Markina, Z., Vedyagin, A., Kuznetsov, V., Gusev, A., Park, S., & Golokhvast, K. (2020). Comparison of the level and mechanisms of toxicity of carbon nanotubes, carbon nanofibers, and silicon nanotubes in bioassay with four marine microalgae. *Nanomaterials, 10*(3), 485. https://doi.org/10.3390/nano10030485
143. Pecot, C. V., Calin, G. A., Coleman, R. L., Lopez-Berestein, G., & Sood, A. K. (2011). RNA interference in the clinic: Challenges and future directions. *Nature Reviews Cancer, 11*(1), 59–67. https://doi.org/10.1038/nrc2966
144. Kah, M., Beulke, S., Tiede, K., & Hofmann, T. (2013). Nanopesticides: State of knowledge, environmental fate, and exposure modeling. *Critical Reviews in Environmental Science and Technology, 43*(16), 1823–1867. https://doi.org/10.1080/10643389.2012.671750
145. Tripathi, D. K., Mishra, R. K., Singh, S., Singh, S., Vishwakarma, K., Sharma, S., Singh, V. P., Singh, P. K., Prasad, S. M., Dubey, N. K., Pandey, A. C., Sahi, S., & Chauhan, D. K. (2017). Nitric oxide ameliorates zinc oxide nanoparticles phytotoxicity in wheat seedlings: Implication of the ascorbate–glutathione cycle. *Frontiers in Plant Science, 8*, 1. https://doi.org/10.3389/fpls.2017.00001
146. Tripathi, D. K., Singh, S., Singh, V. P., Prasad, S. M., Dubey, N. K., & Chauhan, D. K. (2017). Silicon nanoparticles more effectively alleviated UV-B stress than silicon in wheat (*Triticum aestivum*) seedlings. *Plant Physiology and Biochemistry, 110*, 70–81. https://doi.org/10.1016/j.plaphy.2016.06.026
147. Manna, I., & Bandyopadhyay, M. (2019). A review on the biotechnological aspects of utilizing engineered nanoparticles as delivery systems in plants. *Plant Gene, 17*, 100167. https://doi.org/10.1016/j.plgene.2018.100167
148. Mwaanga, P. (2018). Risks, uncertainties, and ethics of nanotechnology in agriculture. *In: New Visions in Plant Science, 22*(3).

Toxicological Implications Associated with Nanomaterials in Agro-ecosystems

Archana Ashok Sharbidre

Contents

A. A. Sharbidre (✉)
Department of Zoology, Savitribai Phule Pune University, Pune, MS, India
e-mail: aasharbidre@unipune.ac.in

R. S. Chaughule, A. V. Shelar (eds.), *Nanotechnology in Agriculture*,
https://doi.org/10.1007/978-3-031-97544-8_12

1 Introduction

Amidst the growing demand for higher agricultural yields and more sustainable farming practices, the scientific community has turned its attention to the promising realm of nanotechnology [1, 2]. Nanotechnology offers a multitude of potential applications in the agricultural sector, from enhancing nutrient delivery and extending the shelf life of food products to detecting pathogens and toxins [3].

Researchers have made significant strides in exploring the use of nanomaterials, such as (nanoparticles) NPs and nano-formulations, in the development of nano-agrochemicals like nano-pesticides and nano-fertilizers [4]. These innovative solutions hold the promise of improving crop productivity and overcoming the limitations of conventional farming methods [1]. However, the unique physicochemical properties of nanometer-scale materials have also raised concerns about their potential toxic effects on environmental and biological systems [5].

One of the key areas of focus has been the impact of nanomaterials on soil microbiota, which play a crucial role in agroecosystems [6]. Studies have shown that the introduction of nanomaterials can lead to varying responses from soil fungi and bacteria, potentially resulting in infertility [5].

Nanotechnology has revolutionized various sectors, including agriculture, with the development of nano-pesticides and nano-fertilizers Nanotechnology has brought significant advancements to various sectors, including agriculture, through the creation of nano-pesticides and nano-fertilizers [7]. These nanomaterials possess unique physical, chemical, and biological properties that can enhance the effectiveness of agricultural inputs [8].

Specifically, nano-pesticides have the potential to provide more targeted and efficient delivery of active ingredients, thus possibly reducing the environmental impact of traditional pesticides [7]. Similarly, nano-fertilizers can enhance nutrient uptake and utilization by plants, leading to increased crop yields while potentially requiring lower resource inputs. Nevertheless, the widespread use of nanomaterials in agriculture raises concerns regarding their potential accumulation in the environment and associated ecotoxicological risks. Nanomaterials may pose risks to biodiversity at much lower concentrations than previously thought, and their effects on non-target organisms and ecosystems are still not fully understood [8].

It is important to carefully consider the development and application of nano-pesticides and nano-fertilizers to balance their potential benefits with potential risks. Rigorous ecological risk assessments and a comprehensive understanding of the environmental fate and behavior of these nanomaterials are crucial to ensure the sustainable and responsible use of these technologies in agriculture [7].

As the scientific community continues to delve deeper into the complex interactions between nanomaterials and environmental systems, the need for a comprehensive understanding of their physiological, biochemical, molecular, and genotoxic effects has become increasingly crucial [2]. The accumulation of nanomaterials in the environment and their probable influence on biodiversity have also emerged as significant concerns [9].

In this context, the scientific community has emphasized the importance of developing a responsible regulatory framework to ensure the safer and more sustainable use of nanotechnology in agriculture [1, 2]. By addressing the challenges and potential risks, researchers can harness the transformative power of nanotechnology to revolutionize the agricultural sector and enhance food security while prioritizing environmental protection and human health.

2 Sources of Nanomaterials in Agro-Ecosystems

2.1 Nano-Pesticides: Exploring the Potential Benefits and Risks

Nano-pesticides, a rapidly emerging field in the agricultural industry, have recently garnered significant attention due to their potential to revolutionize pest management practices. The development of nano-scale formulations has been driven by the need to address the limitations of traditional chemical pesticides, which can have detrimental effects on the environment and human health.

Nanoscale particles, defined as those with dimensions ranging from 1 to 100 nanometers, exhibit unique properties that set them apart from their larger counterparts. The presence of these characteristics, such as increased surface area, improved solubility, and greater penetration, has led to the development of nano-pesticides. They are designed to offer improved efficacy, precise delivery, and reduced environmental impact.

Nano-based pesticides are designed to enhance the delivery and efficacy of active ingredients, while potentially reducing the overall application rates and environmental impact. The advantages of these products may include improved formulation properties, simplified application, targeted pest control, increased effectiveness, reduced application rates, and enhanced environmental safety.

Nano-pesticides have a significant advantage due to their ability to respond to specific biotic or abiotic stimuli, allowing for a more precise and controlled release of the active components. This "smart" delivery mechanism can help minimize the indiscriminate use of pesticides, ultimately reducing the risk of exposure to non-target organisms and the environment [10]. Moreover, nano-based formulations have the potential to improve the stability and penetration of active ingredients, allowing for more effective pest control [11]. The unique properties of nanomaterials, such as their increased surface area and altered physicochemical characteristics, can contribute to enhanced efficacy and targeted delivery, potentially leading to reduced application rates and improved environmental profiles [7]. Nano-pesticides offer promising benefits, but their use raises concerns about their impact on the environment and the need for thorough risk assessment. Regulatory bodies and researchers are actively working to address these issues and ensure the safe and responsible production and use of nano-pesticide products.

The ecological risk assessment of nano-pesticides must consider factors that differ from conventional pesticides, such as the unique behavior and fate of nanomaterials in the environment, their potential interactions with biotic and abiotic components, and the potential for unintended consequences [7]. Additionally, the potential for nano-pesticides to adversely affect non-target species, including beneficial insects and other organisms, is a critical area of concern that requires thorough investigation [12].

As the field of nano-pesticides continues to evolve, it is crucial to balance the potential benefits and the careful management of the associated risks. Ongoing research and collaborative efforts between regulators, scientists, and the agrochemical industry will be instrumental in guiding the responsible development and implementation of nano-enabled pest control solutions that prioritize environmental sustainability and human health.

2.2 Nano-Herbal Insecticides

Nano-herbal insecticides represent a promising development in sustainable pest management. They combine the targeted action of nanotechnology with the environmentally friendly profile of botanical insecticides.

Essentially, these insecticides work by encapsulating plant-derived insecticidal compounds within NP [13]. This nano-encapsulation offers several benefits:

- **Improved Bioavailability:** The nano-sized particles increase the surface area of the active ingredient, leading to better absorption and effectiveness.
- **Enhanced Stability:** Encapsulation protects the active compounds from degradation, extending their shelf life and effectiveness in the field [14].
- **Targeted Delivery:** NPs can be designed to target specific pests or release the insecticide in response to environmental cues [15].

This approach offers a potentially transformative solution for pest management, reducing reliance on synthetic chemicals and minimizing environmental impact [16]. However, further research is needed to ensure the safety and optimize the cost-effectiveness of these novel formulations [17].

2.3 The Potential of Nanofertilizers in Sustainable Agriculture

The use of nano-technology has opened up new possibilities for innovation in agriculture. Nano-fertilizers, in particular, show great promise in addressing the limitations of traditional fertilizers. They offer several benefits such as improved nutrient efficiency, reduced environmental impact, and increased crop productivity [18–20].

One of the main advantages of nanofertilizers is their ability to precisely and efficiently deliver nutrients to crops [21]. The nano-scale materials have a significantly higher surface area to volume ratio, allowing for targeted and controlled release of active ingredients based on environmental triggers and biological demands [19]. This targeted delivery can help reduce nutrient waste and mitigate the environmental pollution caused by excessive fertilizer use.

Given the continuing global population growth and the heightened demand for food production, the potential of nanofertilizers to increase crop yields and improve resource efficiency becomes increasingly important. However, the large-scale adoption of nanofertilizers is still in its early stages, with production and availability on a commercial scale lagging [20, 21].

As the global population continues to grow and the demand for food production intensifies, the potential of nanofertilizers to boost crop yields and improve resource efficiency becomes increasingly important. However, the large-scale adoption of nano-fertilizers is still in its early stages, with production and availability on a commercial scale lagging.

2.3.1 Nanosensors

The emergence of nanosensors has revolutionized the way we monitor and manage agro-ecosystems. Nanotechnology has shown great potential in addressing the challenges faced by the agricultural sector, such as improving crop productivity, enhancing soil quality, and optimizing resource use [22–24].

Nanosensors have played a key role in the sustainable management of agricultural soil. These sensors can provide real-time data on soil parameters, such as moisture, pH, and nutrient levels, enabling farmers to make informed decisions and optimize resource utilization [25]. Furthermore, nanosensors can also detect the presence of plant pathogens, enabling early intervention and prevention of crop diseases [22]. Additionally, nanosensors have been used to monitor the presence of pesticides and other agrochemicals, ensuring the safety and quality of food products [25].

Beyond soil and crop monitoring, nanotechnology has also found applications in precision farming which is facilitated by technologies like satellite positioning systems and remote sensing, which can significantly improve productivity and reduce resource consumption [23, 25].

However, despite the numerous benefits, the widespread adoption of nanotechnology in agriculture faces several challenges. There are concerns regarding the potential environmental and health impacts of nanomaterials, which require thorough investigation and risk assessment [24]. Addressing these challenges will be crucial to unlocking the full potential of nanosensors in agro-ecosystems.

In conclusion, the integration of nanosensors in agro-ecosystems has demonstrated the ability to enhance sustainable agricultural practices, improve crop productivity, and ensure food safety and quality. As the technology continues to evolve, it is essential to address the remaining challenges to fully harness the power of nanosensors in transforming the agricultural landscape [22, 24, 26].

3 Uptake and Transport of Nanomaterials in Plants

3.1 Mechanisms of Uptake

The potential for nanomaterials to be taken up by plants has received significant attention in recent years. These tiny particles have unique physical and chemical properties that may allow them to interact with and become part of plant systems. The mechanisms by which nanomaterials are transported and accumulated within plants are complex and not yet fully understood [15]. Nanomaterials can potentially enter plants through various pathways, including root uptake, foliar uptake, and even direct entry through the plant's vascular system [27]. It is believed that the uptake and movement of nanomaterials depend on factors such as particle size, surface charge, concentration, exposure time, and plant species. The transport of nanomaterials is influenced by their physical and chemical properties, as well as the conditions of the surrounding environment. For instance, the size and shape of NP and their surface properties affect their movement within the plant. Nanomaterials can enter plants through openings such as stomata, root hairs, and cracks on the leaf surface. Once inside, they can move through the plant via diffusion, bulk flow, and phloem loading [15].

Previous research has explored various methods of applying nanomaterials to plants, including leaf spraying, root application, branch injection, and seed treatment [28].

While this technology is still in its early stages, it holds promise for diverse applications, such as targeted delivery of genetic material, improved disease management through antimicrobial agents, and controlled release of nutrients to enhance plant growth [27]. However, the ability of nanomaterials to penetrate living plant tissues also raises concerns about their potential accumulation in the food chain and the need to further understand their impacts [28]. Continued research is necessary to elucidate the complex pathways and implications of nanomaterial uptake and transport within plant systems.

3.2 Translocation Pathways

Research efforts have focused on uncovering the processes involved in the uptake, transportation, and accretion of various NPs inside plant structures and cell organelles [29–33]. The translocation and accrual of NPs in plants are inclined by the physiological and structural characteristics of plant cells, the interaction between nanomaterials and soil, and the properties and stability of the NPs [34, 35].

The plant cell wall serves as a selective barrier, controlling the entry of NPs into the cell and determining their solubility and ability to pass through based on their characteristics [36–38]. The primary restriction for NP entering plant cells is the size of the pore in the cell wall. For NPs to enter plant cells from the surface, they must have a size ranging between 40 and 50 nanometers [39, 40].

The composition of NPs is a key factor that affects their ability to pass through the cell wall and cell membrane [41]. Furthermore, the characteristics of the NPs also influence their ability to attach to the root surface or root exudates [42].

Due to their positive charge, NPs may have a greater capacity to adhere to the cell wall. Furthermore, the shape of these substances and the NP coating they have can significantly impact their behavior in the rhizosphere and their effects on plants. Newly formed seeds are located in the upper part of the plant, where NPs, originally gathered by the plant's roots, migrate [43].

The translocation of NPs is significantly influenced by the properties of both the NPs and the plants. For example, studies have shown that the shoot of the *Oryza sativa* plant can accumulate gold nanoparticles (AuNPs) [44], while they do not build up in the shoots of *Cucurbita pepo* or *Raphanus raphanistrum* [45, 46]. Moreover, plant roots have the highest absorption rate for positively charged AuNPs. In contrast, negatively charged AuNPs are more effectively transferred from the roots to the shoots of plants [47].

Previous studies have demonstrated that titanium dioxide (TiO_2) and silicon dioxide (SiO_2) are the most stable NPs and can be found in their pure state within plant tissues [48, 49]. Based on previous research findings, TiO_2NPs play a role in translocation within *Triticum aestivum* spp. The particles with a diameter larger than 140 nm are unable to accumulate in the root. Additionally, particles with a diameter below 36 nm can accumulate in the root's parenchyma, but cannot reach the stele and cannot be transported into the shoot of *T. aestivum* [50]. Research has shown that TiO_2NPs can speed up the growth of Spinach and Vitex vegetable (*Vitex agnus-*Castus L) [41, 51]. TiO_2NPs also increased plant growth by improving nitrogen metabolism [52]. TiO_2 has shown assurance in directly quashing crop diseases via antimicrobial and insecticidal activities [53].

NPs (NPs) such as copper (II) oxide (CuO), cerium oxide (CeO2), lanthanum oxide (La2O3), and nickel oxide (NiO) can change due to disparity, which can result in accumulating alterations in plant speciation. Researchers used synchrotron X-ray absorption (XAS) spectroscopy to gather information about the changes that zinc oxide (ZnO) NPs went through when they were exposed to different plants [54].

Zea mays has been found to have the highest concentration of zinc in its roots and shoots when exposed to ZnO NPs through hydroponic methods, especially in the form of zinc phosphate. This could be due to the increased solubility of zinc in the rhizosphere, which allows it to be taken up and distributed within the plant in its ionic form [55]. The same patterns of zinc accumulation were also observed in wheat crops [56, 57].

Researchers have observed the movement of ZnO and CeO2 NPs from the soil into *Glycine max* (*G. max*) [58]. The CeO2 NPs were found to be transported as NPs, while Zn underwent biotransformation into Zn-citrus within the plant tissue [59]. CuO NPs were transported from the roots to the shoots through the xylem in *Z. mays*. They may also be transported back from the shoots to the roots through the phloem [60].

3.3 The Interactions between Crops and Nanomaterials: Inherent and External Factors

The complex nature of bio-nano interactions warrants a careful examination of both the inherent and external factors that influence these interactions. Inherent factors, such as the physicochemical properties of nanomaterials (e.g., size, shape, and composition) and the specific plant species, play a vivacious part in determining the uptake, translocation, and nanomaterial's probable toxicity inside the plant system [61, 62].

This investigation focuses on the concept of inherent factors (size, composition, and concentration), which refers to the innate characteristics and properties of nanomaterials that can significantly affect their interactions with crops and their development [63]. Similarly, the distinctive surface chemistry and reactivity of specific nanomaterials can change their bioavailability and uptake by plants, resulting in both positive and negative effects [59].

External factors, such as the soil environment, microbial communities, and climate conditions, can also significantly impact the fate and behavior of nanomaterials in agricultural ecosystems. Understanding these interacting factors is essential for developing effective and sustainable applications of nanomaterials in agriculture while minimizing potential risks to crop productivity and the environment.

The entry of nanomaterials into the soil system can affect soil quality and plant growth through various mechanisms, including nutrient release, impacts on soil biota, and changes in soil organic matter. Furthermore, the mechanisms involved in the uptake and translocation of nanomaterials within plants, as well as the associated defense mechanisms, must be elucidated to warrant the innocuous and efficient use of nanomaterials in agronomy [62].

Ongoing research efforts aim to promote the sustainable development of nano-enabled agriculture by addressing the challenges and identifying future research directions. This includes a shift towards testing and using nanomaterials based on agricultural requirements, rather than a sole focus on the development of nanomaterials [63]. Simultaneously, the transport and ecological toxicity of nanomaterials must be thoroughly investigated to ensure their benign usage in agriculture [64].

As the field of nanotechnology continues to evolve, its integration with agriculture holds great promise for enhancing crop productivity, improving soil health, and supporting sustainable food production. However, a comprehensive understanding of the interactions between crops and nanomaterials, considering both inherent and external factors, is crucial for the successful and responsible implementation of this emerging technology in the agricultural sector.

4 Agricultural Nanoparticle Pollution: Toxicological Effects on Plants

4.1 NPs' Adherence and Accumulation: Implications for Plant Toxicity

NPs have gained significant attention in the field of agriculture due to their unique physical and chemical properties, which can be leveraged for various applications [28]. However, the potential risks associated with nanoparticle accumulation and toxicity in plants have raised growing concerns [65, 66].

The effects and uptake of NPs in plants are complex and multifaceted. NPs can be absorbed by plants and significantly impact their physiological activities [15]. Research has shown that the size, surface charge, and chemical composition of NPs can influence their absorption and transportation within the plant, particularly in the leaf and root systems. Smaller NPs tend to have higher rates of uptake and translocation, while the surface charge can affect their interaction with plant cells and tissues. Moreover, the accumulation of NPs in plants can lead to phytotoxicity, which can ultimately impact the plant's growth, development, and overall health. The specific mechanisms underlying nanoparticle-induced toxicity in plants are not yet fully understood, and further research is needed to elucidate the complex interactions between NPs and plant systems [65].

One of the key challenges in understanding the implications of nanoparticle accumulation in plants is the potential for bioaccumulation and transfer through the food chain. As plants are a fundamental component of ecosystems and the primary source of food for humans and other organisms, it is crucial to evaluate the potential environmental and health risks associated with nanoparticle exposure in plants [15].

In summary, the adherence and accumulation of NPs in plants can lead to significant toxicity, underscoring the need for a comprehensive understanding of the complex interactions between NPs and plant systems.

4.2 *Formation of Reactive Oxygen and Nitrogen Species in Plant Cells*

The potential environmental and toxicological implications of the widespread use of NPs in agriculture have raised concerns among researchers and regulatory bodies [2]. One key aspect that warrants investigation is the impact of nanoparticle pollution on the formation of reactive oxygen and nitrogen species in plant cells, as these can lead to oxidative stress and subsequent cellular damage.

Studies have shown that the exposure of plants to certain types of NPs, such as metal and metal oxide NPs, can induce the generation of reactive oxygen species and reactive nitrogen species within plant cells. These reactive species can disrupt cellular processes, damage cellular components like chloroplasts and vascular bundles, and ultimately lead to impaired plant growth and development [66]. The mechanisms by which NPs can trigger the formation of these reactive species are complex and may involve various pathways, including the disruption of electron transport chains, the activation of enzymatic systems, and the interaction of NPs with cellular components [28] and the oxidative stress response in plants [34]. Given the widespread use of NPs in agricultural practices, such as in the development of nanopesticides, nanofertilizers, and other nano-enabled products, there is a pressing need to understand the potential consequences of nanoparticle pollution on plant health and the broader agroecosystem [64, 68]. Comprehensive risk assessment and the development of appropriate regulatory frameworks are crucial to ensure the safe and responsible use of nanomaterials in agriculture while maximizing their benefits.

Most NPs accumulate toxins at high concentrations within plants, which can alter their biochemical, physiological, morphological, and genetic processes. This can include detrimental cellular and subcellular organelles and reducing productivity [69]. The most well-known indication of engineered NPs (ENP) phytotoxicity in plants includes plummeting nutrient uptake, hydraulic transference and photosynthesis, blockage of the apoplastic pores and barriers, DNA mutilation, and augmented reactive nitrogen species generation and ROS [70–72].

Several studies have shown the damaging effects of ENPs on plants, particularly in the accumulation of toxic metals and metalloids like CdO, CuO, ZnO, TiO_2, and Al_2O_3 NPs [73, 74]. These NPs can induce oxidative stress, cell membrane damage, and root growth inhibition in important food crops like wheat, rice, and cucumber [75]. The oxidative stress response in plants triggered by ENP exposure is a key factor in their toxicity, resulting in the formation of reactive oxygen and nitrogen species that can damage lipids, proteins, and DNA [72]. The extent of ENP phytotoxicity depends on complex interactions between the NP properties, the plant species, and the environmental conditions.

Careful assessment of NP effects on plants is crucial as the widespread use of ENP-containing products could lead to significant agricultural nanopollution that may threaten global food security. To effectively mitigate these risks, a deeper understanding of the molecular mechanisms underlying ENP-plant interactions is needed [76].

4.3 Influence of Agricultural NPs on Plant and Microbial Secondary Metabolism

Agricultural NPs can have a significant impact on the secondary metabolism of both plants and soil microorganisms, which has implications for crop productivity and ecosystem sustainability.

Secondary metabolites are organic compounds produced by plants and microorganisms that do not directly contribute to their primary growth and development, yet often play crucial ecological roles. These secondary metabolites can serve a wide range of functions, such as defense against herbivores and pathogens, attraction of pollinators, and nutrient acquisition. Emerging evidence suggests that agricultural NPs, such as those used in fertilizers or pesticides, may have the potential to alter the production of these secondary metabolites, either by directly influencing the metabolic pathways or indirectly through changes in plant-microbe interactions [77, 78]. This can have significant implications for the ecological balance and the quality of agricultural products.

Certain NPs, such as those containing essential minerals, can enhance the production of secondary metabolites by optimizing plant mineral nutrition and alleviating nutrient deficiencies [63]. Conversely, some NPs can induce oxidative stress and disrupt cellular processes, leading to a decrease in secondary metabolite synthesis [77]. Nanoparticle entry into the soil can occur through various pathways, including the application of nano-agrochemicals, contamination from industrial wastes, and atmospheric deposition and their interactions with soil microbes, including mycorrhizal fungi and rhizobia, can have both stimulatory and inhibitory effects on their secondary metabolite production [78]. In addition to their direct effects on plants, NPs can also influence the secondary metabolism of soil microorganism. Microbes, particularly those involved in plant-microbe symbioses, play a crucial role in the production of various secondary metabolites that contribute to plant health and productivity [77, 78].

4.4 Genotoxic and Growth-Inhibitory Effects of NPs in Plants

NPs have gained significant attention in the agricultural sector due to their potential to enhance crop productivity and nutrient delivery [28, 65, 79]. However, the impact of NPs on plant growth and development, particularly their genotoxic and growth-inhibitory effects, requires careful evaluation [15, 64].

Some nanomaterials have been shown to have beneficial effects on plant growth, such as improved nutrient uptake and enhanced photosynthesis [68]. Conversely, certain NPs can have inhibitive effects, leading to reduced growth, altered germination, and disrupted physiological processes [65]. The transport and ecological toxicity of NPs in plant-soil systems is a significant concern, as they may accumulate in the food chain and impact higher-level organisms [65]. Careful selection and application of NPs in agriculture are necessary to maximize their benefits while minimizing the potential risks [68].

Understanding the complex interactions between NPs, plants, and the surrounding environment is crucial for the sustainable and responsible use of these emerging technologies in agriculture.

5 Toxicological Effects on Soil Microorganisms

The literature indicates that NPs can have complex and varied effects on microbial communities, ranging from neutral to negative to even positive impacts. Inorganic NPs, such as those made of metals or metal oxides, may have a greater toxic potential on soil microorganisms compared to organic NPs. The increasing use of ENPs in various industrial and agricultural applications has raised concerns about their potential environmental impacts [80]. NPs can enter the soil ecosystem through various pathways, including the application of nano-agrochemicals, contaminated amendments, and industrial waste [78].

Numerous studies have investigated the effects of NPs on soil microorganisms and their activities [80–83]. Evidence suggests that NPs can have both positive and negative impacts on soil microbial communities and their functions. On the one hand, some studies have reported that certain NPs can enhance soil enzyme activity and microbial diversity. This is attributed to the potential of NPs to improve nutrient availability, soil structure, and microbial growth [80]. However, other studies have found that NPs can have detrimental effects on soil microorganisms, such as inhibiting microbial Nitrogen fixation, symbiotic relationships, and the production of antifungal compounds. The toxicity of NPs to soil microorganisms is highly dependent on the specific characteristics of the NPs, as well as the properties of the soil.

The disruption of microbial communities by NPs can have far-reaching consequences, particularly for beneficial plant-microbe symbioses like mycorrhizal and rhizobial associations. Some studies have reported that NPs can negatively impact the development and functioning of mycorrhizal and rhizobial symbioses, leading to reduced colonization, nodule formation, and overall plant–microbe interactions [78]. This disruption can result in compromised nutrient cycling, plant growth, and overall ecosystem resilience. These symbioses are crucial for the health and sustainability of both natural and agricultural ecosystems, and their disruption by NPs could potentially compromise ecosystem functioning and plant productivity.

However, the specific effects of NPs on microbial communities and their symbiotic relationships with plants are highly dependent on the type of nanoparticle, the soil properties, and the microbial species involved. Therefore, a more comprehensive understanding of these complex interactions is crucial to mitigate the potential risks posed by nanoparticle contamination in terrestrial ecosystems. Inorganic NPs appear to have a greater toxic potential to soil microorganisms than organic NPs [81]. Furthermore, the fate and behavior of NPs in the soil environment, including their transformation, transport, and interactions with soil biota, are crucial factors in determining their overall impact on soil ecosystems [80].

To ensure the safe and sustainable use of engineered NPs in agriculture and other applications, a better understanding of their interactions with soil microorganisms and their effects on soil quality and ecosystem functions is necessary. This requires a multidisciplinary approach, including ecotoxicological and biosafety studies, to assess the potential risks and benefits of NPs in the soil environment.

5.1 The Impact of NPs on Nutrient Cycling

NPs, which are defined as particles with at least one dimension less than 100 nm, have become increasingly prevalent in various industries and consumer products, from electronics and cosmetics to water treatment and agriculture. As these nanomaterials are increasingly discharged into the environment, understanding their impact on ecological systems, particularly nutrient cycling, is of growing importance.

The application of nanoelements in plant nutrition has been explored as a potential strategy to enhance crop productivity and stress tolerance [82]. However, the environmental fate and effects of nanofertilizers on soil ecosystems remain largely unknown. Nanomaterials may affect the bioavailability and cycling of essential nutrients, potentially altering the delicate balance of the soil-plant-microbe system [62].

The fate and transformation of NPs in municipal wastewater treatment systems have significant implications for their environmental impact. During production and application, NPs can enter the environment through various pathways, and their small size and large surface area confer unique physical and chemical properties that can affect their behavior and interactions with living organisms [84].

Several studies have investigated the potential phytotoxic mechanisms of NPs, indicating that they can disrupt the structure and function of essential plant components, such as chloroplasts and vascular bundles, leading to oxidative stress and alterations in nutrient uptake and transport [67].

Future research should focus on elucidating the mechanisms by which NPs interact with and impact nutrient cycling in terrestrial and aquatic ecosystems. Comprehensive risk assessment and sustainable development of nano-enabled agriculture are crucial to ensure the responsible and environmentally sound use of these emerging technologies.

5.2 *Toxicological Effects on Non-target Organisms*

5.2.1 Effects on Beneficial Insects

The widespread adoption of nano-agrochemicals has raised concerns regarding their potential effects on beneficial insects, which play a crucial role in maintaining the delicate balance of ecosystems [4, 6, 7]. Beneficial insects, such as pollinators, predatory insects, and decomposers, are essential components of agricultural systems, contributing to the overall health and productivity of crops [85].

The unique properties of nano-agrochemicals, including their increased surface area, enhanced bioavailability, and targeted delivery mechanisms, may inadvertently disrupt the intricate interactions between these beneficial insects and their environments [7, 86]. The potential risks associated with nano-agrochemicals include the accumulation of these materials in the soil, water, and plant tissues, which could lead to the bioaccumulation and biomagnification of harmful compounds within the food chain [87–89].

While the benefits of nano-agrochemicals, such as improved pest control and increased crop yields, are well-documented, it is crucial to consider the long-term implications of their use on beneficial insects. Comprehensive risk assessment and ongoing monitoring of the environmental impacts of nano-agrochemicals are essential to ensure the sustainable and responsible development of these technologies, balancing the needs of agricultural productivity with the preservation of crucial ecosystem services provided by beneficial insects [90, 91].

5.2.2 Bioaccumulation in Wildlife

Nano-sized agrochemicals may have a substantial impact on ecosystems and biodiversity. The widespread use of these nano-agrochemicals has raised concerns about their potential impact on the environment, particularly on wildlife and their ecosystems.

Using nano-agrochemicals in agricultural fields can cause insects and other animals to accumulate dangerous amounts of nanoparticles, which can mess up food chains and cause populations to drop [92]. Additionally, these chemicals can accumulate in water sources, further impacting aquatic ecosystems and potentially leading to long-term environmental damage [88, 93]. For example, using nano-agrochemicals in rice paddies can harm aquatic organisms in nearby water bodies due to runoff. This can disrupt the balance of the ecosystem and result in declines in fish populations and other aquatic species. As these chemicals build up in the tissues of animals higher up in the food chain, they can cause reproductive issues, behavioral changes, and even death [87, 94]. To prevent further harm to the environment, researchers and policymakers must monitor the use of nano-agrochemicals and their effects on wildlife.

Nano-pesticides, in particular, have garnered significant attention due to their ability to penetrate biological barriers and their increased toxicity compared to conventional pesticides [95]. This raises concerns about the potential for bioaccumulation, where the nanomaterials or their breakdown products may accumulate in the tissues of wildlife, potentially leading to adverse effects on their health and the overall ecosystem [96]. The increased surface area and unique physicochemical properties of nano-pesticides can enhance their interaction with and uptake by various organisms, including plants, invertebrates, and vertebrates. This heightens the risk of bioaccumulation, as the nanomaterials may be passed up the food chain, potentially leading to magnification of their concentrations in higher trophic levels [87].

Regulating the use of nano-agrochemicals and assessing their ecological risks is a complex challenge, as the traditional risk assessment frameworks may not be adequate in addressing the unique characteristics of nanomaterials [6]. Comprehensive studies are needed to understand the fate, behavior, and potential impacts of nano-agrochemicals on various wildlife species and their ecosystems, to develop effective regulatory strategies and mitigation measures.

In conclusion, the increasing use of nano-agrochemicals in agriculture raises significant concerns about their potential for bioaccumulation in wildlife. Further research and regulatory efforts are necessary to ensure the sustainable and responsible use of these emerging technologies, safeguarding the health of ecosystems and the well-being of wildlife.

5.2.3 Implications for Human Health

One major concern is the potential for nanopesticides to accumulate in the environment and pose risks to biodiversity [8]. The "unsystematic" use of these nanomaterials can lead to the development of pesticide resistance, reduced nitrogen fixation, and the bioaccumulation of pesticides in agricultural products and water bodies, ultimately threatening both ecosystems and human health [13].

While nanotechnology holds promise for more sustainable agricultural practices by enabling targeted delivery and controlled release of agrochemicals [96, 97], the need for a more holistic, "One Health" approach to managing these technologies is critical [98]. This will require a deeper understanding of the complex interactions between human, animal, and environmental exposures in the context of nano-agrochemical use.

Rigorous risk assessment and regulation will be essential to mitigate the potential negative impacts of these emerging technologies on human health and the environment. Ongoing research and collaboration across disciplines will be key to developing safe and sustainable applications of nano-agrochemicals in the future [99].

The novel physicochemical properties of nanometer-scale materials have generated significant research interest across various sectors. However, extensive studies have shown that engineered nanoparticles, even at relatively small scales, may pose threats to biological and environmental systems. As a result, examining the potential

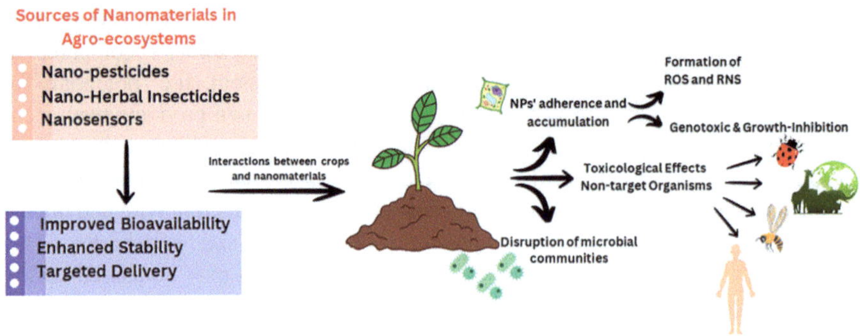

Fig. 1 Toxicological implications associated with nanomaterials in agro-ecosystems

risks associated with nanomaterials has become a critical concern for scientists. The main areas of research in the literature include the genotoxic, biochemical, molecular, and physiological effects of these materials, which have been investigated in both small and large studies. Additionally, Fig. 1 illustrates the negative impacts of nanomaterials on food chains and agroecosystems, highlighting the need for safer and more eco-friendly approaches to their use.

6 Risk Assessment and Regulation of Nanomaterials in Agro-Ecosystems

6.1 Regulatory Framework for NPs in Agriculture

One of the main concerns about using NPs in agriculture is their potential impact on the environment and human health. The unique properties of NPs, such as their small size and high surface-to-volume ratio, can lead to increased reactivity and the potential for bioaccumulation, which may have adverse effects on ecosystems and human health [1, 6]. To address these concerns, regulators have started developing guidelines and regulations to govern the use of NPs in agricultural applications.

Internationally, organizations like the Organization for Economic Cooperation and Development and the United Nations Environment Programme have taken steps to establish frameworks for assessing and managing nanomaterials, including their use in agriculture [9]. In the United States, the Environmental Protection Agency has established regulations for registering and labeling nanoscale pesticides and other nano-enabled agricultural products, and the Food and Drug Administration has issued guidance on evaluating nanomaterials in food and feed applications [100].

Similarly, the European Union has implemented regulations on the use of nanomaterials in food, cosmetics, and other consumer products, which may also have implications for their use in agriculture [2]. In India, the Ministry of Science and Technology has developed guidelines for the safety assessment and regulation of nanomaterials, including their use in agricultural applications [1]. Despite these efforts, the regulatory landscape for NPs in agriculture remains fragmented and incomplete, requiring more coordinated and comprehensive approaches to ensure the safe and responsible use of these technologies. As the use of NPs in agriculture continues to expand, policymakers, scientists, and stakeholders will need to work together to develop robust and flexible regulatory frameworks that can keep pace with the rapidly evolving field of nanotechnology.

The use of NPs in agricultural applications has garnered significant attention in recent years, as these microscopic materials have the potential to enhance crop yields, improve pest control, and promote sustainable farming practices. However, the regulatory landscape surrounding the use of NPs in agriculture remains a complex and evolving landscape, with a need for comprehensive guidelines to ensure the safe and responsible deployment of these technologies. The use of NPs in agricultural applications has gained significant attention in recent years due to their potential to enhance crop yields, improve pest control, and promote sustainable farming practices. However, the regulations governing their use in agriculture are complex and evolving, and there is a need for comprehensive guidelines to ensure their safe and responsible deployment.

6.2 Challenges in Assessing Agricultural Nanomaterial Toxicity

Assessing the toxicity of nanomaterials in agriculture presents unique challenges due to their novel properties and the complexity of agricultural ecosystems which are listed in Fig. 2.

1. **Diversity and Variability of Nanomaterials**

There is a vast diversity of engineered nanomaterials used in agriculture, each with unique physicochemical properties that influence their toxicity [2]. These properties, including size, shape, surface charge, and composition, can change over time in the environment, making it difficult to predict their long-term effects [64].

2. **Complexity of Agroecosystems**

Agricultural ecosystems are complex, with interactions between various biotic and abiotic factors. Nanomaterials can interact with soil components, water, air, and living organisms, leading to unpredictable transformations and effects [104].

3. **Fate and Transport of Nanomaterials**

Understanding the fate and transport of nanomaterials in soil, water, and air is crucial for assessing their potential exposure routes and impacts [5]. Their small size and high surface area can lead to increased mobility and uptake by plants and animals [65]. The bioavailability of nanomaterials, or the extent to which they are absorbed

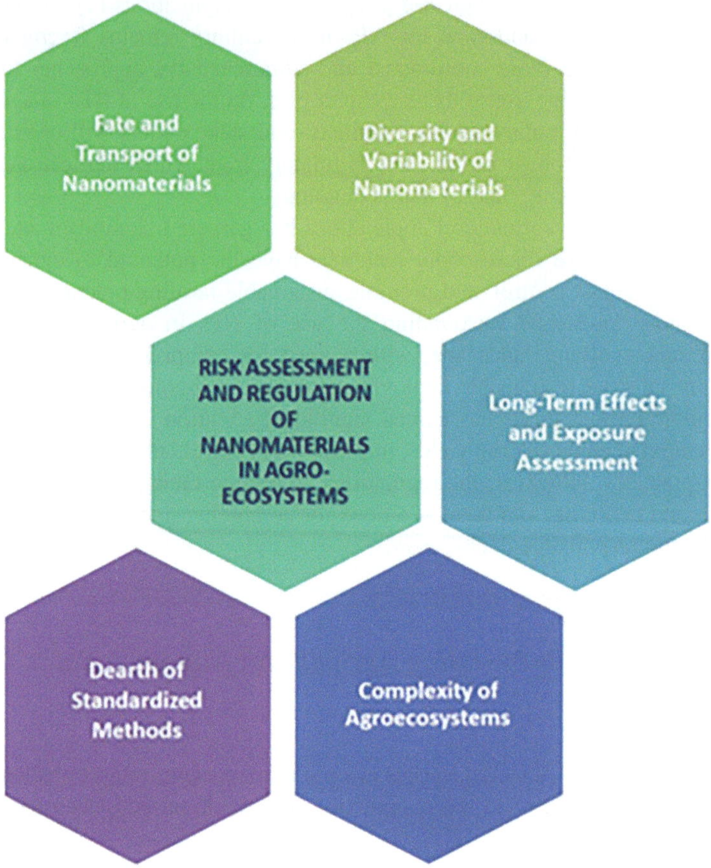

Fig. 2 Unique challenges in assessing the toxicity of agro-nanomaterials

by organisms, is a key factor determining their toxicity. Nanomaterials can accumulate in organisms over time, potentially leading to biomagnification in food chains [101].

4. **Long-Term Effects and Exposure Assessment**

Assessing the long-term effects of nanomaterials is challenging due to their persistence in the environment and the potential for delayed toxicity. Accurately assessing exposure to nanomaterials in the field is difficult due to their low concentrations and the complexity of environmental matrices [5].

5. **Dearth of Standardized Methods**

There is a dearth of standardized methods for assessing the toxicity of nanomaterials in agricultural settings. This makes it difficult to compare results across studies and develop reliable risk assessments.

6. **Regulatory Challenges**

The regulatory landscape for nanomaterials in agriculture is still evolving. Clear guidelines and regulations are needed to ensure the safe and responsible use of nanotechnology in agriculture [2].

6.3 Future Research Directions of Agronanochemicals

Agronanochemicals, which utilize nanotechnology for improved delivery and efficacy of agrochemicals, hold immense potential for sustainable agriculture. However, to fully realize their benefits while mitigating potential risks, several key research directions need to be explored which are listed in Fig. 3.

6.3.1 Targeted Delivery and Controlled Release

Enhancing targeting efficiency is crucial for maximizing the benefits of nano-agrochemicals while minimizing their potential risks. We can lower the overall amount of chemicals used, minimize side effects, and improve treatment effectiveness by allowing site-specific application. This can be done by creating nanocarriers that can deliver pesticides, fertilizers, or other agricultural inputs precisely where they are needed with controlled release rates. By delivering higher concentrations of active ingredients directly to the target pests, this targeted approach can be particularly beneficial in combating pesticide resistance [87]. By developing stimuli-responsive nanomaterials that release agrochemicals based on environmental cues (e.g., pH, moisture, temperature) or plant physiological needs, we can optimize release profiles [88, 94].

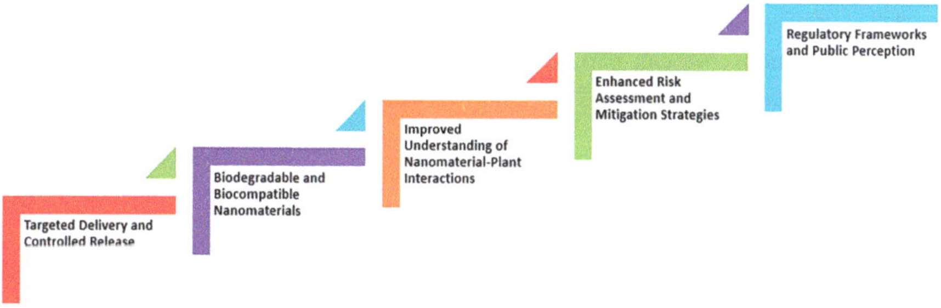

Fig. 3 Future research directions of agronanochemicals

6.3.2 Biodegradable and Biocompatible Nanomaterials

By exploring bio-based materials, nanocarriers can be synthesized from biodegradable and biocompatible materials like chitosan, cellulose, and other biopolymers to minimize environmental persistence. To assess their long-term impacts, comprehensive life-cycle assessments of agronanochemicals can be conducted to evaluate their long-term environmental fate, biodegradability, and potential accumulation in food chains [102, 103].

6.3.3 Improved Understanding of Nanomaterial-Plant Interactions

Investigating the mechanisms can help us understand how plants absorb, translocate, and metabolize different nanomaterials. Also, the impact of agronanochemicals on plant physiology can be explored to understand their effects on plant growth, development, yield, and nutritional quality [104, 105].

6.3.4 Enhanced Risk Assessment and Mitigation Strategies

Developing standardized toxicity tests by establishing standardized protocols for evaluating the toxicity of agronanochemicals to various organisms, including soil microbes, beneficial insects, and aquatic life. Modeling environmental fate can be done by developing predictive models to assess the transport, transformation, and potential exposure risks of agronanochemicals in different environmental compartments [106].

6.3.5 Regulatory Frameworks and Public Perception

Harmonizing regulations to develop clear and science-based regulatory frameworks for the safe and responsible use of agronanochemicals. By promoting transparent communication and public engagement initiatives to address concerns and build trust regarding the use of nanotechnology in agriculture [107]. By focusing on these research directions, we can harness the potential of agronanochemicals to enhance agricultural productivity while ensuring the long-term health of our environment and food systems.

References

1. Mishra, M., Dashora, K., Srivastava, A., Fasake, V. D., & Nag, R. H. (2019). Prospects, challenges and need for regulation of nanotechnology with special reference to India. *Ecotoxicology and Environmental Safety, 171*, 677–682. https://doi.org/10.1016/j.ecoenv.2018.12.085

2. Iavicoli, I., Leso, V., Beezhold, D. H., & Shvedova, A. A. (2017). Nanotechnology in agriculture: Opportunities, toxicological implications, and occupational risks. *Toxicology and Applied Pharmacology, 329*, 96–111. https://doi.org/10.1016/j.taap.2017.05.025
3. Mitter, N., & Hussey, K. (2019). Moving policy and regulation forward for nanotechnology applications in agriculture. *Nature Nanotechnology, 14*(6), 508–510. https://doi.org/10.1038/s41565-019-0464-4
4. Li, L., Xu, Z., Kah, M., Lin, D., & Filser, J. (2019). Nanopesticides: A comprehensive assessment of environmental risk is needed before widespread agricultural application. *Environmental Science & Technology, 53*(14), 7923–7924. https://doi.org/10.1021/acs.est.9b03146
5. Mishra, S., Keswani, C., Abhilash, P. C., Fraceto, L. F., & Singh, H. B. (2017). Integrated approach of agri-nanotechnology: Challenges and future trends. *Frontiers in Plant Science, 8*. https://doi.org/10.3389/fpls.2017.00471
6. Tripathi, S., Mahra, S., J, V, Tiwari, K., Rana, S., Tripathi, D. K., Sharma, S., & Sahi, S. (2023). Recent advances and perspectives of nanomaterials in agricultural management and associated environmental risk: A review. *Nanomaterials, 13*(10), 1604. https://doi.org/10.3390/nano13101604
7. Walker, G. W., Kookana, R. S., Smith, N. E., Kah, M., Doolette, C. L., Reeves, P. T., Lovell, W., Anderson, D. J., Turney, T. W., & Navarro, D. A. (2017). Ecological risk assessment of Nano-enabled pesticides: A perspective on problem formulation. *Journal of Agricultural and Food Chemistry, 66*(26), 6480–6486. https://doi.org/10.1021/acs.jafc.7b02373
8. Singh, R. P., Handa, R., & Manchanda, G. (2021). Nanoparticles in sustainable agriculture: An emerging opportunity. *Journal of Controlled Release, 329*, 1234–1248. https://doi.org/10.1016/j.jconrel.2020.10.051
9. Agathokleous, E., Feng, Z., Iavicoli, I., & Calabrese, E. J. (2020). Nano-pesticides: A great challenge for biodiversity? The need for a broader perspective. *Nano Today, 30*, 100808. https://doi.org/10.1016/j.nantod.2019.100808
10. Camara, M. C., Campos, E. V. R., Monteiro, R. A., Pereira, A. D. E. S., De Freitas Proenca, P. L., & Fraceto, L. F. (2019). Development of stimuli-responsive nano-based pesticides: Emerging opportunities for agriculture. *Journal of Nanobiotechnology, 17*(1), 100. https://doi.org/10.1186/s12951-019-0533-8
11. Hayles, J., Johnson, L., Worthley, C., & Losic, D. (2017). Nanopesticides: A review of current research and perspectives. In *Elsevier eBooks* (pp. 193–225). https://doi.org/10.1016/b978-0-12-804299-1.00006-0
12. Jasrotia, P., Nagpal, M., Mishra, C. N., Sharma, A. K., Kumar, S., Kamble, U., Bhardwaj, A. K., Kashyap, P. L., Kumar, S., & Singh, G. P. (2022). Nanomaterials for postharvest management of insect pests: Current state and future perspectives. *Frontiers in Nanotechnology, 3*. https://doi.org/10.3389/fnano.2021.811056
13. De Oliveira, J. L., Campos, E. V. R., Bakshi, M., Abhilash, P., & Fraceto, L. F. (2014). Application of nanotechnology for the encapsulation of botanical insecticides for sustainable agriculture: Prospects and promises. *Biotechnology Advances, 32*(8), 1550–1561. https://doi.org/10.1016/j.biotechadv.2014.10.010
14. Isman, M. B. (2006). Botanical insecticides, deterrents, and repellents in modern agriculture and an increasingly regulated world. *Annual Review of Entomology, 51*(1), 45–66. https://doi.org/10.1146/annurev.ento.51.110104.151146
15. Wang, D., Saleh, N. B., Byro, A., Zepp, R., Sahle-Demessie, E., Luxton, T. P., Ho, K. T., Burgess, R. M., Flury, M., White, J. C., & Su, C. (2022). Nano-enabled pesticides for sustainable agriculture and global food security. *Nature Nanotechnology, 17*(4), 347–360. https://doi.org/10.1038/s41565-022-01082-8
16. Meng, F. L. (2011). Research on the development of botanical pesticide based on analysis of scientific materials. *Advanced Materials Research, 282–283*, 509–513. https://doi.org/10.4028/www.scientific.net/amr.282-283.509

17. Kookana, R. S., Boxall, A. B. A., Reeves, P. T., Ashauer, R., Beulke, S., Chaudhry, Q., Cornelis, G., Fernandes, T. F., Gan, J., Kah, M., Lynch, I., Ranville, J., Sinclair, C., Spurgeon, D., Tiede, K., & Van Den Brink, P. J. (2014). Nanopesticides: Guiding principles for regulatory evaluation of environmental risks. *Journal of Agricultural and Food Chemistry, 62*(19), 4227–4240. https://doi.org/10.1021/jf500232f
18. Bhardwaj, A. K., Arya, G., Kumar, R., Hamed, L., Pirasteh-Anosheh, H., Jasrotia, P., Kashyap, P. L., & Singh, G. P. (2022). Switching to nanonutrients for sustaining agroecosystems and environment: The challenges and benefits in moving up from ionic to particle feeding. *Journal of Nanobiotechnology, 20*(1), 19. https://doi.org/10.1186/s12951-021-01177-9
19. Preetha, P. S., & Balakrishnan, N. (2017b). A review of nano fertilizers and their use and functions in soil. *International Journal of Current Microbiology and Applied Sciences, 6*(12), 3117–3133. https://doi.org/10.20546/ijcmas.2017.612.364
20. Sivarethinamohan, R., & Sujatha, S. (2021). Unlocking the potentials of using nanotechnology to stabilize agriculture and food production. *AIP Conference Proceedings*. https://doi.org/10.1063/5.0039418
21. Sagar, D. S. (2020). Nanofertilizers: Perspective to enhance growth, yield and NUE of crops. *Indian Journal of Pure & Applied Biosciences, 8*(6), 339–349. https://doi.org/10.18782/2582-2845.8405
22. Kim, D., Kadam, A., Shinde, S., Saratale, R. G., Patra, J., & Ghodake, G. (2017). Recent developments in nanotechnology transforming the agricultural sector: A transition replete with opportunities. *Journal of the Science of Food and Agriculture, 98*(3), 849–864. https://doi.org/10.1002/jsfa.8749
23. Fraceto, L. F., Grillo, R., De Medeiros, G. A., Scognamiglio, V., Rea, G., & Bartolucci, C. (2016). Nanotechnology in agriculture: Which innovation potential does it have? *Frontiers in Environmental Science, 4*. https://doi.org/10.3389/fenvs.2016.00020
24. Gupta, A., Rayeen, F., Mishra, R., Tripathi, M., & Pathak, N. (2023). Nanotechnology applications in sustainable agriculture: An emerging eco-friendly approach. *Plant Nano Biology, 4*, 100033. https://doi.org/10.1016/j.plana.2023.100033
25. Sekhon, B. (2014). Nanotechnology in agri-food production: An overview. *Nanotechnology Science and Applications, 31*. https://doi.org/10.2147/nsa.s39406
26. Prasad, R., Gupta, N., Kumar, M., Kumar, V., Wang, S., & Abd-Elsalam, K. A. (2017). Nanomaterials act as plant defense mechanism. In *Springer eBooks* (pp. 253–269). https://doi.org/10.1007/978-981-10-4678-0_14
27. Banerjee, K., Pramanik, P., Maity, A., Joshi, D., Wani, S., & Krishnan, P. (2019). Methods of using nanomaterials to plant systems and their delivery to plants (mode of entry, uptake, translocation, accumulation, biotransformation and barriers). In *Elsevier eBooks* (pp. 123–152). https://doi.org/10.1016/b978-0-12-815322-2.00005-5
28. Ruttkay-Nedecky, B., Krystofova, O., Nejdl, L., & Adam, V. (2017). Nanoparticles based on essential metals and their phytotoxicity. *Journal of Nanobiotechnology, 15*(1), 33. https://doi.org/10.1186/s12951-017-0268-3
29. Raliya, R., Franke, C., Chavalmane, S., Nair, R., & Reed, N. (2016). Quantitative understanding of nanoparticle uptake in watermelon plants. *Frontiers in Plant Science, 7*. https://doi.org/10.3389/fpls.2016.01288
30. Anjum, N. A., Rodrigo, M. A. M., Moulick, A., Heger, Z., Kopel, P., Zítka, O., Adam, V., Lukatkin, A. S., Duarte, A. C., Pereira, E., & Kizek, R. (2016). Transport phenomena of nanoparticles in plants and animals/humans. *Environmental Research, 151*, 233–243. https://doi.org/10.1016/j.envres.2016.07.018
31. Tripathi, D. K., Shweta, N., Singh, S., Singh, S., Pandey, R., Singh, V. P., Sharma, N. C., Prasad, S. M., Dubey, N. K., & Chauhan, D. K. (2017). An overview on manufactured nanoparticles in plants: Uptake, translocation, accumulation and phytotoxicity. *Plant Physiology and Biochemistry, 110*, 2–12. https://doi.org/10.1016/j.plaphy.2016.07.030

32. Chen, H. (2018). Metal based nanoparticles in agricultural system: Behavior, transport, and interaction with plants. *Chemical Speciation and Bioavailability, 30*(1), 123–134. https://doi.org/10.1080/09542299.2018.1520050
33. Bashri, G., Parihar, P., Singh, R., Patel, A., & Prasad, S. M. (2018). Plant and nanoparticle Interface at the molecular level. In *Elsevier eBooks* (pp. 325–344). https://doi.org/10.1016/b978-0-12-811487-2.00015-3
34. Rastogi, A., Zivcak, M., Sytar, O., Kalaji, H. M., He, X., Mbarki, S., & Brestic, M. (2017). Impact of metal and metal oxide nanoparticles on plant: A critical review. *Frontiers in Chemistry, 5*. https://doi.org/10.3389/fchem.2017.00078
35. Gao, M., Chang, J., Wang, Z., Zhang, H., & Wang, T. (2023). Advances in transport and toxicity of nanoparticles in plants. *Journal of Nanobiotechnology, 21*(1), 75. https://doi.org/10.1186/s12951-023-01830-5
36. Schwab, F., Zhai, G., Kern, M., Turner, A., Schnoor, J. L., & Wiesner, M. R. (2015). Barriers, pathways and processes for uptake, translocation and accumulation of nanomaterials in plants—Critical review. *Nanotoxicology, 10*(3), 257–278. https://doi.org/10.3109/17435390.2015.1048326
37. Chichiriccò, G., & Poma, A. (2015). Penetration and toxicity of nanomaterials in higher plants. *Nanomaterials, 5*(2), 851–873. https://doi.org/10.3390/nano5020851
38. Kurczyńska, E., Godel-Jędrychowska, K., Sala, K., & Milewska-Hendel, A. (2021). Nanoparticles—Plant interaction: What we know, where we are? *Applied Sciences, 11*(12), 5473. https://doi.org/10.3390/app11125473
39. Etxeberria, E., Gonzalez, P., Bhattacharya, P., Sharma, P., & Ke, P. C. (2016). Determining the size exclusion for nanoparticles in citrus leaves. *HortScience, 51*(6), 732–737. https://doi.org/10.21273/hortsci.51.6.732
40. Hubbard, J. D., Lui, A., & Landry, M. P. (2020). Multiscale and multidisciplinary approach to understanding nanoparticle transport in plants. *Current Opinion in Chemical Engineering, 30*, 135–143. https://doi.org/10.1016/j.coche.2020.100659
41. Jeyasubramanian, K., Thoppey, U. U. G., Hikku, G. S., Selvakumar, N., Subramania, A., & Krishnamoorthy, K. (2016). Enhancement in growth rate and productivity of spinach grown in hydroponics with iron oxide nanoparticles. *RSC Advances, 6*(19), 15451–15459. https://doi.org/10.1039/c5ra23425e
42. Nair, R., Varghese, S. H., Nair, B. G., Maekawa, T., Yoshida, Y., & Kumar, D. S. (2010). Nanoparticulate material delivery to plants. *Plant Science, 179*(3), 154–163. https://doi.org/10.1016/j.plantsci.2010.04.012
43. Cifuentes, Z., Custardoy, L., De La Fuente, J. M., Marquina, C., Ibarra, M. R., Rubiales, D., & Pérez-De-Luque, A. (2010). Absorption and translocation to the aerial part of magnetic carbon-coated nanoparticles through the root of different crop plants. *Journal of Nanobiotechnology, 8*(1). https://doi.org/10.1186/1477-3155-8-26
44. Koelmel, J., Leland, T., Wang, H., Amarasiriwardena, D., & Xing, B. (2013). Investigation of gold nanoparticles uptake and their tissue level distribution in rice plants by laser ablation-inductively coupled-mass spectrometry. *Environmental Pollution, 174*, 222–228. https://doi.org/10.1016/j.envpol.2012.11.026
45. Gonnelli, C., Cacioppo, F., Giordano, C., Capozzoli, L., Salvatici, C., Salvatici, M. C., Colzi, I., Del Bubba, M., Ancillotti, C., & Ristori, S. (2015). Cucurbita pepoL. Extracts as a versatile hydrotropic source for the synthesis of gold nanoparticles with different shapes. *Green Chemistry Letters and Reviews, 8*(1), 39–47. https://doi.org/10.1080/17518253.2015.1027288
46. Malejko, J., Godlewska-Żyłkiewicz, B., Vanek, T., Landa, P., Nath, J., Dror, I., & Berkowitz, B. (2021). Uptake, translocation, weathering and speciation of gold nanoparticles in potato, radish, carrot and lettuce crops. *Journal of Hazardous Materials, 418*, 126219. https://doi.org/10.1016/j.jhazmat.2021.126219
47. Zhu, Z. J., Wang, H., Yan, B., Zheng, H., Jiang, Y., Miranda, O. R., Rotello, V. M., Xing, B., & Vachet, R. W. (2012). Effect of surface charge on the uptake and distribution of gold nanoparticles in four plant species. *Environmental Science & Technology, 46*(22), 12391–12398. https://doi.org/10.1021/es301977w

48. Sabaghnia, N., Javanmard, A., Janmohammadi, M., & Nouraein, M. (2018). The influence of nano-TiO2 and Nano-silica particles effects on yield and morphological traits of sunflower. *Helia, 41*(69), 213–225. https://doi.org/10.1515/helia-2018-0010
49. Bovand, F., Chavoshi, S., & Ghorbanpour, M. (2022). Titanium dioxide and silicon dioxide nanoparticles differentially affect germination and biochemical traits in marigold (Calendula officinalis L.) and lemon balm (Melissa officinalis L.). *Nanotechnology for Environmental Engineering, 8*(1), 281–295. https://doi.org/10.1007/s41204-022-00279-5
50. Larue, C., Laurette, J., Herlin-Boime, N., Khodja, H., Fayard, B., Flank, A. M., Brisset, F., & Carriere, M. (2012). Accumulation, translocation and impact of TiO2 nanoparticles in wheat (Triticum aestivum spp.): Influence of diameter and crystal phase. *The Science of the Total Environment, 431*, 197–208. https://doi.org/10.1016/j.scitotenv.2012.04.073
51. Farahi, S. M. M., Yazdi, M. E. T., Einafshar, E., Akhondi, M., Ebadi, M., Azimipour, S., Mahmoodzadeh, H., & Iranbakhsh, A. (2023). The effects of titanium dioxide (TiO2) nanoparticles on physiological, biochemical, and antioxidant properties of Vitex plant (*Vitex agnus* - Castus L). *Heliyon, 9*(11), e22144. https://doi.org/10.1016/j.heliyon.2023.e22144
52. Kamal, R., & Mogazy, A. M. (2021). Effect of doping on TiO2 nanoparticles characteristics: Studying of fertilizing effect on cowpea plant growth and yield. *Journal of Soil Science and Plant Nutrition, 23*(1), 325–337. https://doi.org/10.1007/s42729-021-00648-0
53. Prakashraj, R., Vijayakumar, S., Punitha, V. N., Vidhya, E., Nilavukkarasi, M., & Praseetha, P. K. (2021). Fabricated TiO2 nanofertilizers for foliar assimilation to enhance yield and disease resistance in Capsicum annuum L. *Journal of Plant Growth Regulation, 41*(8), 3387–3394. https://doi.org/10.1007/s00344-021-10520-1
54. Lin, D., & Xing, B. (2008). Root uptake and phytotoxicity of ZnO nanoparticles. *Environmental Science & Technology, 42*(15), 5580–5585. https://doi.org/10.1021/es800422x
55. Lv, Z., Sun, H., Du, W., Li, R., Mao, H., & Kopittke, P. M. (2021). Interaction of different-sized ZnO nanoparticles with maize (*Zea mays*): Accumulation, biotransformation and phytotoxicity. *The Science of the Total Environment, 796*, 148927. https://doi.org/10.1016/j.scitotenv.2021.148927
56. Srivastav, A., Ganjewala, D., Singhal, R. K., Rajput, V. D., Minkina, T., Voloshina, M., Srivastava, S., & Shrivastava, M. (2021). Effect of ZnO nanoparticles on growth and biochemical responses of wheat and maize. *Plants, 10*(12), 2556. https://doi.org/10.3390/plants10122556
57. Solanki, P., & Laura, J. (2018). Effect of ZnO nanoparticles on seed germination and seedling growth in wheat (Triticum aestivum). *Journal of Pharmacognosy and Phytochemistry, 7*(5), 2048–2052. https://www.phytojournal.com/archives/2018/vol7issue5/PartAI/7-5-265-970.pdf
58. López-Moreno, M. L., De La Rosa, G., Hernández-Viezcas, J., Castillo-Michel, H., Botez, C. E., Peralta-Videa, J. R., & Gardea-Torresdey, J. L. (2010). Evidence of the differential biotransformation and genotoxicity of ZnO and CeO2 nanoparticles on soybean (*Glycine max*) plants. *Environmental Science & Technology, 44*(19), 7315–7320. https://doi.org/10.1021/es903891g
59. Hernandez-Viezcas, J., Castillo-Michel, H., Servin, A., Peralta-Videa, J., & Gardea-Torresdey, J. (2011). Spectroscopic verification of zinc absorption and distribution in the desert plant *Prosopis juliflora-velutina* (velvet mesquite) treated with ZnO nanoparticles. *Chemical Engineering Journal, 170*(2–3), 346–352. https://doi.org/10.1016/j.cej.2010.12.021
60. Wang, Z., Xie, X., Zhao, J., Liu, X., Feng, W., White, J. C., & Xing, B. (2012). Xylem- and phloem-based transport of CuO nanoparticles in maize (Zea maysL.). *Environmental Science & Technology, 46*(8), 4434–4441. https://doi.org/10.1021/es204212z
61. Mukherjee, A., Peralta-Videa, J. R., Gardea-Torresdey, J., & White, J. C. (2016). Effects and uptake of nanoparticles in plants, 386–408. https://doi.org/10.1002/9781119275855.ch20
62. Usman, M., Farooq, M., Wakeel, A., Nawaz, A., Cheema, S. A., Rehman, H. U., Ashraf, I., & Sanaullah, M. (2020). Nanotechnology in agriculture: Current status, challenges and future opportunities. *The Science of the Total Environment, 721*, 137778. https://doi.org/10.1016/j.scitotenv.2020.137778

63. Achari, G. A., & Kowshik, M. (2018). Recent developments on nanotechnology in agriculture: Plant mineral nutrition, health, and interactions with soil microflora. *Journal of Agricultural and Food Chemistry, 66*(33), 8647–8661. https://doi.org/10.1021/acs.jafc.8b00691
64. Saleem, H., & Zaidi, S. J. (2020). Recent developments in the application of nanomaterials in agroecosystems. *Nanomaterials, 10*(12), 2411. https://doi.org/10.3390/nano10122411
65. Maghsoodi, M. R., Lajayer, B. A., Hatami, M., & Mirjalili, M. H. (2019d). Challenges and opportunities of nanotechnology in plant-soil mediated systems: Beneficial role, phytotoxicity, and phytoextraction. In *Elsevier eBooks* (pp. 379–404). https://doi.org/10.1016/b978-0-12-815322-2.00018-3
66. Mukherjee, A., Peralta-Videa, J. R., Gardea-Torresdey, J., & White, J. C. (2016c). *Effects and uptake of nanoparticles in plants* (pp. 386–408). https://doi.org/10.1002/9781119275855.ch20
67. Van Nhan, L., Ma, C., Rui, Y., Liu, S., Li, X., Xing, B., & Liu, L. (2015). Phytotoxic mechanism of nanoparticles: Destruction of chloroplasts and vascular bundles and alteration of nutrient absorption. *Scientific Reports, 5*(1). https://doi.org/10.1038/srep11618
68. Baazaoui, N., Sghaier-Hammami, B., Hammami, S. B. M., Khefacha, R., Chaari, S., Elleuch, L., Messaoud, M., & Abdelly, C. (2021). A handbook guide to better use of nanoparticles in plants. *Communications in Soil Science and Plant Analysis, 52*(4), 287–321. https://doi.org/10.1080/00103624.2020.1836198
69. El-Moneim, D. A., Dawood, M. F. A., Moursi, Y. S., Farghaly, A. A., Afifi, M., & Sallam, A. (2021). Positive and negative effects of nanoparticles on agricultural crops. *Nanotechnology for Environmental Engineering, 6*(2). https://doi.org/10.1007/s41204-021-00117-0
70. Ma, X., Geisler-Lee, J., Deng, Y., & Kolmakov, A. (2014). Corrigendum to "Interactions between engineered nanoparticles (ENPs) and plants: Phytotoxicity, uptake and accumulation" [Sci. Total Environ. 408 (2010) 3053–3061]. *The Science of the Total Environment, 481*, 635. https://doi.org/10.1016/j.scitotenv.2014.02.044
71. Verma, S. K., Das, A. K., Patel, M. K., Shah, A., Kumar, V., & Gantait, S. (2018). Engineered nanomaterials for plant growth and development: A perspective analysis. *The Science of the Total Environment, 630*, 1413–1435. https://doi.org/10.1016/j.scitotenv.2018.02.313
72. Marslin, G., Sheeba, C. J., & Franklin, G. (2017). Nanoparticles alter secondary metabolism in plants via ROS burst. *Frontiers in Plant Science, 8*. https://doi.org/10.3389/fpls.2017.00832
73. Tolaymat, T., Genaidy, A., Abdelraheem, W., Dionysiou, D., & Andersen, C. (2017b). The effects of metallic engineered nanoparticles upon plant systems: An analytic examination of scientific evidence. *The Science of the Total Environment, 579*, 93–106. https://doi.org/10.1016/j.scitotenv.2016.10.229
74. Eduok, S., & Coulon, F. (2017). Microbiological toxicity of nanoparticles. In *Elsevier eBooks* (pp. 97–117). https://doi.org/10.1016/b978-0-323-42980-1.00006-6
75. Rizwan, M., Ali, S., Qayyum, M. F., Ok, Y. S., Adrees, M., Ibrahim, M., Zia-Ur-Rehman, M., Farid, M., & Abbas, F. (2017). Effect of metal and metal oxide nanoparticles on growth and physiology of globally important food crops: A critical review. *Journal of Hazardous Materials, 322*, 2–16. https://doi.org/10.1016/j.jhazmat.2016.05.061
76. Lv, J., Christie, P., & Zhang, S. (2019). Uptake, translocation, and transformation of metal-based nanoparticles in plants: Recent advances and methodological challenges. *Environmental Science Nano, 6*(1), 41–59. https://doi.org/10.1039/c8en00645h
77. Mahawar, H., & Prasanna, R. (2018). Prospecting the interactions of nanoparticles with beneficial microorganisms for developing green technologies for agriculture. *Environmental Nanotechnology Monitoring & Management, 10*, 477–485. https://doi.org/10.1016/j.enmm.2018.09.004
78. Tian, H., Kah, M., & Kariman, K. (2019). Are nanoparticles a threat to mycorrhizal and Rhizobial symbioses? A critical review. *Frontiers in Microbiology, 10*. https://doi.org/10.3389/fmicb.2019.01660

79. Barkataki, M. P., & Singh, T. (2019). Plant-nanoparticle interactions: Mechanisms, effects, and approaches. In *Comprehensive analytical chemistry* (pp. 55–83). https://doi.org/10.1016/bs.coac.2019.09.007
80. Hermes, P. H., Gabriela, M. P., Ileana, V. R., Fusaro, C., Fernando, L. V., Mariana, M. A., Padilla-Rodríguez, C., & Fabián, F. L. (2020). Carbon nanotubes as plant growth regulators: Prospects. In *Nanotechnology in the life sciences* (pp. 77–115). https://doi.org/10.1007/978-3-030-39246-8_4
81. Karimi, E., & Fard, E. M. (2017). Nanomaterial effects on soil microorganisms. In *Soil biology* (pp. 137–200). https://doi.org/10.1007/978-3-319-46835-8_5
82. Morales-Díaz, A. B., Ortega-Ortíz, H., Juárez-Maldonado, A., Cadenas-Pliego, G., González-Morales, S., & Benavides-Mendoza, A. (2017). Application of nanoelements in plant nutrition and its impact in ecosystems. Advances in Natural Sciences Nanoscience and Nanotechnology, 8(1), 013001. https://doi.org/10.1088/2043-6254/8/1/013001
83. Hegde, K., Brar, S. K., Verma, M., & Surampalli, R. Y. (2016). Current understandings of toxicity, risks and regulations of engineered nanoparticles with respect to environmental microorganisms. *Nanotechnology for Environmental Engineering, 1*(1). https://doi.org/10.1007/s41204-016-0005-4
84. Wang, Y., Sun, C., Zhao, X., Cui, B., Zeng, Z., Wang, A., Liu, G., & Cui, H. (2016). The application of Nano-TiO2 photo semiconductors in agriculture. *Nanoscale Research Letters, 11*(1), 529. https://doi.org/10.1186/s11671-016-1721-1
85. Prasad, R., Bhattacharyya, A., & Nguyen, Q. D. (2017). Nanotechnology in sustainable agriculture: Recent developments, challenges, and perspectives. *Frontiers in Microbiology, 8*. https://doi.org/10.3389/fmicb.2017.01014
86. Kumar, S., Nehra, M., Dilbaghi, N., Marrazza, G., Hassan, A. A., & Kim, K. H. (2019). Nano-based smart pesticide formulations: Emerging opportunities for agriculture. *Journal of Controlled Release, 294*, 131–153. https://doi.org/10.1016/j.jconrel.2018.12.012
87. Chaud, M., Souto, E. B., Zielinska, A., Severino, P., Batain, F., Oliveira-Junior, J., & Alves, T. (2021). Nanopesticides in agriculture: Benefits and challenge in agricultural productivity, toxicological risks to human health and environment. *Toxics, 9*(6), 131. https://doi.org/10.3390/toxics9060131
88. Chhipa, H. (2017). Nanopesticide: Current status and future possibilities. *Agricultural Research & Technology Open Access Journal, 5*(1). https://doi.org/10.19080/artoaj.2017.05.555651
89. Mwaanga, P. (2018). Risks, uncertainties, and ethics of nanotechnology in agriculture. In *InTech eBooks*. https://doi.org/10.5772/intechopen.76590
90. Pradhan, S., Durgam, M., & Mailapalli, D. R. (2020). Urea loaded hydroxyapatite nanocarrier for efficient delivery of plant nutrients in rice. *Archives of Agronomy and Soil Science, 67*(3), 371–382. https://doi.org/10.1080/03650340.2020.1732940
91. De Oliveira, J. L., Campos, E. V., Camara, M. C., De Carvalho, L. B., Monteiro, R. A., De Lima, R., & Fraceto, L. F. (2019). Nanotechnology-based delivery systems: Highlights in agricultural applications. *Journal of Siberian Federal University Biology*, 311–328. https://doi.org/10.17516/1997-1389-0305
92. Pullagurala, V. L. R., Adisa, I. O., Rawat, S., White, J. C., Zuverza-Mena, N., Hernandez-Viezcas, J. A., Peralta-Videa, J. R., & Gardea-Torresdey, J. L. (2019). Fate of engineered nanomaterials in agroenvironments and impacts on agroecosystems. In *Elsevier eBooks* (pp. 105–142). https://doi.org/10.1016/b978-0-12-814835-8.00004-2
93. Zhang, Y., & Goss, G. G. (2022). Nanotechnology in agriculture: Comparison of the toxicity between conventional and nano-based agrochemicals on non-target aquatic species. *Journal of Hazardous Materials, 439*, 129559. https://doi.org/10.1016/j.jhazmat.2022.129559
94. Kah, M. (2015). Nanopesticides and nanofertilizers: Emerging contaminants or opportunities for risk mitigation? *Frontiers in Chemistry, 3*. https://doi.org/10.3389/fchem.2015.00064
95. Garg, D., Sridhar, K., Inbaraj, B. S., Chawla, P., Tripathi, M., & Sharma, M. (2023). Nano-biofertilizer formulations for agriculture: A systematic review on recent advances and prospective applications. *Bioengineering, 10*(9), 1010. https://doi.org/10.3390/bioengineering10091010

96. Paramo, L. A., Feregrino-Pérez, A. A., Guevara, R., Mendoza, S., & Esquivel, K. (2020). Nanoparticles in agroindustry: Applications, toxicity, challenges, and trends. *Nanomaterials, 10*(9), 1654. https://doi.org/10.3390/nano10091654
97. Pavan, M. V. R., & Khabashesku, V. N. (2015). Environmental consequences of engineered nanomaterials: An awareness campaign to promote safe nanotechnology and dispel related misconceptions. *All Days.* https://doi.org/10.4043/26039-ms
98. Lombi, E., Donner, E., Dusinska, M., & Wickson, F. (2019). A one health approach to managing the applications and implications of nanotechnologies in agriculture. *Nature Nanotechnology, 14*(6), 523–531. https://doi.org/10.1038/s41565-019-0460-8
99. Maharramov, A. M., Hasanova, U. A., Suleymanova, I. A., Osmanova, G. E., & Hajiyeva, N. E. (2019). The engineered nanoparticles in food chain: Potential toxicity and effects. *SN Applied Sciences, 1*(11). https://doi.org/10.1007/s42452-019-1412-5
100. Advancing EPA's Understanding of the Next Generation of Pesticides. (2022, May 9). https://www.epa.gov/sciencematters/advancing-epas-understanding-next-generation-pesticides
101. Guha, T., Gopal, G., Chatterjee, R., Mukherjee, A., & Kundu, R. (2020). Differential growth and metabolic responses induced by nano-scale zero valent iron in germinating seeds and seedlings of Oryza sativa L. cv. Swarna. *Ecotoxicology and Environmental Safety, 204*, 111104. https://doi.org/10.1016/j.ecoenv.2020.111104
102. Vinzant, K., Rashid, M., & Khodakovskaya, M. V. (2023). Advanced applications of sustainable and biological nano-polymers in agricultural production. *Frontiers in Plant Science, 13.* https://doi.org/10.3389/fpls.2022.1081165
103. Xi, L., Zhang, M., Zhang, L., Lew, T. T. S., & Lam, Y. M. (2022). Novel materials for urban farming (Adv. Mater. 25/2022). *Advanced Materials, 34*(25). https://doi.org/10.1002/adma.202270190
104. Aslani, F., Bagheri, S., Julkapli, N. M., Juraimi, A. S., Hashemi, F. S. G., & Baghdadi, A. (2014). Effects of engineered nanomaterials on plants growth: An overview. *The Scientific World Journal, 2014*, 1–28. https://doi.org/10.1155/2014/641759
105. Zuverza-Mena, N., Martínez-Fernández, D., Du, W., Hernandez-Viezcas, J. A., Bonilla-Bird, N., López-Moreno, M. L., Komárek, M., Peralta-Videa, J. R., & Gardea-Torresdey, J. L. (2017). Exposure of engineered nanomaterials to plants: Insights into the physiological and biochemical responses-A review. *Plant Physiology and Biochemistry, 110*, 236–264. https://doi.org/10.1016/j.plaphy.2016.05.037
106. Nowak, K. M., Miltner, A., & Kästner, M. (2020). Environmental fate assessment of chemicals and the formation of Biogenic Non-extractable Residues (bioNER). In *The handbook of environmental chemistry* (pp. 81–111). https://doi.org/10.1007/698_2020_515
107. Shelar, A., Nile, S. H., Singh, A. V., Rothenstein, D., Bill, J., Xiao, J., Chaskar, M., Kai, G., & Patil, R. (2023). Recent advances in Nano-enabled seed treatment strategies for sustainable agriculture: Challenges, risk assessment, and future perspectives. *Nano-Micro Letters, 15*(1), 54. https://doi.org/10.1007/s40820-023-01025-5

Risk Assessment and Registration Policies for the Safety of Nanomaterials in Agriculture

Kishor Raut and Archana A. Sharbidre

Contents

K. Raut
Entomology Division, Ross Lifescience Ltd., Pune, MS, India

A. A. Sharbidre (✉)
Department of Zoology, Savitribai Phule Pune University, Pune, MS, India
e-mail: aasharbidre@unipune.ac.in

R. S. Chaughule, A. V. Shelar (eds.), *Nanotechnology in Agriculture*,
https://doi.org/10.1007/978-3-031-97544-8_13

1 Introduction

The need for progress in agriculture and biosciences is underscored by the continuing challenges of climate change and the introduction of new pathogens and diseases. In the midst of the twenty-first century, these fields must be able to rise to the challenge of meeting the growing demands of a continually expanding global population that increasingly competes with one another for increasingly scarce natural resources. These problems have been of critical importance as humanity wrestles with these issues, and the development of systems that optimize resource utilization and minimize waste has been an important challenge. But environmental and biological warfare, along with the socio-political resistance these efforts often face, make the task more difficult. Such complexities make clear the necessity for innovative solutions to development and resilience to these multifaceted threats are addressed.

In the context of these global challenges the notion of "nano" is a powerful and promising tool. The term "nano" itself is derived from the Greek word nanos meaning "dwarf" and the prefix "nano," representing one-billionth or 10^{-9}, has since become synonymous with nanoparticles (NPs), which refer to objects with dimensions ranging from 1 to 100 nm. These tiny particles are notable for their unique properties and potential applications in various fields [1–4].

Nanoparticles exist in all three dimensions within this size range and have drawn significant attention in recent years for their innovative applications. Despite being a relatively recent scientific focus, the phenomenon of nanoscale structures has long existed in nature. Examples include volcanic ash, mineral composites, forest fire particulates, and even the microscopic structures responsible for the vibrant colors in butterfly wings. The natural prevalence of these nanoscale phenomena emphasizes their fundamental role in both biological and environmental processes. Researchers have increasingly recognized their potential to contribute to solutions for the modern challenges of sustainability, efficiency, and environmental protection [5, 6]. By integrating insights from nanotechnology with advancements in agri- and biosciences, humanity stands a better chance of addressing the critical demands of a changing world. From improving agricultural productivity to developing bioscience innovations, nanoparticles could play a pivotal role in shaping a more sustainable and resilient future. The prefix "nano" comes from the Greek word nano which

Table 1 Different terminologies used in nanotechnology with their definitions

Terminology	Definition and its reference
Nanomaterial	It is a natural, incidental, or manufactured material containing particles, in an unbound state or as an agglomerate and where, for 50%, or more of particles in the number size distribution, one or more external dimensions are in the size range 1 nm to 100 nm [8]
Nanopesticide	Any pesticide formulation that involves either very small particles of a pesticide's active ingredients or other small, engineered structures with useful pesticide properties [9]
Nano-coating	Surface coatings or nanocoatings of nanomaterials, on fertilizers or pesticides active ingredients, can hold the material more strongly due to higher surface tension than the conventional surfaces and thus help in controlled release [10]
Nanofertilizers and nanomaterials enhance fertilizers	Nanomaterials can supply one or more nutrients to plants and enhance their growth and yields. Nanomaterial-enhanced fertilizers, when combined with conventional fertilizers, can enhance their effectiveness without directly supplying crops with nutrients [11]
Nano-sensor	Nanoscale devices capable of detecting and responding to physio-chemical and biological stimuli. It may provide a larger surface area for the immobilization of the target-recognition elements or confer the sensor their own optical-physical and electrochemical properties [12]
Nano-enabled formulation	It encompasses those emulsions made of a smaller number of surfactants, or microcapsules with well-defined nanopore networks [13]

means dwarf, something extremely small. Scientifically, it means 1 billionth of a meter, or 10^{-9} meters. At this scale, it also offers a basis for distinguishing nanoscience from nanotechnology. Nanoscience studies (typically 1 to 100 nanometers) structures and molecules at the nanoscale. But nanotechnology is the practical application of this knowledge, which helps to develop new devices, materials, and solutions. This scale is put in context by the fact that one human hair is 60,000 nanometers thick, and the radius of the DNA double helix is just 1 nm [7]. The different terminologies used in nanotechnology are listed in Table 1 with definition & their references.

Nanoparticles (NPs) have received a great deal of attention in the last several years because of the tremendous differences in properties between the NPs and their bulk material analogs. Distinctive characteristics of nanoparticles, such as higher surface area to volume ratio, differences in surface charge distribution, and patterns of light scattering, produce these distinctions. In addition to altering the chemical and optical behavior of nanoparticles, these attributes make them more reactive. Particle entropy is increased by the irregular distribution of atomic constituents within nanoparticles, thus increasing their Gibbs free energy and their reactivity. Nanoparticles as devices have also been shown by comparative studies to have very potent effects on the development, growth, and vitality of plants and represent transformative tools for agricultural science [14]. Agrochemical solutions, micronutrient formulations, and nanofertilizers based on nanoparticles have also been developed and successfully used to increase plant growth and vigor.

The interaction of nanofertilizers with the rhizosphere is one of the important advantages of nanofertilizers in improving the availability and uptake of nutrients by plants. They release their active compounds slowly and constantly, unlike traditional fertilizers, and they maintain effectiveness in the plant system for a longer period of time. Fertilizers based on nanoporous zeolite have been developed recently in nanotechnology that provide significantly better absorption by plants of nitrogen. For example, natural zeolite treated with aminopropyl trimethoxy silane is very attractive for the binding of nitrogen in urea. Increased urea adsorption and controlled release from the treated zeolite were achieved due to its amine groups on the surface of the treated zeolite, creating more efficient and sustained fertilizer delivery [15]. This lays bare the spurring potential of nanoparticles in agriculture to improve resource efficiency and sustainability. But by using the specific properties of nanoparticles, researchers and agricultural practitioners can open up a more productive and environmentally friendly future.

The field of Nanotechnology has emerged as a transformative technological space having far-reaching potential in Applications ranging from Medicine, Catalysis, Energy, Materials, and Agriculture. This has enabled new routes to groundbreaking applications, owing to its ability to manipulate matter at the nanoscale (1–100 nm) characterized by small particle sizes and large surface areas. As the final stage in the development of these nanoparticles (NPs), they have unique physical, chemical, and biological properties that make them appropriate for many industrial, medical, and agricultural applications. Research has focused on the synthesis of nanoparticles as a focus of research and these methods are classified into physical, chemical, and biological approaches [16].

Nevertheless, usual methods suffer from scalability issues, the difficulty of separating nanoparticles from emulsions, and a negative environmental impact due to the large amounts of surfactants needed. In recent years, plant extract-based green synthesis methods have become increasingly important due to simplicity, greenness, and reduced reaction times. Not only do these methods work, they are sustainable and further, they do not allow the release of harmful chemicals into the environment, as per international efforts to cut down pollution [17]. In particular, green synthesis of nanomaterials has proven to be advantageous in agriculture where they improve the fertilization process, act as plant growth regulators, and improve pesticide efficiency [18]. Nanotechnology integration in agriculture is considered a vital step towards increased productivity at minimal environmental damage.

The vast potential of nanotechnology to fundamentally change the way farming is conducted has attracted considerable public funding and considerable interest over the last decade. Nanotechnology integration into agriculture, as an open system exchanging freely energy and matter, poses special challenges. While farming activities are more diverse, they also involve large-scale inputs and operate under various, and often uncontrollable, environmental conditions, unlike industrial systems. However, despite this, nanotechnology provides innovative solutions that will enable productivity increase, resource wastage minimization, and environmental impact [19].

2 Applications of Nanotechnology

The main applications of nanotechnology in the agricultural field comprise:

2.1 *Plant Disease Diagnostics*

Diagnoses based on nanotechnology include accurate and rapid detection of plant pathogens thus facilitating timely intervention. Nanotechnology provides cutting-edge tools for the precise, sensitive, and rapid detection of plant pathogens, including bacteria, viruses, fungi, and nematodes, ensuring timely intervention. Traditional diagnostic methods often rely on time-intensive and labor-intensive techniques, such as culturing pathogens or using polymerase chain reactions (PCR), which may delay actionable responses. Nanotechnology overcomes these limitations by integrating nanoscale materials and systems that enhance sensitivity and reduce detection time.

By enabling early and accurate detection of plant pathogens, nanotechnology plays a critical role in mitigating crop losses, improving agricultural productivity, and supporting sustainable farming practices. This revolutionary approach aligns with modern precision agriculture techniques, empowering farmers with actionable insights to combat plant diseases effectively.

2.2 *Nanosensors for Crop Protection*

These devices can measure plant diseases early on, monitor soil health, and identify residues of agrochemicals to achieve precision crop management.

- Postharvest Management: Nanotechnology helps in the storage, packaging, and transportation of agricultural produce increases its shelf life, and diminishes losses.
- Nanodevices for Genetic Engineering: Gene transfer and plant genetic modification at the nanoscale tools for making insect-resistant and drought-tolerant plants are provided.
- Animal Health and Poultry Production: Nanoscale Innovations. Support better animal health monitoring, breeding, and disease.
- Postharvest Management: Nanotechnology improves the storage, packaging, and transportation of agricultural produce, extending shelf life, and reducing losses.

2.3 Nanoformulations for Agrochemicals

Fertilizers and pesticides are being advanced into more advanced formulations for targeted delivery to crops which reduce the overall quantity of chemicals used while improving the efficacy.

With the aid of nanotechnology, precision agriculture holds promise to reduce nitrogen loss through controlled release mechanisms, reduce soil and water contamination, and maintain microbial diversity. For example, nanoparticle-mediated gene or DNA transfer has been tried for the development of insect-resistant crop varieties. The biomass-to-fuel production process is also being advanced by nanotechnology which is providing sustainable energy solutions.

The advantages of nanotechnology for agriculture are great, but the environmental impact of nanotechnology and the health effects of working with nanomaterials must be dealt with. If nanoparticles are to be used safely in open farming systems where the possibility of uncontrolled exposure to soil and water ecosystems exists, risk assessments will be required. Biodegradable nanoparticle attractants being developed from proteins and carbohydrates are being explored to minimize risks to human health and the environment.

All stages of production, processing, storage, packaging, and transportation of agricultural products are integrated into the use of nanotechnology in agriculture. For example, nanocapsules control the release of agrochemicals; nanosensors monitor plant health, and nanodevices promote genetic improvement. Nurtured by these technologies, plants will be able to better absorb nutrients, diseases will be detected well, and pests will be controlled better. Its transformative potential also applies to the field of food technology and aquaculture, where its use can be to improve food safety, extend the shelf life of products, and develop sustainable fisheries. Although progress has been made, it is not far enough to fully realize nanotechnology's potential in agriculture and there remain challenges, e.g. regulatory hurdles, public perception, and long-term ecological impacts. However, with ongoing research and development, nanotechnology will change the agricultural and food industries in ways to make them grow sustainably and secure global food security [20, 21]. Some of the other nanoparticles or formulations are listed in Table 2 with their application usage.

3 Risk Assessment of Nanomaterials in Agriculture

Rapidly expanding is the incorporation of nanomaterials (NMs) in agricultural science, stimulated by growing global demand for higher agricultural yields and the inability to obtain environmentally efficient practices to counter traditional agricultural methods that also have steep negative environmental and socioeconomic impacts [36]. With Crop Synthesis Nanotechnology can revolutionize Precision farming and Sustainable agriculture by increasing crop yields while reducing Inputs

Table 2 Nanoparticles or formulations application with functions and benefits

Nanoparticles type	Used concentration	Mode of application	Physiological function	Reference
Gold nanoparticles (Au NPs)	5–11 mg/L and 500–1000 μM	Soil drenching	Enhanced seed germination and vegetative growth	[22]
	5–11 mg/L	Soil drenching	Increased seed germination rate	
	5–15 ppm	Seed priming	Improved germination of aged seeds and enhanced growth	[23]
	10, 25, 50 and 100 mg / L	Spray by Foliar	10 ppm boosted leaf count and seed yield, while 25 ppm increased sugar content	[24]
	20 μg /mL	Seed priming	Improving plant growth and photosynthesis	
Sliver nanoparticles (Ag NPs)	5 & 10 mg/L	Seed priming	Enhances seed germination, starch metabolism, and water intake	[25]
	100 & 1000 mg/L		Decreased the germination and growth of rice seedlings	[26]
	60 mg / L		Enhanced agro-morphology, biochemistry, and enzyme activity	
	150 mg / L	Soil drenching	Increase of antioxidants, lipid peroxidation, and reduced contentsof chlorophyll, carot-enoids, total carbohydrate, and total soluble proteins	
	200 mg / L	Vegetative growth stage foliar application	Increase of antioxidants, lipid peroxidation, and reduced contentsof chlorophyll, carot-enoids, total carbohydrate, and total soluble proteins	[27]
Copper nanoparticles (Cu NPs)	4.44 mg / L	Seed priming	Enhance the plant biomass in both conditions i.e. normal and drought	[28]
	250 mg / L	Seed priming	Boosted bioactive compounds (vitamin C, lycopene, phenols, flavonoids) and antioxidant enzymes (SOD, CAT)	[29]
	0–20 mg / L	Hydroponic culture	Reduced root length in lettuce and alfalfa with dose-dependent	[30]
	Cu @ CuO and nCu SO4.5H_2O 10 and 20 mg / L	Hydroponic culture	Decreased water content, root length, and biomass. Altered defense metabolites.	[31]
	nCu (OH)2-b 1050 mg / L to 2100 mg / L	Foliar spray	Enhanced leaf biomass with metabolite changes indicating a defensive response	[30]

(continued)

Table 2 (continued)

Nanoparticles type	Used concentration	Mode of application	Physiological function	Reference
Zinc nanoparticles (Zn ONPs)	500 mg / mL	Soil drenching	Improving the growth, chlorophyll contents, Zn contents	[32]
	1600 mg / L	Seed priming	Reduced germination: Alfalfa 40%, tomato 20%, increased germination in cucumber	[33]
	100 mg / L	Seed soaking	Enhances germination rate	[34]
	400 mg / L	Foliar application	Improve oxidative stress	[35]
	500 mg / L	Mixed in soil	Translocation of Zn to aerial parts; root elongation; increased Zn in roots	[28]

like fertilizers, pesticides, and herbicides. The focus of this approach is to monitor environmental variables and adopt targeted actions to meet desired outcomes to promote resource efficiency and sustainability [37, 38]. Yet the increasing use of nanotechnology in agriculture causes concern over its potential detriment to people's health, the environment, and global economy [39]. Incorporation of NMs intentionally into agricultural processes may pose risks of bioaccumulation in the environment and food chain and residue uptake by humans, animals, and plants [40]. High stability and resistance to degradation in environmental matrices such as soil, water, and air have been shown for nanostructured materials, including lead and tin nanoparticles. Together, these properties, with their toxicological effects on tissues and organs, highlight the environmental risk of maintaining these compounds in the environment [41].

The ecological risks associated with nanomaterials are exemplified by silver nanoparticles, used in numerous consumer goods. They are found to disrupt aquatic ecosystems and disrupt fish, algae, bacteria, and other aquatic organisms [42]. Factors including particle size, mass, structure, and crystalline configuration influence the toxicological profile of nano-agrochemicals. For example, nanocomposites with crystallinity are more hazardous than their amorphous counterparts because they are more reactive. Adverse effects have been noted with smaller nanoparticles, where increasing toxicity is shown for decreasing particle size. Certain nanofertilizers, such as copper nanoparticles (Cu NPs), can overdose a plant such that molecular pathways are altered, which can decrease the levels of critical plant hormones, such as indole-3 acetic acid (IAA) and abscisic acid (ABA). Second-generation crops treated with the same dose of iron oxide nanoparticles (Fe_3O_4) as are beneficial for first-generation crops have been observed to suffer significant physiological damage from iron accumulation [43].

Nano-agrochemical ecotoxicological risk assessments should include organism-level dose-response analyses, but also studies of cellular and subcellular damage mechanisms [44]. While nanoparticles have been widely employed in sustainability-focused applications, there is limited understanding of their long-term impact on human health and environmental systems. Therefore, integrated risk analyses are needed to consider the life cycle of nanomaterials. Exposure pathway, hazard potential, and toxicological effects should be assessed using standardized testing and monitoring frameworks such as analyses. Since carbon-based nanoparticles, such as fullerenes, carbon nanotubes (CNTs), and graphene, are fewer toxic alternatives to mitigate the risks associated with nanomaterials [45], these materials have been explored. Potential avenues to reduce the adverse effects of nanotechnology in agriculture are provided by these materials. Although significant research, comprehensive risk assessments, and sustainable nanomaterial management should be developed to achieve a balance between using the advantages of nanotechnology and protecting human and environmental health.

4 Registration Policies for Nanomaterials in Agriculture

The application of nanotechnology in agriculture has received much consideration over the past decade, to promote sustainable practices. Nanopesticides, nanobiosensors, and nanomaterials for soil decontamination aimed to mitigate the biotic and abiotic stresses on crops have been developed by various nanotechnology-based innovations [46, 47]. These developments intend to maximize crop yields, improve nutrient use efficiency, and reduce reliance on classic plant protection agents. Therefore, a large amount of attention has been put on the regulation and safe use of nanomaterials (NMs) and engineered nanomaterials (ENMs) in agriculture across different levels of authority, including national and international.

However, the delivery of NMs and ENMs into plant cells is complicated by the physical barrier formed by the cell wall with a size exclusion limit of 5–20 nm [48]. Typical means of delivering biomolecules into plants are hampered by limited transgene efficiency, limited scope due to species, limited types of cargo that can be transferred, and risk of tissue damage. But nano-carrier systems have shown promise as a solution to successfully genetically modify crops, like rice, maize, tobacco, wheat, and cotton [49]. As a result, it is now recognized that regulatory frameworks should be developed for evaluating and reducing potential unintended impacts of nano-based agricultural products on humans, animals, and the environment.

5 Regulatory Frameworks Across Countries

Regulatory regimes for nanotechnology use in agriculture have been developed in several nations, including the United States, European Union member states, India, China, Canada, and Australia. The regulations relate to the development, commercialization, and safety assessment of nano-based agriproducts (nanofertilizers, nanopesticides, and genetically modified crops) using nanocarriers.

5.1 Key Regulatory Elements and Oversight and Governance

In the United States, responsibility for pesticides is with the Environmental Protection Agency (EPA), for biotechnology products with the U.S. Department of Agriculture (USDA), and for food safety with the Food and Drug Administration (FDA). The European Union also trusts the European Chemicals Agency (ECHA) and European Food Safety Authority (EFSA) to oversee the use of nano-based agri products. Each nanotechnology application ensures compliance with safety, labeling, and environmental standards by each regulatory body [50].

5.2 Risk Assessment and Safety Evaluation

Before being approved, nanomaterials must undergo an in-depth risk assessment of their toxicity, environmental impact, and potential health hazards, on behalf of regulatory authorities. They are assessing bioaccumulation, exposure routes, and long-term effects of NMs on ecosystems and human health.

5.3 Labeling Requirements

Because of transparency, and informed consumer choice, the regulatory framework demands labeling of agricultural products with nanomaterials. It allows better awareness and responsible use of nano-enhanced products.

5.4 International Standards

The International Organization for Standardization (ISO) developed standards for the safe use and characterization of nanomaterials in global organizations. For example, for nanomaterials, ISO/TS 80004-1 establishes standard terminology.

5.5 Encouragement of Research and Development

Research initiatives to develop safer and more efficient nano-based agriproducts with environmental sustainability are supported by Governments.

5.5.1 Country-Specific Regulations

Some examples are as follows:

1. **India**

Agriculture nanotechnology regulation is done by many agencies like the Food Safety and Standards Authority of India (FSSAI), the Department of Biotechnology (DBT), the Ministry of Environment, Forest and Climate Change (MoEFCC), etc.

Key legislations include:

The Environment Protection Act, 1986: Governs the handling of hazardous substances, including NMs.
The Food Safety and Standards Act, 2006: Ensures food safety, with provisions for nanotechnology in food products.
The Insecticides Act, 1968: Regulates nano-based insecticides.

2. **United States**

Regulatory frameworks include:

The Toxic Substances Control Act (TSCA): Is responsible for overseeing the handling of chemicals, in general, and NMs in particular, in agricultural products.
The Federal Insecticide, Fungicide, and Rodenticide Act (FIFRA): Registration should be managed and usage of nano-enhanced pesticides.

3. **European Union**

The EU has implemented several regulations specific to nano-agriculture, including:

Regulation (EC) No 1107/2009: Authorizes plant protection products containing NMs.
Regulation (EU) No 2019/1009: Establishes safety standards for fertilizers containing nanomaterials.

5.6 *Addressing Safety and Sustainability*

While nanotechnology may offer a lot of potentials, there are still many concerns as to how any NMs will affect human health and the environment over the long term. Risk studies have identified nanoparticle bioaccumulation and toxicity. As an example, some nanofertilizers have been associated with hormonal imbalances in plants and damage to the physiology of crops [43]. These risks must be addressed by regular updating of regulatory frameworks based on the most up-to-date scientific research.

Safe and sustainable use of nanotechnology agriculture requires globally harmonized robust regulations. Working in collaboration among researchers, manufacturers, regulatory bodies, and policymakers, nanotechnology can be unlocked to its full potential with respect to human and environmental health.

5.7 Importance of Safety in Nanomaterials in Agriculture

Nanotechnology has many applications for the development of new agricultural technologies by introducing nanomaterials (NMs) into agriculture. Although nanotechnology is a known application in biomedicine, food, energy, defense, textiles, and consumer goods, its use in agriculture is being increasingly investigated. These include the development of nano-priming for rapid seed germination, nanofertilizers, nanopesticides, nanoweedicides, and others as transformative solutions to sustainable agriculture. These advances promise to alleviate these stressor challenges related to abiotic and biotic stressors that impede agricultural productivity.

Factors of extreme temperature fluctuations due to climate change are examples of abiotic stressors while examples of biotic stressors include pathogens such as bacteria, viruses, fungi, and insect infestation. These stressors are under study as tools to improve plant tolerance and thus crop resilience and productivity, using NMs. At low concentrations, NMs can support plant adaptation to abiotic stresses through two primary mechanisms: First, they activate plant cell signaling pathways by generation of reactive oxygen species (ROS) and reactive nitrogen species (RNS) and stimulate the plant's defense mechanisms. The enzymatic and nonenzymatic antioxidant activities, as well as the accumulation of osmolytes, amino acids, and other essential nutrients [51], are included in these defense systems. Because of their inherent redox properties, metal-based nanomaterials, including metal oxides and salts, have shown potential in addressing both abiotic and biotic stressors. For example, nanofertilizers have been applied successfully to crops under adverse environmental conditions to enhance crop yields. A well-known example is urea-modified hydroxyapatite (HA) nanoparticles whose release of nitrogen is controlled over 60 days versus the 30-day release profile of conventional fertilizers [52]. Mesoporous silica nanoparticles have also been shown to encapsulate urea to achieve a fivefold increase in release duration, thus enhancing nutrient efficiency in the soil [53]. In addition to providing nutrients, NMs including nano-silicon dioxide (SiO_2) have been shown to increase seed germination, seedling weight, and stress tolerance in crops such as tomatoes and wheat. For example, nano-SiO_2 alleviated UV-B-induced stress in wheat by triggering NO-mediated activation of antioxidant pathways and thus decreasing oxidative damage while preserving plant health [54]. As with other metal/metal oxide nanoparticles such as silver (Ag), copper oxide (CuO), zinc oxide (ZnO), and magnesium oxide (MgO), these have been used to fight soil-borne and plant diseases [55–57]. Research into nanocomposites further underscores the versatility of NMs in agriculture. Silver-incorporated chitosan nanocomposites (Ag@CS) show synergistic antifungal activity to conventional fungicides against *Phytophthora capsici*, a causal agent of pepper blight. The results indicate that lower fungicide dosages will be possible resulting in reduced environmental toxicity without decreased efficacy [58]. *Bacillus thuringiensis* (Bt) nanocomposites have been shown to increase pesticide shelf life and decrease dosage requirements, although their mode of action still needs further study before commercialization [59]. Mitigation of abiotic stresses such as salinity and heavy metal toxicity is the further beneficial application of NMs. For example, nano-iron sulfate ($FeSO_4$) foliar application in sunflower plants under saline conditions increased biomass, chlorophyll content, photosynthetic rates, and decreased sodium ion accumulation in leaves. The nano-$FeSO_4$ also exceeded its conventional counterpart in increasing chlorophyll *a* content and biomass production [60]. Like, silicon nanoparticles (SiNPs) increased UV-B stress tolerance in wheat seedlings with enhanced antioxidant enzyme activity, which is more effective than traditional silicon treatments [54]. Along with that, biogenic NMs including AgNPs have been also revealed to be able to alleviate drought stress in lentils and other crops by improving growth parameters as well as antioxidant activities and by regulating metabolic processes [61]. It has also been reported that titanium dioxide (TiO_2) nanoparticles can abate oxidative stress in plants challenged by heavy metals, increasing photosynthetic efficiency and enzyme activities in crops, corn and soybean [62]. CeO_2 Nanoparticles improved maize salt tolerance by maintaining ionic homeostasis, increasing photosynthetic efficiency, and decreasing ROS [63]. These advancements demonstrate the important role of NMs in the counteraction of both biotic and abiotic stress factors in agriculture. Consistently experimental evidence is produced that shows the potential of NMs to increase crop yield, improve plant stress tolerance, and promote sustainable agricultural practices. Although these are encouraging results, there are still many things that need to be watched for the long-term safety, environmental impact, and regulatory frameworks regarding nanomaterials use in agriculture. Additional research is needed to learn what the specific mechanisms of NM action are, optimize these formulations, and ensure safe and effective integration of NM into agricultural systems.

6 Advancements in Nanotechnology for Sustainable Agriculture: Nanopesticides, Nanoherbicides, and Nanocarriers

Recently, nanotechnology has been progressing in a transformative way concerning the challenges confronting sustainable agriculture. Effective delivery of macronutrients, micronutrients, plant growth regulators, pesticides, and fertilizers are being intensively developed in smart nanoscale carriers. The results presented here have demonstrated great potential in terms of improving the efficiency of agricultural inputs and at the same time, limiting the environmental impacts. In the following sections, innovations and applications of nanopesticides, nanoherbicides, and nanocarriers are elaborated on and their mechanisms, benefits, and progress in research are explained.

6.1 *Nanopesticides: Addressing Limitations of Traditional Pesticides*

Synthetic pesticides and synthetic herbicides have helped raise yields on crops, which are needed to meet the growing global demand for food. Nevertheless, the use of pesticides and herbicides in the traditional way is limited by the low water solubility of active agents and their unpredictable release dynamics. Challenges in the transport and delivery of active ingredients in pesticides have been addressed through the design of the nanoformulation of pesticides.

1. **Mechanism of Action**

Pesticide nanoformulations are controlled by controlling the transport potential of active agents. These nanocarriers not only improved the bioavailability of active ingredients to plants but also reduced chemical discharge and environmental pollution. For instance:

2. **Improvements in Efficacy**

Systemic insecticide nano-imidacloprid has shown superior photo degradability and consequently decreased environmental persistence [64]. Reduced Non-target toxicity: nanoforms such as nano-acephate are less toxic to non-target organisms when compared to their conventional counterparts [65].

3. **Enhanced Stability and Transport**

Bifenthrin polymeric nanocapsules (nCAP4-BIF) had improved release profiles and transport potential under agricultural conditions [66]. There are very few studies that have monitored nanopesticide effects over long periods. Mitter et al. [67] reported that bioClay RNAi/nanoparticle platform provided 20 days of protection against plant viruses post-application. Extended Activity: Stored grain experiments showed that insecticidal formulations were effective for 5 months [68].

6.2 *Nanocarriers for Agrochemical Delivery*

Nanocarriers provide novel approaches for the efficient delivery of fungicides, herbicides, and insecticides. To minimize wastage, these carriers are made to integrate nicely with the soil and plant structures so that they provide optimal delivery.

Soil and Ecosystem Integration: Plant roots are secured by nanocarriers to organic materials and soil structure is stabilized via agrochemical delivery. Improved Elution: In loamy sand columns saturated with artificial pore water, nanocarriers have enhanced mobility and elution of pyrethroid nanoformulations when co-applied with fertilizer [66]. Metallic nanoparticles such as Ag, Cu, and Al, incorporated into nanoformulations, are very effective against bacterial, fungal, and viral pathogens; and are also less toxic to plants [69].

6.3 Nanoherbicides- Advancements and Challenges

The development of nanoherbicides has mirrored progress in nanopesticides, with a primary focus on reducing the non-target toxicity of herbicides.

6.4 Innovations in Herbicide Delivery

1. **Enhanced Bioavailability**

Glyphosate nanoemulsions have demonstrated higher efficacy than commercial formulations due to increased bioavailability [70].

2. **Targeted Action**

Nanoformulations of herbicides such as atrazine exert their effects directly through plant vascular tissues, as observed in Indian mustard (*Brassica juncea*) [71].

3. **Reducing Soil Leaching**

Nanoparticle-based carriers have been shown to reduce the leaching of herbicides, preserving soil health. Some of the examples include:

(i) **Rice Husk-Based Carriers**
Nano-sized rice husk particles loaded with 2,4-D herbicide demonstrated superior herbicidal action while minimizing leaching effects [72].

(ii) **Alginate and Chitosan Carriers**
Paraquat herbicide encapsulated in alginate or chitosan showed reduced leaching and soil sorption [73].

7 Environmental and Pathogen Control with Nanotechnology

Nanoparticles have been shown to counter bacterial, fungal, and viral pathogens effectively in in-vitro settings. Highly toxic inorganic nanoparticles, including silver, copper, and aluminum are attractive as agents for pest and disease control.

7.1 Balancing Pathogen Control and Soil Health

These nano-carried herbicides (e.g., montmorillonite clay layers with pH-responsive polymer coatings) have been shown not to negatively impact soil bacterial populations [74].

7.2 Challenges and Future Directions

Despite promising advances in sustainable agriculture, many barriers remain to the application of nanotechnology:

1. **Environmental Persistence**

To understand fully the long-term effects of nanoparticles on soil and ecosystem health more research is needed.

2. **Field Application**

To be valid for real-world situations, laboratory findings must be optimized, especially concerning nanoparticle mobility, stability, and aggregation under environmental conditions.

3. **Public Acceptance**

Transparent risk assessments and relevant risk regulatory oversight are especially important in the case of nanopesticides and nanoherbicides.

4. **Lower Dosage Requirements**

As a result, nanoformulations enable the lowering of the dosage requirements, thus water and energy saving, urging for further work on scaling and commercialization. Agricultural innovation driven by nanotechnology represents a paradigm shift offering tools for precision delivery, enhanced efficacy, and reduced environmental impact of agrochemicals. By tackling the challenges of today, while investing in long-term research, nanoformulations can transform the way sustainable agriculture is conducted, and greatly benefit global food security.

8 Advanced Applications and Implications of Nanotechnology in Agriculture and Environmental Defense

Nanoparticles (NPs) interacting with plants affect a large number of physiological and morphological changes in plants, the effectiveness of which varies depending on NP type, chemical composition, size, surface charge, reactivity, dosage, frequency of application, and plant species. Transformative tools in agricultural biotechnology are emerging through nano-enabled delivery systems. For instance, nanotechnology makes it possible to deliver engineered DNA into plant cells and to cause them to express desirable traits. Gene editing at the cellular level becomes feasible through NP-mediated delivery systems of DNA, which allows for the incorporation of desired traits into organisms rapidly. The use of this approach allows seed characteristics, such as color, yield, and growth cycle, to be customized. The potential for external and internal crop modifications via nanoparticles is great. Externally, they

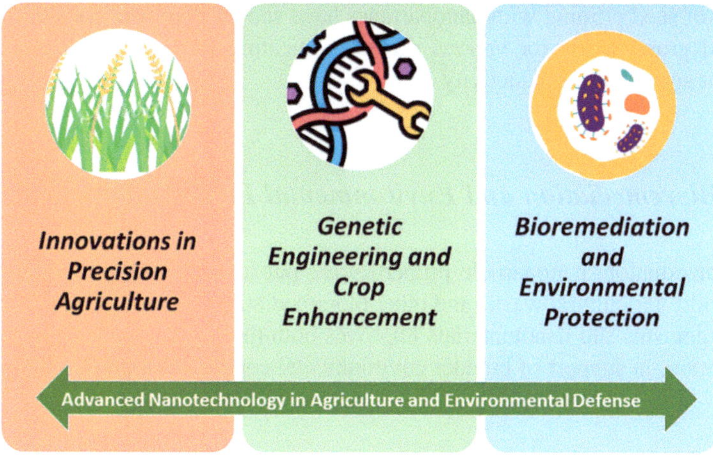

Fig. 1 Advanced applications and implications of Nano-agrotechnology

optimize nutrient delivery and external crop protection. Rapid modifications of gene expression patterns are enabled internally. In addition, NPs have been applied for the efficacy delivery of agrochemicals with new strategies to enhance crop protection and minimize environmental damage which are displayed in Fig. 1. Nanoparticle-based fertilizers and pesticides target specific areas and allow for the delivery of nutrients and chemicals with an efficiency that minimizes the ecological footprint of conventional agricultural strategies.

8.1 Innovations in Precision Agriculture and Nanotechnology

Nano-biosensors which are nanoscale-based systems integrated into precision agriculture combine biological and nanoscale structures to detect environmental conditions and monitor issues in real-time. Actionable data from these tools is essential for sustainable farming and optimizing resource use. The eco-friendly approach of green synthesis of nanomaterials using microorganisms like fungi and bacteria has attracted attention. For example, endophytic fungi can make silver nanoparticles, as sustainable alternatives to traditional production methods.

8.2 Genetic Engineering and Crop Enhancement

Plant genetic engineering has been revolutionized by nanodevices that permit the transfer and manipulation of DNA to improve crop traits, e.g., insect resistance.

Results of seed priming with nanoparticles have shown improved rates of germination and growth rates for various species, indicating that nano-primed seed may augment agricultural productivity.

8.3 Bioremediation and Environmental Protection

In bioremediation, nanoparticle properties are put to use for detoxifying contaminated soils, agricultural waste, and other bioremediation purposes. Combined use of microorganisms and nanomaterials improves both the efficacy and sustainability of remediation, in support of broader environmental conservation goals.

9 Challenges and Future Perspectives

The promise of nanotechnology has not been realized yet due to challenges in the understanding and mitigation of the environmental impacts of nanomaterials (NMs). Key issues include the need for developing advanced analytical technologies to detect and monitor transformed NMs in environmental matrices, and the requirement to evaluate the bioavailable fractions for toxicity assessment. The increasing volume of the nanomaterial market demands that these regulations on production, handling, and disposal be stringent.

It is well known that the environmental behavior and toxicity of engineered nanomaterials (ENMs) are related to their physicochemical properties including type, size, surface charge, coatings, and crystal chemistry but also to the exposure conditions. A comprehensive characterization of these properties is needed to understand their terrestrial ecosystem effects. Moreover, the interaction of NMs with soil physicochemical properties and their transformation in the environment is still an ongoing research priority.

A rational design of nanomaterials specifically tailored for agricultural and ecological applications is a promising way to overcome the limitations. Improving the stability, mobility, and selectivity of NMs under real-world conditions is essential if the full potential of NMs is to be realized. However, efforts to advance the state of the science on NM risks to ecosystems and human health need to also include the development of standardized approaches to quantify NM risks. The molecular-level impact of NMs could be advanced by advanced omics techniques and bioinformatics tools to inform the development of safer, more effective NMs.

10 Conclusion

Agricultural innovation and ecological defense are a frontier of nanotechnology that could bring about a "second green revolution." The wide applications of nanomaterials such as precision farming, and bioremediation, represent the revolutionary potentials of nanomaterials in supporting environmentally friendly agriculture and environmental protection. However, to reach widespread adoption, gaps in known knowledge of environmental safety and commercial scalability need to be addressed. Nanotechnology can bridge these gaps, moving from a promising scientific concept to the centerpiece of modern agricultural and ecological practice.

References

1. Drexler, K. E. (2022). Nanosystems: Molecular machinery, manufacturing, and computation | Wiley, Wiley.com.
2. Gong, X., Liu, Y., Huang, D., Zeng, G., Liu, S., Tang, H., Zhou, L., Hu, X., Zhou, Y., & Tan, X. (2016). Effects of exogenous calcium and spermidine on cadmium stress moderation and metal accumulation in Boehmeria nivea (L.) Gaudich. *Environmental Science and Pollution Research International, 23*(9), 8699–8708. https://doi.org/10.1007/s11356-016-6122-6
3. Ariga, K. (2024). Materials Nanoarchitectonics at dynamic interfaces: Structure formation and functional manipulation. *Materials (Basel, Switzerland), 17*(1), 271. https://doi.org/10.3390/ma17010271
4. Ashraf, S. A., Siddiqui, A. J., Elkhalifa, A. E. O., Khan, M. I., Patel, M., Alreshidi, M., Moin, A., Singh, R., Snoussi, M., & Adnan, M. (2021). Innovations in nanoscience for the sustainable development of food and agriculture with implications on health and environment. *The Science of the Total Environment, 768*, 144990. https://doi.org/10.1016/j.scitotenv.2021.144990
5. Lv, W., Geng, H., Zhou, B., Chen, H., Yuan, R., Ma, C., Liu, R., Xing, B., & Wang, F. (2022). The behavior, transport, and positive regulation mechanism of ZnO nanoparticles in a plant-soil-microbe environment, *Environmental Pollution (Barking, Essex : 1987), 315*, 120368. https://doi.org/10.1016/j.envpol.2022.120368
6. Wu, H., & Li, Z. (2022). Nano-enabled agriculture: How do nanoparticles cross barriers in plants? *Plant Communications, 3*(6), 100346. https://doi.org/10.1016/j.xplc.2022.100346
7. Ma, L. Q., Guan, D. X., Bradford, S., Letcher, R. J., Rinklebe, J., Ok, Y. S., & Whittle, M. (2022). New measures in 2022 to enhance the quality and reputation of critical reviews in environmental science and technology journal. *Critical Reviews in Environmental Science and Technology, 52*(22), 3943–3946. https://doi.org/10.1080/10643389.2022.2052587
8. Potocnik, J. (2011). Commission recommendation of 18 October 2011 on the definition of nanomaterial. *Journal of the European Communities Legislation, 275*, 38–40.
9. Bergeson, L. L. (2010). Nanosilver: US EPA's pesticide office considers how best to proceed. *Environmental Quality Management, 19*(3), 79.
10. Ghormade, V., Deshpande, M. V., & Paknikar, K. M. (2011). Perspectives for nano-biotechnology enabled protection and nutrition of plants. *Biotechnology Advances, 29*(6), 792–803.
11. Liu, R., & Lal, R. (2015). Potentials of engineered nanoparticles as fertilizers for increasing agronomic productions. *Science of the Total Environment, 514*, 131–139.
12. Omanović-Mikličanina, E., & Maksimović, M. (2016). Nanosensors applications in agriculture and food industry. *Bulletin of the Chemists and Technologists of Bosnia Herzegovina, 47*, 59–70.

13. Kah, M. (2015). Nanopesticides and nanofertilizers: Emerging contaminants or opportunities for risk mitigation? *Frontiers in Chemistry, 3*, 64.
14. Liu, J., Yue, J., Wang, H., Xie, L., Zhao, Y., Zhao, M., & Zhou, H. (2023). Strategies for engineering virus resistance in potato. *Plants, 12*(9), 1736.
15. Singh, D., Sharma, A., Verma, S. K., Pandey, H., & Pandey, M. (2024). Impact of nanoparticles on plant physiology, nutrition, and toxicity: A short review. *Next Nanotechnology, 6*, 100081.
16. Ghidan, A. Y., Al-Antary, T. M., Salem, N. M., & Awwad, A. M. (2017). Facile green synthetic route to the zinc oxide (ZnONPs) nanoparticles: Effect on green peach aphid and antibacterial activity. *Journal of Agricultural Science, 9*(2), 131–138.
17. Huang, S., Wang, L., Liu, L., Hou, Y., & Li, L. (2015). Nanotechnology in agriculture, livestock, and aquaculture in China. A review. *Agronomy for Sustainable Development, 35*, 369–400.
18. Ghidan, A. Y., Al-Antary, T. M., Awwad, A. M., & Akash, M. W. (2017). Aphidicidal potential of green synthesized magnesium hydroxide nanoparticles using Olea europaea leaves extract. *ARPN Journal of Agricultural and Biological Science, 2017b, 12*(10), 293–301.
19. Khot, L. R., Sankaran, S., Maja, J. M., Ehsani, R., & Schuster, E. W. (2012). Applications of nanomaterials in agricultural production and crop protection: A review. *Crop Protection, 35*, 64–70.
20. Bayda, S., Adeel, M., Tuccinardi, T., Cordani, M., & Rizzolio, F. (2019). The history of nanoscience and nanotechnology: From chemical–physical applications to nanomedicine. *Molecules, 25*(1), 112.
21. Kah, M., & Hofmann, T. (2014). Nanopesticide research: Current trends and future priorities. *Environment International, 63*, 224–235.
22. Gopinath, K., Gowri, S., Karthika, V., & Arumugam, A. (2014). Green synthesis of gold nanoparticles from fruit extract of Terminalia arjuna, for the enhanced seed germination activity of Gloriosa superba. *Journal of Nanostructure in Chemistry, 4*, 1–11.
23. Mahakham, W., Theerakulpisut, P., Maensiri, S., Phumying, S., & Sarmah, A. K. (2016). Environmentally benign synthesis of phytochemicals-capped gold nanoparticles as nanopriming agent for promoting maize seed germination. *Science of the Total Environment, 573*, 1089–1102.
24. Lovecká, P., Macůrková, A., Záruba, K., Hubáček, T., Siegel, J., & Valentová, O. (2021). Genomic damage induced in Nicotiana tabacum L. plants by colloidal solution with silver and gold nanoparticles. *Plants, 10*(6), 1260.
25. Mahakham, W., Sarmah, A. K., Maensiri, S., & Theerakulpisut, P. (2017). Nanopriming technology for enhancing germination and starch metabolism of aged rice seeds using phytosynthesized silver nanoparticles. *Scientific Reports, 7*(1), 8263.
26. Saleeb, N., Robinson, B., Cavanagh, J., Hossain, A. K. M., & Gooneratne, R. (2019). Biochemical changes in sunflower plant exposed to silver nanoparticles/silver ions.
27. Cui, D., Zhang, P., Ma, Y. H., He, X., Li, Y. Y., Zhao, Y. C., & Zhang, Z. Y. (2014). Phytotoxicity of silver nanoparticles to cucumber (Cucumis sativus) and wheat (Triticum aestivum). *Journal of Zhejiang University Science A, 15*(8), 662–670.
28. Elshoky, H. A., Yotsova, E., Farghali, M. A., Farroh, K. Y., El-Sayed, K., Elzorkany, H. E., & Apostolova, E. (2021). Impact of foliar spray of zinc oxide nanoparticles on the photosynthesis of Pisum sativum L. under salt stress. *Plant Physiology and Biochemistry, 167*, 607–618.
29. Mukherjee, A., Peralta-Videa, J. R., Bandyopadhyay, S., Rico, C. M., Zhao, L., & Gardea-Torresdey, J. L. (2014). Physiological effects of nanoparticulate ZnO in green peas (Pisum sativum L.) cultivated in soil. *Metallomics, 6*(1), 132–138.
30. Trujillo-Reyes, J., Majumdar, S., Botez, C. E., Peralta-Videa, J. R., & Gardea-Torresdey, J. L. (2014). Exposure studies of core–shell Fe/Fe3O4 and Cu/CuO NPs to lettuce (Lactuca sativa) plants: Are they a potential physiological and nutritional hazard? *Journal of Hazardous Materials, 267*, 255–263.

31. Jiang, M., Song, Y., Kanwar, M. K., Ahammed, G. J., Shao, S., & Zhou, J. (2021). Phytonanotechnology applications in modern agriculture. *Journal of Nanobiotechnology, 19*, 1–20.
32. Adrees, M., Khan, Z. S., Hafeez, M., Rizwan, M., Hussain, K., Asrar, M., & Ali, S. (2021). Foliar exposure of zinc oxide nanoparticles improved the growth of wheat (Triticum aestivum L.) and decreased cadmium concentration in grains under simultaneous Cd and water deficient stress. *Ecotoxicology and Environmental Safety, 208*, 111627.
33. García-López, J. I., Niño-Medina, G., Olivares-Sáenz, E., Lira-Saldivar, R. H., Barriga-Castro, E. D., Vázquez-Alvarado, R., et al. (2019). Foliar application of zinc oxide nanoparticles and zinc sulfate boosts the content of bioactive compounds in habanero peppers. *Plants, 8*(8), 254.
34. de la Rosa, G., López-Moreno, M. L., de Haro, D., Botez, C. E., Peralta-Videa, J. R., & Gardea-Torresdey, J. L. (2013). Effects of ZnO nanoparticles in alfalfa, tomato, and cucumber at the germination stage: Root development and X-ray absorption spectroscopy studies. *Pure and Applied Chemistry, 4th, 85*(12), 2161–2174.
35. Kareem, H. A., Hassan, M. U., Zain, M., Irshad, A., Shakoor, N., Saleem, S., & Wang, Q. (2022). Nanosized zinc oxide (n-ZnO) particles pretreatment to alfalfa seedlings alleviate heat-induced morpho-physiological and ultrastructural damages. *Environmental Pollution, 303*, 119069.
36. Eivazi, F., Afrasiabi, Z., & Elizabeth, J. O. S. E. (2018). Effects of silver nanoparticles on the activities of soil enzymes involved in carbon and nutrient cycling. *Pedosphere, 28*(2), 209–214.
37. Sampathkumar, K., Tan, K. X., & Loo, S. C. J. (2020). Developing nano-delivery systems for agriculture and food applications with nature-derived polymers. *Iscience, 23*(5), 101055.
38. Mishra, A., Saini, R. K., & Bajpai, A. K. (2020). Polymer formulations for pesticide release. In *Controlled release of pesticides for sustainable agriculture* (pp. 185–206). Springer.
39. Scott, N., & Chen, H. (2012). Nanoscale science and engineering for agriculture and food systems. *Industrial Biotechnology, 8*(6), 340–343.
40. Singh, J., Vishwakarma, K., Ramawat, N., Rai, P., Singh, V. K., Mishra, R. K., et al. (2019). Nanomaterials and microbes' interactions: a contemporary overview. *3 Biotech, 9*, 1–14.
41. Singh, K., & Sharma, S. (2021). Quantum dots: Fabrication, functionalization, and applications. In *Functionalized nanomaterials II* (pp. 129–164). CRC Press.
42. Navarro, E., Piccapietra, F., Wagner, B., Marconi, F., Kaegi, R., Odzak, N., & Behra, R. (2008). Toxicity of silver nanoparticles to Chlamydomonas reinhardtii. *Environmental Science & Technology, 42*(23), 8959–8964.
43. Le Van, N., Ma, C., Shang, J., Rui, Y., Liu, S., & Xing, B. (2016). Effects of CuO nanoparticles on insecticidal activity and phytotoxicity in conventional and transgenic cotton. *Chemosphere, 144*, 661–670.
44. Xiong, T., Zhang, S., Kang, Z., Zhang, T., & Li, S. (2021). Dose-dependent physiological and transcriptomic responses of lettuce (Lactuca sativa L.) to copper oxide nanoparticles—Insights into the phytotoxicity mechanisms. *International Journal of Molecular Sciences, 22*(7), 3688.
45. Yang, S., Yin, R., Wang, C., Yang, Y., & Wang, J. (2023). Phytotoxicity of zinc oxide nanoparticles and multi-walled carbon nanotubes, alone or in combination, on Arabidopsis thaliana and their mutual effects on oxidative homeostasis. *PLoS One, 18*(2), e0281756.
46. Sonawane, H., Arya, S., Math, S., & Shelke, D. (2021). Myco-synthesized silver and titanium oxide nanoparticles as seed priming agents to promote seed germination and seedling growth of Solanum lycopersicum: A comparative study. *International Nano Letters, 11*(4), 371–379.
47. An, C., Sun, C., Li, N., Huang, B., Jiang, J., Shen, Y., et al. (2022). Nanomaterials and nanotechnology for the delivery of agrochemicals: Strategies towards sustainable agriculture. *Journal of Nanobiotechnology, 20*(1), 11.
48. Arya, S. S., Lenka, S. K., Cahill, D. M., & Rookes, J. E. (2021). Designer nanoparticles for plant cell culture systems: Mechanisms of elicitation and harnessing of specialized metabolites. *BioEssays, 43*(11), 2100081.

49. Demirer, G. S., Zhang, H., Goh, N. S., González-Grandío, E., & Landry, M. P. (2019). Carbon nanotube–mediated DNA delivery without transgene integration in intact plants. *Nature Protocols, 14*(10), 2954–2971.
50. Kumari, R., Suman, K., Karmakar, S., Mishra, V., Lakra, S. G., Saurav, G. K., & Mahto, B. K. (2023). Regulation and safety measures for nanotechnology-based agri-products. *Frontiers in Genome Editing, 5*, 1200987.
51. Khan, M. N., Mobin, M., Abbas, Z. K., AlMutairi, K. A., & Siddiqui, Z. H. (2017). Role of nanomaterials in plants under challenging environments. *Plant Physiology and Biochemistry, 110*, 194–209.
52. Kottegoda, N., Munaweera, I., Madusanka, N., & Karunaratne, V. (2011). A green slow-release fertilizer composition based on urea-modified hydroxyapatite nanoparticles encapsulated wood. *Current Science*, 73–78.
53. Wanyika, H., Gatebe, E., Kioni, P., Tang, Z., & Gao, Y. (2012). Mesoporous silica nanoparticles carrier for urea: Potential applications in agrochemical delivery systems. *Journal of Nanoscience and Nanotechnology, 12*(3), 2221–2228.
54. Tripathi, D. K., Singh, S., Singh, V. P., Prasad, S. M., Dubey, N. K., & Chauhan, D. K. (2017). Silicon nanoparticles more effectively alleviated UV-B stress than silicon in wheat (Triticum aestivum) seedlings. *Plant Physiology and Biochemistry, 110*, 70–81.
55. Giannousi, K., Avramidis, I., & Dendrinou-Samara, C. (2013). Synthesis, characterization and evaluation of copper based nanoparticles as agrochemicals against Phytophthora infestans. *RSC Advances, 3*(44), 21743–21752.
56. Imada, K., Sakai, S., Kajihara, H., Tanaka, S., & Ito, S. I. (2016). Magnesium oxide nanoparticles induce systemic resistance in tomato against bacterial wilt disease. *Plant Pathology, 65*(4), 551–560.
57. Malandrakis, A. A., Kavroulakis, N., & Chrysikopoulos, C. V. (2019). Use of copper, silver and zinc nanoparticles against foliar and soil-borne plant pathogens. *Science of the Total Environment, 670*, 292–299.
58. Le, V. T., Bach, L. G., Pham, T. T., Le, N. T. T., Ngoc, U. T. P., Tran, D. H. N., & Nguyen, D. H. (2019). Synthesis and antifungal activity of chitosan-silver nanocomposite synergize fungicide against Phytophthora capsici. *Journal of Macromolecular Science, Part A, 56*(6), 522–528.
59. Devi, P. V., Duraimurugan, P., & Chandrika, K. S. V. P. (2019). Bacillus thuringiensis-based nanopesticides for crop protection. In *Nano-biopesticides today and future perspectives* (pp. 249–260). Academic Press.
60. Torabian, S., Zahedi, M., & Khoshgoftar, A. H. (2017). Effects of foliar spray of nano-particles of FeSO4 on the growth and ion content of sunflower under saline condition. *Journal of Plant Nutrition, 40*(5), 615–623.
61. Hojjat, S. S., & Kamyab, M. (2017). The effect of silver nanoparticle on Fenugreek seed germination under salinity levels. *Russian Agricultural Sciences, 43*, 61–65.
62. Singh, J., & Lee, B. K. (2016). Influence of nano-TiO2 particles on the bioaccumulation of Cd in soybean plants (Glycine max): A possible mechanism for the removal of Cd from the contaminated soil. *Journal of Environmental Management, 170*, 88–96.
63. Liu, Y., Cao, X., Yue, L., Wang, C., Tao, M., Wang, Z., & Xing, B. (2022). Foliar-applied cerium oxide nanomaterials improve maize yield under salinity stress: Reactive oxygen species homeostasis and rhizobacteria regulation. *Environmental Pollution, 299*, 118900.
64. Guan, H., Chi, D., Yu, J., & Li, X. (2008). A novel photodegradable insecticide: Preparation, characterization and properties evaluation of nano-Imidacloprid. *Pesticide Biochemistry and Physiology, 92*(2), 83–91.
65. Pradhan, S., Roy, I., Lodh, G., Patra, P., Choudhury, S. R., Samanta, A., & Goswami, A. (2013). Entomotoxicity and biosafety assessment of PEGylated acephate nanoparticles: A biologically safe alternative to neurotoxic pesticides. *Journal of Environmental Science and Health, Part B, 48*(7), 559–569.

66. Petosa, A. R., Rajput, F., Selvam, O., Öhl, C., & Tufenkji, N. (2017). Assessing the transport potential of polymeric nanocapsules developed for crop protection. *Water Research, 111*, 10–17.
67. Mitter, N., Worrall, E. A., Robinson, K. E., Li, P., Jain, R. G., Taochy, C., et al. (2017). Clay nanosheets for topical delivery of RNAi for sustained protection against plant viruses. *Nature Plants, 3*(2), 1–10.
68. Yang, F. L., Li, X. G., Zhu, F., & Lei, C. L. (2009). Structural characterization of nanoparticles loaded with garlic essential oil and their insecticidal activity against Tribolium castaneum (Herbst)(Coleoptera: Tenebrionidae). *Journal of Agricultural and Food Chemistry, 57*(21), 10156–10162.
69. Gogos, A., Knauer, K., & Bucheli, T. D. (2012). Nanomaterials in plant protection and fertilization: Current state, foreseen applications, and research priorities. *Journal of Agricultural and Food Chemistry, 60*(39), 9781–9792.
70. Jiang, L. C., Basri, M., Omar, D., Rahman, M. B. A., Salleh, A. B., Rahman, R. N. Z. R. A., & Selamat, A. (2012). Green nano-emulsion intervention for water-soluble glyphosate isopropylamine (IPA) formulations in controlling Eleusine indica (E. indica). *Pesticide Biochemistry and Physiology, 102*(1), 19–29.
71. Bombo, A. B., Pereira, A. E. S., Lusa, M. G., de Medeiros Oliveira, E., de Oliveira, J. L., Campos, E. V. R., et al. (2019). A mechanistic view of interactions of a nanoherbicide with target organism. *Journal of Agricultural and Food Chemistry, 67*(16), 4453–4462.
72. Chidambaram, R. (2016). Application of rice husk nanosorbents containing 2, 4-dichlorophenoxyacetic acid herbicide to control weeds and reduce leaching from soil. *Journal of the Taiwan Institute of Chemical Engineers, 63*, 318–326.
73. dos Santos Silva, M., Cocenza, D. S., Grillo, R., de Melo, N. F. S., Tonello, P. S., de Oliveira, L. C., et al. (2011). Paraquat-loaded alginate/chitosan nanoparticles: Preparation, characterization and soil sorption studies. *Journal of Hazardous Materials, 190*(1–3), 366–374.
74. Maruyama, C. R., Guilger, M., Pascoli, M., Bileshy-José, N., Abhilash, P. C., Fraceto, L. F., & De Lima, R. (2016). Nanoparticles based on chitosan as carriers for the combined herbicides imazapic and imazapyr. *Scientific Reports, 6*(1), 19768.

Index

R. S. Chaughule, A. V. Shelar (eds.), *Nanotechnology in Agriculture*, https://doi.org/10.1007/978-3-031-97544-8

MIX
Papier aus verantwortungsvollen Quellen
Paper from responsible sources
FSC® C105338

If you have any concerns about our products,
you can contact us on
ProductSafety@springernature.com

In case Publisher is established outside the EU,
the EU authorized representative is:
Springer Nature Customer Service Center GmbH
Europaplatz 3, 69115 Heidelberg, Germany

Printed by Libri Plureos GmbH
in Hamburg, Germany